Investigation on Different Properties of Bitumen and Asphalt Mixtures Using Advanced Techniques

Investigation on Different Properties of Bitumen and Asphalt Mixtures Using Advanced Techniques

Guest Editors

Songtao Lv
Xinghai Peng

MDPI
Basel • Beijing • Wuhan • Barcelona • Belgrade • Novi Sad • Cluj • Manchester

Guest Editors
Songtao Lv
School of Traffic & Transportation Engineering
Changsha University of Science & Technology
Changsha
China

Xinghai Peng
Institute of Highway Engineering
RWTH Aachen University
Aachen
Germany

Editorial Office
MDPI AG
Grosspeteranlage 5
4052 Basel, Switzerland

This is a reprint of the Special Issue, published open access by the journal *Buildings* (ISSN 2075-5309), freely accessible at: www.mdpi.com/journal/buildings/special_issues/58WT6XMY04.

For citation purposes, cite each article independently as indicated on the article page online and using the guide below:

Lastname, A.A.; Lastname, B.B. Article Title. *Journal Name* **Year**, *Volume Number*, Page Range.

ISBN 978-3-7258-3204-0 (Hbk)
ISBN 978-3-7258-3203-3 (PDF)
https://doi.org/10.3390/books978-3-7258-3203-3

Contents

Preface

This reprint presents a collection of cutting-edge research focusing on the multifaceted properties, performance, and applications of bitumen and asphalt mixtures under various conditions. The reprint explores advanced techniques in multi-scale analysis, fatigue characterization, and innovative material modifications.

This work aims to deepen our understanding of the mechanical behaviors and failure mechanisms of asphalt mixtures, offering valuable insights for researchers, engineers, and policymakers. We express our sincere gratitude to the contributing authors, the reviewers, and all those who supported the compilation of this reprint.

Songtao Lv and Xinghai Peng
Guest Editors

buildings

MDPI

Article

Performance Evaluation of Soybean Oil/SBR Reclaimed Asphalt and Mixtures

Yu Chen [1], Xiao Li [1,*] and Xiaoge Tian [2]

1 Faculty of Civil Engineering, Changsha University of Science and Technology, Changsha 410114, China; 202202110136@stu.csust.edu.cn
2 School of Traffic and Transportation Engineering, Changsha University of Science and Technology, Changsha 410114, China; tianxiaoge@126.com
* Correspondence: lixiao@stu.csust.edu.cn

Abstract: This study evaluated the properties of soybean oil/SBR reclaimed asphalt (SSRA). The optimal preparation method for SSRA was determined. Additionally, the feasibility of the optimal SSRA scheme was verified through asphalt mixture performance tests. With the soybean oil dosage enhanced, the penetration and low-temperature rheological performance of SSRA were improved. The incorporation of soybean oil lowered the softening point, viscosity, and rutting index of aged asphalt. The softening points of SBR-4%+Oil-7.5% and SBR-6%+Oil-7.5% were 79.4 °C and 82.9 °C, respectively. The stiffness modulus of SBR-6%+oil-10% decreased by 35.37%. When the soybean oil dosage was 10% and the SBR dosage was 6% (SBR-6%+oil-10%), the properties of RTFOT+PAV aged asphalt were restored to those of its original state. The splitting tensile strength ratio of the SBR-6%+oil-10% mixture was 89%, with a decrease of 1.5% compared to the original asphalt mixture. The SBR-6%+oil-10% mixture exhibited improved high-temperature and low-temperature service properties. The total deformation of the SBR-6%+oil-10% mixture decreased by 8.43%, while its dynamic stability increased by 22.21%. This degree of improvement compared to the original asphalt mixture was not significant. The rejuvenation of the aged asphalt and mixture performance can mainly be attributed to the soybean oil supplementing the lost lightweight components of the aged asphalt, while SBR supplemented the degraded polymers. Utilizing soybean oil as a rejuvenating asphalt agent facilitates waste material recycling. Furthermore, this study provides a new idea for the recycling of polymer-modified asphalt and reclaimed asphalt pavement.

Keywords: soybean oil; SBR; rejuvenate aged asphalt; rheological properties; mixture

Citation: Chen, Y.; Li, X.; Tian, X. Performance Evaluation of Soybean Oil/SBR Reclaimed Asphalt and Mixtures. *Buildings* **2024**, *14*, 2085. https://doi.org/10.3390/buildings14072085

Academic Editor: Bjorn Birgisson

Received: 29 May 2024
Revised: 23 June 2024
Accepted: 4 July 2024
Published: 8 July 2024

1. Introduction

During its service life, asphalt pavement is exposed to ultraviolet radiation, thermal cycles, oxygen, and water, inducing both physical and chemical changes that lead to aging. Asphalt aging results in the hardening and embrittlement of the mixture. Over time, the volatilization of light fractions within the asphalt makes the material more rigid, reducing its elasticity and flexibility. This increased rigidity makes the pavement more prone to cracking, especially in low-temperature environments. Asphalt aging can severely impact the durability of pavements, significantly shortening their service life [1]. The increasing amount of recycled asphalt mixtures generated due to aging leads to the wastage of aggregates and petroleum and environmental pollution. Therefore, effectively recycling aged asphalt is essential.

1.1. Bio-Oil Recycled Asphalt and Recycled Asphalt Pavement

As a renewable material, bio-oil exhibits an excellent compatibility with asphalt [2]. Bio-oil has shown significant potential in the field of asphalt rejuvenation. Ji utilized waste cooking oil to restore UV-aged asphalt, which decreased the rutting factor of the asphalt [3]. However, bio-oil only replenishes the light fractions in aged asphalt and does not effectively restore its viscoelastic properties. Ye investigated the rejuvenating effect of bio-oil, revealing that the incorporation of 7% bio-oil could reduce the complex shear modulus by approximately 90% [4]. The temperature sensitivity of the rejuvenated asphalt binder was also reduced. Additionally, the feasibility of bio-oil-rejuvenated asphalt was confirmed through molecular dynamics simulations. Lv investigated the rejuvenation of asphalt at various degrees of aging using bio-oil [5]. Bio-oil increased the irrecoverable creep compliance and improved the low-temperature behavior of aged asphalt binder. Muhammad's research demonstrated that bio-oil reduced the carbonyl functional group area in asphalt, validating the rejuvenating effect of bio-oil [6]. Nizamuddin investigated the rejuvenation effect of plastic bio-oil on aged asphalt [7]. The bio-oil induced the softening of the aged asphalt, thereby enhancing its fatigue performance and reducing its stiffness [8].

Peng investigated the rejuvenation effects of different types of bio-oil [2], which indicated that bio-oil decreased the temperature sensitivity of asphalt. Additionally, the rejuvenated asphalt maintained good high-temperature behavior. The microstructure of the rejuvenated asphalt closely matched that of the original asphalt. Chen summarized the rejuvenation effects of vegetable oil on aged asphalt [9]. The good compatibility between the vegetable oil and asphalt ensured the feasibility of rejuvenation. The vegetable oil improved the thermal stability and service performance of the aged asphalt. Chen pointed out that the composite rejuvenation of asphalt with bio-oil and polymers could increase the properties of aged asphalt [10].

Bio-oil can be used not only for asphalt rejuvenation, but also for the rejuvenation of recycled asphalt mixtures, as explored by researchers. Saeed utilized rubber seed oil to regenerate RAP materials, and the results indicated that bio-oil exhibited a favorable viscosity reduction effect on aged asphalt [11]. With an increase in bio-oil dosage, the strength and stiffness of the recycled asphalt mixture gradually decreased. Costa chose cottonseed oil for the regeneration of RAP. The cottonseed oil tended to reduce the fatigue life of its mixture [12]. However, it exhibited significant improvements in resistance to brittle failure and stiffness reduction.

In summary, numerous researchers are currently focused on studying the performance of bio-oil in rejuvenating aged asphalt. Related studies have proven the feasibility of using bio-oil for asphalt rejuvenation. Similarly, bio-oil can also be used directly for the rejuvenation of RAP. By replenishing its light components, bio-oil can rejuvenate the service performance of aged asphalt. Moreover, bio-oil effectively increases aged asphalt's low-temperature performance and fatigue properties. However, the current lack of research on the properties of bio-oil-rejuvenated asphalt mixtures significantly limits their widespread application.

1.2. *Polymer Recycled Asphalt and Recycled Asphalt Pavement*

The superior performance and environmental advantages of polymers have led to their increasing emphasis regarding reclaimed asphalt and mixtures. Hu investigated the role of polymers in rejuvenated asphalt [13]. Polymers enhanced the toughness and anti-aging properties of the asphalt and reinforced the polymer network structure in aged asphalt. Li investigated the effects of Styrene Butadiene Rubber (SBR) on asphalt [14]. The results indicated that SBR can effectively enhance the low-temperature service properties and fatigue cracking performance of asphalt. Yi utilized epoxy resin polymers to rejuvenate aged asphalt [15]. The results demonstrated that the polymers formed a stable network structure within the aged asphalt, thereby improving its fatigue performance and low-temperature properties. Ren used SBS and rubber to rejuvenate aged asphalt [16]. The polymers significantly enhanced the viscoelastic properties and anti-aging behavior of the aged asphalt. Additionally, the polymer-rejuvenated asphalt mixtures exhibited an excellent road performance. Almusawi utilized polymers as rejuvenators for RAP, resulting in significant improvements in the moisture stability and strength of the rejuvenated asphalt mixtures [17]. Viscione employed the dry method to incorporate polymers into rejuvenated asphalt mixtures [18]. The polymers enhanced the moisture stability and crack resistance of the mixtures.

In summary, polymers enhance the low-temperature service behavior and fatigue resistance by forming new cross-linked network structures within aged asphalt. Moreover, aging reduces the adhesion between asphalt and aggregates, severely impacting moisture sensitivity. Polymers significantly increase the high-temperature properties of aged asphalt binder. In addition, polymers can enhance the adhesion between aged asphalt and aggregates, thereby protecting reclaimed asphalt mixtures from water damage.

1.3. Objective

The aging process of SBS-modified asphalt includes two parts: neat asphalt aging and polymer degradation. In this study, soybean oil and SBR were utilized to reclaim aged SBS-modified asphalt. On the one hand, soybean oil, as a waste product, has essentially the same composition as neat asphalt and contains sufficient lightweight components. Soybean oil can supplement the volatilized lightweight components. On the other hand, SBR can supplement the deteriorated polymers. The feasibility and effects of soybean oil and SBR on aged SBS-modified asphalt and its mixture are worth investigating.

This study evaluated the properties of reclaimed asphalt with different dosages of soybean oil and SBR. The reclaimed asphalt properties mainly include conventional and rheological properties, with indicators such as penetration, softening point, viscosity, rutting index, m-value, and stiffness modulus. The optimal preparation content for SSRA was determined. Additionally, the road properties of the SSRA mixture were tested.

2. Materials and Tests Design

2.1. Materials

The SBS-modified asphalt was purchased from Baoli Bitumen Co., Ltd. in Changsha, China. The dosage of the SBS modifier was 5%. The property indicators of the original asphalt are displayed in Table 1. SBS-modified asphalt was utilized to prepare the aged asphalt. The SBR was purchased from Shanghai Jinlang Rubber & Plastic Technology Co., Ltd., in Shanghai, China. The bio-oil was purchased from Shandong Fenghui New Energy Co., Ltd., in Jinan, China. Soybean oil and SBR were used as regenerating agents, and the performance indicators are described in Tables 2 and 3.

Table 1. Performance indicators of original asphalt.

Index	Results	Technical Requirements
Solubility (%)	99.8	≥99.5
Penetration (0.1 mm)	58.6	60~80
Ductility (cm)	126.4	≥100
Viscosity (60 °C) (Pa·s)	206.5	≥160
Softening point (°C)	79.6	≥46
Flashpoint (°C)	256.2	≥230
	After RTFOT aged	
Mass loss (%)	0.06	≤±0.8
Penetration ratio (%)	65	≥61
Ductility (5 cm/min, 10 °C)	7.9	≥6

Table 2. Performance indicators of soybean oil.

Index	Results
PH	2.76
Flashpoint (°C)	239
Ash dosage (%)	0.07
Aliphatic acid dosage (%)	65.8
Alcohol dosage (%)	5.62
Density (g/cm^3)	0.941

Table 3. Performance indexes of SBR.

Index	Results
Solid dosage (%)	58.4
Viscosity (MPa)	74.8
PH	7.2
Tensile strength (MPa)	24.5
Styrene dosage (%)	26.3

2.2. Preparation of Soybean Oil/SBR Reclaimed Asphalt

The asphalt was then heated at 163 °C for 85 min with a specified air flow rate, preparing rolling thin-film oven test (RTFOT) aged asphalt. Subsequently, the RTFOT aged asphalt was subjected to the PAV aging procedure. The test conditions for PAV were an air pressure of 2.1 MPa ± 0.1 MPa, a test temperature of 100 °C ± 0.5 °C, and an aging time of 20 h. Then, the RTFOT+PAV aged asphalt was prepared.

The soybean oil and RTFOT+PAV aged asphalt were blended first. The soybean oil dosages were 5%, 7.5%, 10%, and 12.5% (mass ratio of aged asphalt) [5]. During this process, the shear temperature was 150 °C. The shear rate was 1500 r/min for 30 min [19]. Soybean oil reduced the viscosity of the aged asphalt, facilitating a better blend of the SBR and aged asphalt. Next, the SBR was blended with the soybean oil reclaimed asphalt. The SBR dosages were 4% and 6% (mass ratio of aged asphalt) [19]. The shear temperature was 160 °C, and the shear rate was 3000 r/min for 30 min [5]. To ensure that the SBR and soybean oil reclaimed asphalt fully dispersed, the asphalt was developed at 160 °C for 60 min. The abbreviations for the soybean oil/SBR reclaimed asphalt (SSRA) are shown in Table 4.

Table 4. Abbreviations of soybean oil/SBR reclaimed asphalt.

SBR Dosage	Soybean Oil	Abbreviation
4%	5%	SBR-4%+Oil-5%
	7.5%	SBR-4%+Oil-7.5%
	10%	SBR-4%+Oil-10%
6%	7.5%	SBR-6%+Oil-7.5%
	10%	SBR-6%+Oil-10%
	12.5%	SBR-6%+Oil-12.5%

2.3. Test Methods

2.3.1. Conventional Properties Tests of Asphalt

Penetration, softening point, viscosity, and segregation tests were performed to assess the conventional performance of the recycled asphalt. The penetration tests were performed at 25 °C [20]. The softening point tests were conducted according to the specification requirements [21]. The viscosity tests were performed at 160 °C. The segregation test was performed to assess the storage stability of the reclaimed asphalt. The compatibility of the soybean oil, SBR, and aged asphalt was investigated. To ensure the accuracy of the test results, three parallel tests were conducted for each conventional properties test.

2.3.2. Rheological Behavior Tests of Asphalt

Temperature sweep and bending beam rheometer (BBR) tests were utilized to reveal the rheological properties of the reclaimed asphalt. The temperature sweep test was conducted at 46–82 °C. The temperature sweep test was conducted in strain-controlled mode. The gap distance and diameter of the specimens were 1 mm and 25 mm, respectively. The tests were performed at a constant loading frequency of 10 rad/s. The RTFOT+PAV aged asphalt samples were also subjected to BBR tests. The BBR tests were performed at −24 °C [22]. To ensure the accuracy of the test results, three parallel tests were conducted for each rheological behavior test.

2.3.3. Performance Tests of Asphalt Mixture

This study conducted road performance tests on the recycled asphalt mixtures. To ensure the accuracy of the test results, three parallel tests were conducted for each mixture test. The wheel tracking test (T 0719-2011) was utilized to verify the high-temperature stability of the mixtures [23,24]. In the laboratory, asphalt mixture specimens with dimensions of 300 mm × 300 mm × 50 mm were prepared. The specimens were kept at 60 °C for 8 h. The dynamic stability of the specimens was calculated using Equation (1).

$$Dynaamic\ stability = \frac{(t_2 - t_1) \times N}{d_2 - d_1} \times C_1 \times C_2 \quad (1)$$

where: t_2 is 60 min, t_1 is 45 min, N represents the reciprocating rolling speed (42 times/min), C_1 is the equipment factor (C_1 is 1.0), C_2 is the asphalt mixture size factor (C_2 is 1.0), d_1 is the deformation of the specimen at 45 min, and d_2 is the deformation of the specimen at 60 min.

The freeze–thaw indirect tension test can simulate the process experienced by asphalt roads in freeze–thaw regions, accelerating the damage caused by water to the mixture. After subjecting the specimens to freeze–thaw treatment, the indirect tension strength was tested according to the specifications (T 0729-2011) [24].

The low-temperature service performance of the mixture was tested through the low-temperature bending test (T 0715-2011) [24]. The specimens were prismatic with dimensions of 250 mm × 30 mm × 35 mm. Before testing, the beams were conditioned at −10 °C for 3 h. The data were processed using Equation (2).

$$S_B = \frac{R_B}{\varepsilon_B} \quad (2)$$

where R_B represents the bending tensile strength of the specimen (MPa), ε_B represents the maximum strain of the specimen ($\mu\varepsilon$), and S_B represents the flexural modulus of the specimen (MPa).

2.4. Flow Chart

Figure 1 shows the test flow chart of this study.

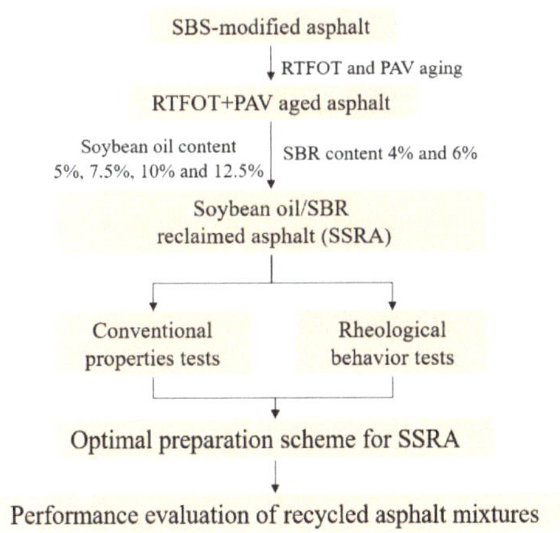

Figure 1. Test flow chart.

3. Soybean Oil/SBR Reclaimed Asphalt Performance Test Results

3.1. Penetration

Penetration is an indicator that reveals the consistency of asphalt. The penetration test results are described in Figure 2.

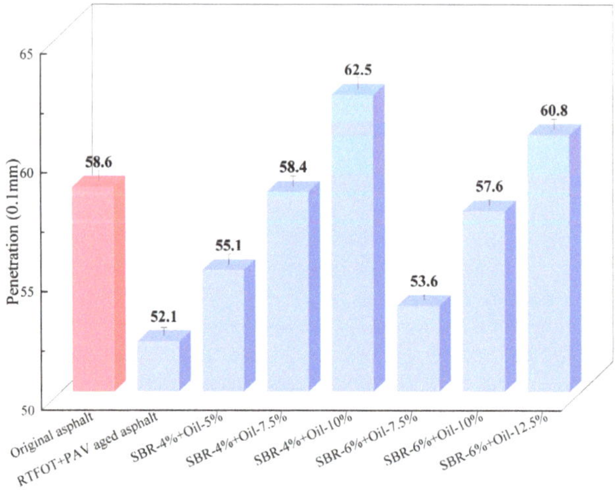

Figure 2. Penetration test results.

After thermo-oxidative aging, the penetration of the original asphalt was reduced from 58.6 to 52.1. Aging causes the lighter components to volatilize and increases the stiffness of the asphalt [25]. From the results of Figure 2, it can be seen that the addition of SBR and soybean oil enhanced its penetration.

When the SBR dosage was 4% and the soybean oil dosage was 5%, 7.5%, and 10%, the penetration of SSRA was 55.1, 58.4, and 62.5, respectively. There was an obvious increase in the penetration of SSRA with an increase in the bio-oil dosage. This was mainly due to the high content of light fractions in the soybean oil, which reduced the stiffness of aged asphalt. This indicated that soybean oil effectively restored the aged asphalt [19].

The penetration of SBR-4%+Oil-5% was still lower than that of the original asphalt. When the SBR dosage was 4% and the soybean oil dosage was 7.5%, the penetration of the RTFOT+PAV aged asphalt could be reclaimed to its original state. However, with the soybean oil dosage increased to 10%, the penetration of SBR-4%+Oil-10% increased by 6.66% compared to the original asphalt.

When the SBR dosage was 6% and the soybean oil dosage was 10%, the penetration of SSRA was 57.6. This showed that the consistency of SBR-6%+Oil-10% was the same as that of the original asphalt. Moreover, SBR-6%+Oil-10% and SBR-4%+Oil-7.5% provided the same degree of penetration recovery for the RTFOT+PAV aged asphalt. When the soybean oil dosage was 10%, the penetration of SBR-6%+Oil-10% and SBR-4%+Oil-10% was 57.6 and 62.5. This indicated that SBR had a reduction effect on the penetration of the reclaimed asphalt.

The analysis of the penetration indicator indicated that the penetration of SBR-4%+Oil-7.5% and SBR-6%+Oil-10% was basically the same as that of the original asphalt samples. As the soybean oil dosage continued to increase, the SSRA penetration would exceed that of the original asphalt. This is due to the fact that the lighter components of bio-oil would reduce the consistency of the aged asphalt [26].

3.2. Softening Point

The softening point, as one of the essential performance indicators of asphalt, can reflect the temperature sensitivity of asphalt, the results of which are illustrated in Figure 3.

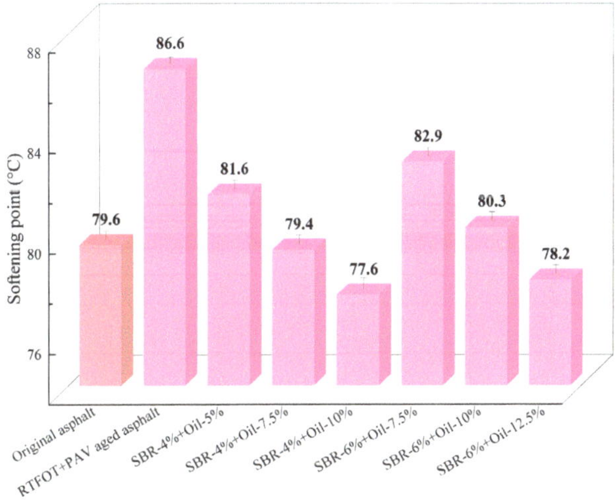

Figure 3. Softening point test results.

The softening point of the RTFOT+PAV aged asphalt increased by 7 °C compared to the original asphalt. The aging process led to the volatilization of lighter components and the degradation of polymers, resulting in an increase in the molecular weight of the asphalt and a higher softening point [27]. Figure 3 illustrates a gradual decrease in the softening point of SSRA with an increasing soybean oil dosage. Specifically, at an SBR dosage of 4%, the softening point of SSRA decreased by 5.8%, 8.31%, and 10.4% with soybean oil dosages of 5%, 7.5%, and 10%, respectively. Soybean oil supplementation replenished the lighter components of the aged asphalt, while bio-oil reduced the temperature sensitivity of the aged asphalt. Notably, at a soybean oil dosage of 7.5%, the softening points of SBR-4%+Oil-7.5% and SBR-6%+Oil-7.5% were 79.4 °C and 82.9 °C, respectively. SBR, acting as a polymer, contributed to enhancing the high-temperature behavior of the RTFOT+PAV aged asphalt [28].

The analysis of the softening point of SSRA demonstrated that the synergistic action of SBR and soybean oil partially restored the softening point. Specifically, when the SBR dosage was 4% and the soybean oil dosage ranged from 7.5% to 10% or when the SBR dosage was 6% and the soybean oil dosage ranged from 10% to 12.5%, the softening point of SSRA closely approximated that of the original asphalt.

3.3. Viscosity

Asphalt viscosity refers to the flowability or consistency of asphalt at a certain temperature. It plays a crucial role in determining the mixing and construction temperatures of asphalt mixtures. Figure 4 displays the viscosity results of the original asphalt, RTFOT+PAV aged asphalt, and various types of SSRA asphalt.

The viscosity of the RTFOT+PAV aged asphalt exhibited a significant increase compared to that of the original asphalt. Thermal oxidative aging resulted in a decrease in the dosage of light components and an increase in the dosage of heavy components in the asphalt [29]. Thermal oxidative aging caused the asphalt to become harder and resulted in an increase in its viscosity. The effect of SBR and soybean oil significantly decreased the viscosity of the RTFOT+PAV aged asphalt. With a continued enhancement in the bio-oil dosage, the soybean oil significantly reduced the viscosity of SSRA. This was consistent with the results of the softening point test. The viscosities of SBR-4%+Oil-10% and SBR-6%+Oil-12.5% were lower than that of the original asphalt.

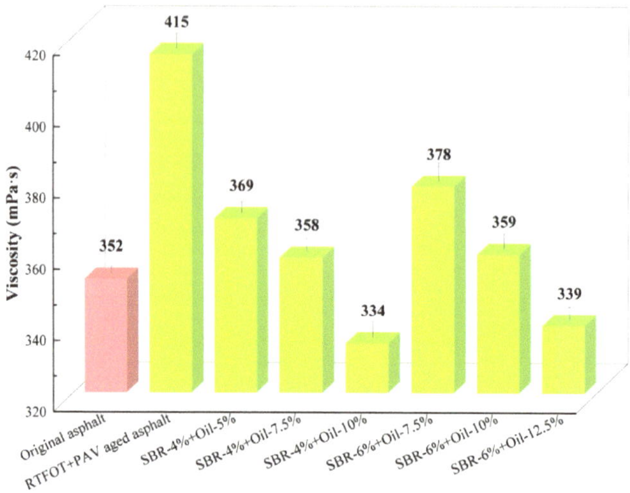

Figure 4. Viscosity test results at 160 °C.

The viscosities of the six types of SSRA prepared in this study did not differ significantly from that of the original asphalt. This finding also suggested that the selected combination of SBR and soybean oil dosage in this study was appropriate for restoring the viscosity of the RTFOT+PAV aged asphalt. The viscosities of SBR-4%+Oil-7.5% and SBR-6%+Oil-10% were essentially consistent with that of the original asphalt.

3.4. Storage Stability Tests

Compatibility reactions between bio-oil and asphalt may occur, resulting in changes in its chemical properties. Certain bio-oils may partially dissolve in asphalt, leading to modifications in its viscosity, flowability, and stability [30]. Therefore, storage stability tests on SSRA must be conducted, and the experimental results are shown in Table 5. Previous studies have indicated that asphalt maintains a good storage stability when the softening point difference is less than 2.5 °C [31].

Table 5. Softening point difference of soybean oil/SBR reclaimed asphalt.

Asphalt Types	Softening Point Difference (°C)	Standard Deviation
Original asphalt	1.3	0.16
PAV aged asphalt	1.8	0.15
SBR-4%+Oil-5%	1.5	0.09
SBR-4%+Oil-7.5%	1.9	0.13
SBR-4%+Oil-10%	2.4	0.18
SBR-6%+Oil-7.5%	1.5	0.13
SBR-6%+Oil-10%	2.1	0.19
SBR-6%+Oil-12.5%	2.9	0.18

The findings presented in Table 5 revealed that, as the soybean oil dosage increased in SSRA, so did the difference in the softening point [30]. Notably, when the soybean oil dosage exceeded 10%, the difference in the softening point of SSRA exceeded 2.5 °C. This suggested that an excessive soybean oil dosage degraded the storage stability of SSRA. However, with a constant soybean oil dosage, the difference in the softening point decreased with an increase in the SBR dosage. The softening point difference between SBR-6%+Oil-10% and SBR-4%+Oil-10% decreased by 0.5 °C. This occurred due to the formation of a cross-linked network structure between the SBR and asphalt, enhancing the storage stability of SSRA [32]. Based on the analysis of the softening point difference of SSRA, it is recommended that the soybean oil dosage should not exceed 10%, as it may otherwise affect the storage stability of the asphalt.

3.5. Temperature Sweep

The rutting index of asphalt indicates its tendency to develop ruts under the load of vehicle traffic and heavy loads. The rutting index has frequently been employed to evaluate the properties of asphalt in resisting deformation. Figure 5 depicts the rutting indexes of the various asphalt binders.

Figure 5 illustrates that PAV aging increased the rutting index of the original asphalt. When the temperature was 64 °C, the RTFOT+PAV aged asphalt showed an increase of 6.693 kPa compared to the original asphalt. After aging, the asphalt's molecular chains may experience breaking, cross-linking, or increased rigidity, leading to an enhancement in the rutting factor [33].

The addition of SBR and soybean oil resulted in a reduction in the rutting index of SSRA. This suggested that the SBR and soybean oil contributed to partially restoring the aged asphalt. At a test temperature of 64 °C, with an SBR dosage at 4% and soybean oil dosage at 5%, 7.5%, and 10%, the rutting index of SSRA increased by 1.92 kPa, 4.5 kPa, and −0.33 kPa, respectively. Moreover, with an increase in the dosage of the soybean oil, there was a significant decrease in the rutting index of SSRA. This indicated that the soybean oil significantly restored the performance of the aged asphalt, bringing its high-temperature performance back to the level of the original asphalt [3].

As illustrated in Figure 5a, when the SBR dosage was 4% and the soybean oil dosage was 10%, the rutting index of SSRA was lower than that of the original asphalt. Similarly, as depicted in Figure 5b, the rutting index of SBR-6%+oil-10% was slightly higher than that of the original asphalt, while the rutting index of SBR-6%+oil-12.5% was lower than that of the original asphalt. This observation indicates that a significant amount of soybean oil contributed to a marked reduction in the rutting index of SSRA.

Comparing the rutting indexes of the original asphalt and SSRA, it became apparent that the synergistic influence of soybean oil and SBR played a crucial part in restoring the rutting index of the RTFOT+PAV aged asphalt. The rutting indexes of SBR-4%+oil-10% and SBR-6%+oil-10% were largely consistent with those of the base asphalt.

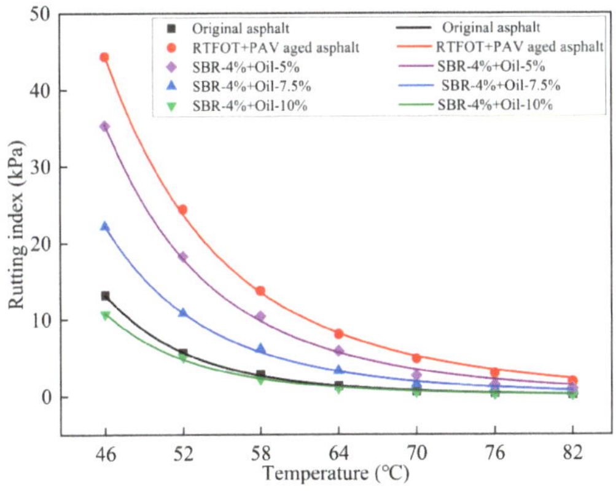

(**a**) SBR dosage 4%

Figure 5. *Cont.*

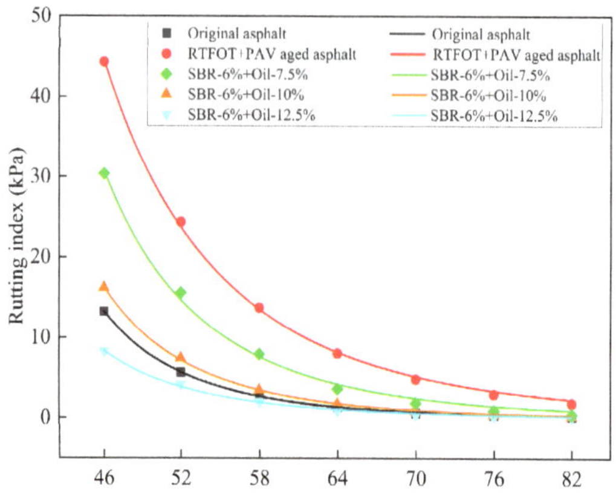

(**b**) SBR dosage 6%

Figure 5. Temperature sweep test results.

3.6. BBR Test

The BBR is a widely utilized test to assess the deformation performance of asphalt under low-temperature conditions [34]. This test has commonly been employed to assess the low-temperature embrittlement characteristics of asphalt binders in order to evaluate their performance in cold climates. The BBR tests were performed at −24°C, and the results are presented in Figure 6.

According to the specifications, when asphalt maintains good low-temperature rheological properties at a certain temperature, the m-value must be higher than 0.3 and the stiffness modulus must be lower than 300 MPa [22]. The findings in Figure 6 suggested that the original asphalt retained a better low-temperature performance at −24 °C. However, after PAV aging, the m-value reduced to 0.248, while the stiffness modulus increased to 376 MPa. Under the influence of oxidation, the asphalt experienced oxidation, hardening, and degradation, resulting in an increased brittleness and susceptibility to cracking [35]. PAV aging markedly deteriorated the low-temperature behavior [36].

The findings in Figure 6 describe that the SBR and soybean oil blended with the aged asphalt and enhanced its low-temperature performance. Moreover, with an increase in the soybean oil dosage, the m-value of SSRA gradually rose while the stiffness modulus declined. At an SBR dosage of 4% and soybean oil dosages of 5%, 7.5%, and 10%, the m-value of SSRA increased by 14.5%, 31%, and 42.74%, respectively, compared to the RTFOT+PAV aged asphalt. Likewise, the stiffness moduli of SBR-6%+oil-7.5%, SBR-6%+oil-10%, and SBR-6%+oil-12.5% decreased by 21.8%, 35.37%, and 42.55%, respectively, compared to the RTFOT+PAV aged asphalt. Soybean oil enhanced the low-temperature toughness of the aged asphalt, restoring its low-temperature performance.

Among the six types of SSRA, SBR-4%+oil-7.5%, SBR-4%+oil-10%, SBR-6%+oil-10%, and SBR-6%+oil-12.5% exhibited favorable resistance to low-temperature cracking at −24 °C. This indicated that, when the SBR dosage was 4% and the soybean oil dosage was 7.5% and 10% or when the SBR dosage was 6% and the soybean oil dosage was 10% and 12.5%, the low-temperature performance of the RTFOT+PAV aged asphalt could be restored to its original state.

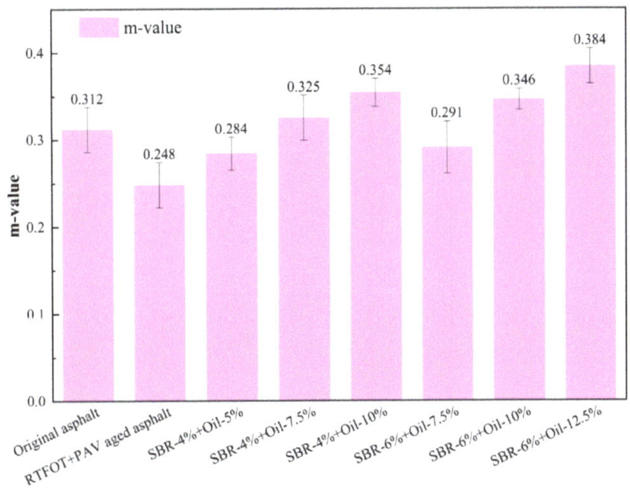

(**a**) m-value

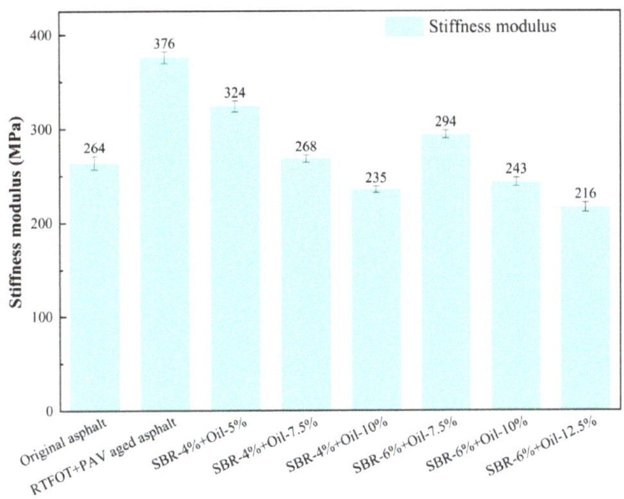

(**b**) Stiffness modulus

Figure 6. BBR test results.

4. Determining the Optimal Preparation Scheme for Soybean Oil/SBR Reclaimed Asphalt

The optimal preparation method of SSRA was determined by analyzing the conventional and rheological properties of various types of SSRA. Table 6 presents the optimal preparation scheme for SSRA, determined through different asphalt tests.

An analysis of various tests revealed that, when the SBR dosage was 6% and the soybean oil dosage was 10%, the PAV aged asphalt could be restored to a performance level equivalent to that of the original asphalt. Thus, the optimal preparation scheme for SSRA, SBR-6%+oil-10%, was determined.

Table 6. SSRA optimal preparation scheme.

Tests	SSRA Optimal Preparation Scheme
Penetration	SBR-4%+Oil-7.5% and SBR-6%+Oil-10%
Softening point	SBR-4%+Oil-7.5%, SBR-4%+Oil-10%, SBR-6%+Oil-10%, and SBR-6%+Oil-12.5%
Viscosity	SBR-4%+Oil-7.5% and SBR-6%+Oil-10%
Storage stability	SBR-4%+Oil-5%, SBR-4%+Oil-7.5%, SBR-4%+Oil-10%, SBR-6%+Oil-7.5%, and SBR-6%+Oil-10%
Temperature sweep	SBR-4%+oil-10% and SBR-6%+oil-10%
BBR	SBR-4%+Oil-7.5%, SBR-4%+Oil-10%, SBR-6%+Oil-10%, and SBR-6%+Oil-12.5%

5. Soybean Oil/SBR Reclaimed Asphalt Mixture Performance Test Results

5.1. Asphalt Mixture Gradation Design

To verify the rejuvenation effects of soybean oil and SBR, this study conducted asphalt mixture tests on both the original asphalt mixture and the SBR-6%+oil-10% mixture. Moreover, the feasibility of the SBR-6%+oil-10% scheme was further validated. The aggregate gradation type of the asphalt mixture was AC-16, as illustrated in Figure 7. Limestone was selected as the aggregate. The limestone originated from the Shengxing Stone Quarry in Guangxi Province, China. The properties of limestone are shown in Table 7. The asphalt–stone ratio for the different asphalt mixtures was established through the Marshall test [37]. The results regarding the mixing temperature, compaction temperature, and asphalt–stone ratio for the various asphalt mixtures are presented in Table 8.

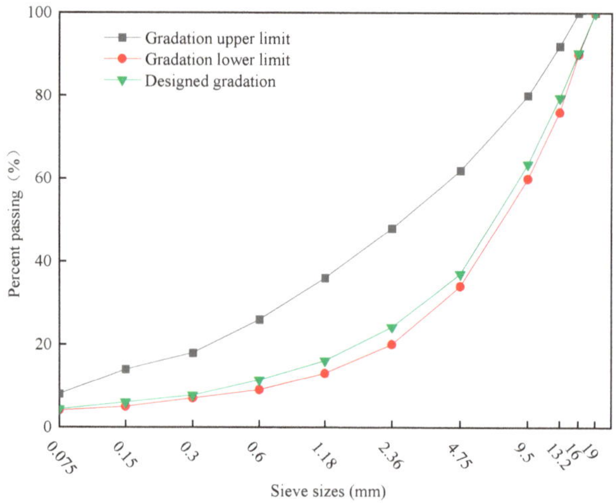

Figure 7. AC-16 mixture gradation.

Table 7. The properties of limestone.

Properties	Test Results	Requirements
Needle-flake particle dosage	3.2%	≤15
Abrasion value	16.3%	≤30
Polished value	59.6%	≥45
Crush value	17.6%	≤26

Table 8. The preparation conditions for different types of mixtures.

Asphalt Types of Mixture	Mixing Temperature	Compaction Temperature	Asphalt-Stone Ratio
Original asphalt	173 °C–175 °C	164 °C–168 °C	4.97
SBR-6%+oil-10%	174 °C–177 °C	165 °C–169 °C	4.93

5.2. Wheel Tracking Test

As traffic loads increase, asphalt pavement becomes more susceptible to plastic deformation, resulting in an increase in rut depth [38]. The rutting resistance of both the SBR-6%+oil-10% and original asphalt mixtures was tested, and the total deformation and dynamic stability are illustrated in Figure 8.

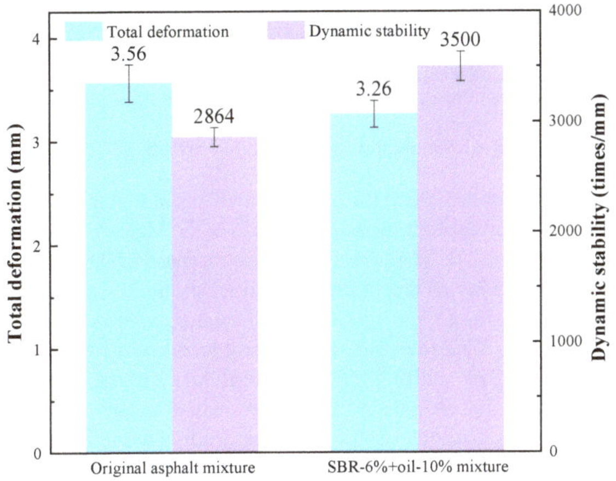

Figure 8. The total deformation and dynamic stability.

As illustrated in Figure 8, the original asphalt mixture exhibited a total deformation of 3.56 mm and a dynamic stability of 2864 times/mm. On the other hand, the SBR-6%+oil-10% mixture demonstrated a total deformation of 3.26 mm and dynamic stability of 3500 times/mm. In comparison, the total deformation decreased by 8.43%, while the dynamic stability increased by 22.21%.

The SBR-6%+oil-10% mixture exhibited improved high-temperature properties compared to the original asphalt mixture. The difference in the high-temperature performance between these two types of asphalt mixtures was minimal. The conclusions drawn from wheel tracking tests on the asphalt mixtures aligned with those obtained from the rheological tests on the asphalt. The synergistic effect of SBR and soybean oil led to an enhanced rejuvenation of the RTFOT+PAV aged asphalt [39,40]. Soybean oil increased the proportion of lighter components in the RTFOT+PAV aged asphalt, while SBR supplemented the polymers that underwent degradation due to the aging process [19].

5.3. Freeze–Thaw Indirect Tension Test

When water infiltrates into asphalt pavement, it can cause a decrease in the adhesion behavior of asphalt, leading to a reduced pavement strength and stiffness [41]. Furthermore, the aging process can also decrease the adhesion property of asphalt. Therefore, evaluating the water stability of recycled asphalt mixtures is of practical significance. Freeze–thaw splitting tests were utilized to assess the moisture sensitivity of the mixture, with the results depicted in Figure 9.

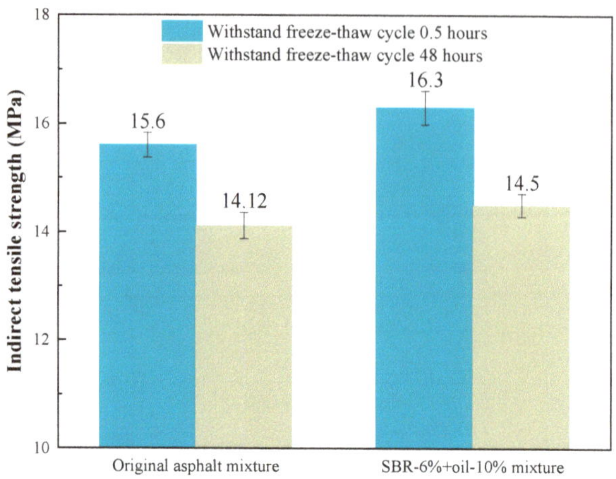

Figure 9. Freeze–thaw indirect tension test results.

After immersing the asphalt mixtures in water for 0.5 h, the indirect tension strength of the original asphalt mixture was 15.6 MPa. Meanwhile, the strength of the SBR-6%+oil-10% mixture was 16.3 MPa, reflecting an increase of 0.8 MPa compared to the original asphalt mixture. After 48 h of immersion in water, there was a notable decrease in the strength [42]. The strength of the original asphalt mixture was 14.12MPa, with a splitting tensile strength ratio of 90.5%. However, the splitting tensile strength ratio of the SBR-6%+oil-10% mixture was 89%. The splitting tensile strength ratios of these two asphalt mixtures were essentially the same. This suggested that the moisture sensitivity of the SBR-6%+oil-10% mixture was essentially equivalent to that of the original asphalt mixture.

5.4. Low-Temperature Bending Test

When there is a sudden drop in temperature, if the thermal stress generated within the asphalt surface layer does not relax promptly, exceeding the tensile strength of the asphalt layer, it may lead to fracture. Low-temperature bending tests were performed at −10 °C to assess the low-temperature performance of the mixtures, and the results are presented in Table 9.

Table 9. The bending test results.

Mixture Types	Parallel Specimens	R_B/MPa	Average Value /MPa	$\varepsilon_B/\mu\varepsilon$	Average Value /$\mu\varepsilon$	S_B/MPa	Average Value /MPa
Original asphalt mixture	1	7.16	7.23	2367	2423	3025	2984
	2	7.23		2446		2956	
	3	7.29		2456		2968	
SBR-6%+oil-10% mixture	1	8.19	8.25	2648	2724	3093	3028
	2	8.21		2759		2976	
	3	8.34		2765		3016	

The failure flexural tensile strength of the original asphalt mixture was 7.23 MPa, while that of the SBR-6%+oil-10% mixture was 8.25 MPa, representing an increase of 14.1% in contrast with the original asphalt mixture. The failure strains of the original asphalt mixture and the SBR-6%+oil-10% mixture were 2423 με and 2724 με, respectively. The flexural moduli of bending of the original asphalt mixture and the SBR-6%+oil-10% mixture were 2984 MPa and 3028 MPa, respectively. Both the failure flexural tensile strength and failure strain of the SBR-6%+oil-10% mixture were improved compared to the original asphalt mixture, indicating a significant increase in low-temperature behavior. This could be attributed to the soybean oil increasing the lighter components proportion in the recycled asphalt, thereby increasing the oil dosage ratio in the SBR-6%+oil-10% mixture [43]. Changes in the asphalt components enhanced the low-temperature service properties of the SBR-6%+oil-10% mixture [4,44].

6. Conclusions

This study explored the regeneration effects of soybean oil and SBR. The optimal rejuvenation method for RTFOT+PAV aged asphalt was determined based on performance test results. Moreover, mixture road properties tests were conducted to validate the feasibility of the optimal rejuvenation method for aged asphalt.

1. With an increase in the soybean oil dosage, the penetration of SSRA gradually increased, while high-temperature performance indicators such as the softening point, viscosity, and rutting index gradually decreased. The softening points of SBR-4%+Oil-7.5% and SBR-6%+Oil-7.5% were 79.4 °C and 82.9 °C, respectively. Soybean oil supplemented the aged asphalt with lighter components, significantly restoring its low-temperature performance. The stiffness modulus of SBR-6%+oil-10% decreased by 35.37%.
2. By comparing the performances of six different kinds of SSRA, it was found that when the soybean oil dosage was 10% and the SBR dosage was 6%, the PAV aged asphalt could be rejuvenated to its original state. Soybean oil supplemented the volatilized lightweight components. SBR supplemented the deteriorated polymers.
3. Mixture tests were conducted on SBR-6%+oil-10% and original asphalt. The results indicated that the SBR-6%+oil-10% mixture exhibited improved high-temperature and low-temperature service properties. In comparison, the total deformation of the SBR-6%+oil-10% mixture decreased by 8.43%, while its dynamic stability increased by 22.21%. However, this degree of improvement compared to the original asphalt mixture was not significant. The splitting tensile strength ratio of the SBR-6%+oil-10% mixture was 89%, which was reduced by 1.5% compared to the original asphalt mixture.
4. The utilization of soybean oil as a waste material in rejuvenating asphalt enables waste utilization. Moreover, this study will offer new ideas for polymer-aged asphalt regeneration.
5. The medium properties and fatigue behavior of SSRA will be revealed in future studies. The types of bio-oils are diverse, and their effects on the performance of asphalt are different. In the future, we will evaluate the rejuvenation effects of different types of bio-oils on aged asphalt to determine the optimal type and dosage of bio-oil for asphalt rejuvenation.

Author Contributions: Conceptualization, X.L.; Methodology, X.T.; Writing—original draft, Y.C. All authors have read and agreed to the published version of the manuscript.

Funding: This research received no external funding.

Data Availability Statement: Data is contained within the article.

Conflicts of Interest: The author declare no conflicts of interest.

References

1. Lv, S.; Tan, L.; Peng, X.; Hu, L.; Borges Cabrera, M. Experimental investigation on the performance of bone glue and crumb rubber compound modified asphalt. *Constr. Build. Mater.* **2021**, *305*, 124734. [CrossRef]
2. Peng, X.; Xie, N.; Xia, C.; Zhou, X.; Zhao, P.; Ma, S.; Zhang, C.; Lv, S. Laboratory evaluation of different bio-oil recycled aged asphalts: Conventional performances and microscopic characteristics. *J. Clean. Prod.* **2023**, *428*, 139442. [CrossRef]
3. Ji, H.; Li, B.; Li, X.; Han, J.; Liu, D.; Dou, H.; Fu, M.; Yao, T. Waste cooking oil as a sustainable solution for UV-aged asphalt binder: Rheological, chemical and molecular structure. *Constr. Build. Mater.* **2024**, *420*, 135149. [CrossRef]
4. Ye, Q.; Yang, Z.; Lv, S.; Jin, J.; Zhang, S. Study on components selection and performance of bio-oil based asphalt rejuvenator based on softening and asphaltene deagglomeration effect. *J. Clean. Prod.* **2023**, *419*, 138238. [CrossRef]

5. Lv, S.; Liu, J.; Peng, X.; Liu, H.; Hu, L.; Yuan, J.; Wang, J. Rheological and microscopic characteristics of bio-oil recycled asphalt. *J. Clean. Prod.* **2021**, *295*, 126449. [CrossRef]
6. Zahoor, M.; Nizamuddin, S.; Madapusi, S.; Giustozzi, F. Recycling asphalt using waste bio-oil: A review of the production processes, properties and future perspectives. *Process Saf. Environ. Prot.* **2021**, *147*, 1135–1159. [CrossRef]
7. Nizamuddin, S.; Baloch, H.A.; Jamal, M.; Madapusi, S.; Giustozzi, F. Performance of waste plastic bio-oil as a rejuvenator for asphalt binder. *Sci. Total Environ.* **2022**, *828*, 154489. [CrossRef] [PubMed]
8. Girimath, S.; Singh, D. Effects of bio-oil on performance characteristics of base and recycled asphalt pavement binders. *Constr. Build. Mater.* **2019**, *227*, 116684. [CrossRef]
9. Chen, C.; Lu, J.; Ma, T.; Zhang, Y.; Gu, L.; Chen, X. Applications of vegetable oils and their derivatives as Bio-Additives for use in asphalt binders: A review. *Constr. Build. Mater.* **2023**, *383*, 131312. [CrossRef]
10. Peng, X.; Yuan, J.; Wu, Z.; Lv, S.; Zhu, X.; Liu, J. Investigation on strength characteristics of bio-asphalt mixtures based on the time–temperature equivalence principle. *Constr. Build. Mater.* **2021**, *309*, 125132. [CrossRef]
11. Saeed, S.M.; Sutanto, M.H.; Napiah, M.; Usman, A.; Batari, A.; Aman, M.Y.; Aliyu Yaro, N.S. Optimization of rubber seed oil content as bio-oil rejuvenator and total water content for cold recycled asphalt mixtures using response surface methodology. *Case Stud. Constr. Mater.* **2021**, *15*, e00561. [CrossRef]
12. da Costa, L.F.; de Medeiros Melo Neto, O.; de Macêdo, A.L.F.; de Figueiredo Lopes Lucena, L.C.; de Figueiredo Lopes Lucena, L. Optimizing recycled asphalt mixtures with zeolite, cottonseed oil, and varied RAP content for enhanced performance and circular economy impact. *Case Stud. Constr. Mater.* **2024**, *20*, e02707. [CrossRef]
13. Hu, M.; Sun, D.; Hofko, B.; Sun, Y.; Mirwald, J.; Xu, L. Multiscale optimization on polymer-based rejuvenators for the efficient recycling of aged high-viscosity modified asphalt: Molecular dynamics simulation and experimental analysis. *J. Clean. Prod.* **2024**, *449*, 141736. [CrossRef]
14. Li, Q.; Zhang, H.; Chen, Z. Improvement of short-term aging resistance of styrene-butadiene rubber modified asphalt by Sasobit and epoxidized soybean oil. *Constr. Build. Mater.* **2021**, *271*, 121870. [CrossRef]
15. Yi, X.; Wong, Y.D.; Chen, H.; Fan, Y.; Yang, J.; Huang, W.; Wang, H. Influence of epoxy resin polymer on recycled asphalt binder properties. *Constr. Build. Mater.* **2023**, *398*, 132549. [CrossRef]
16. Ren, S.; Liu, X.; Wang, H.; Fan, W.; Erkens, S. Evaluation of rheological behaviors and anti-aging properties of recycled asphalts using low-viscosity asphalt and polymers. *J. Clean. Prod.* **2020**, *253*, 120048. [CrossRef]
17. Almusawi, A.; Shoman, S.; Lupanov, A.P. Assessment of the effectiveness and the initial cost efficiency of hot recycled asphalt using polymer modified bitumen. *Case Stud. Constr. Mater.* **2023**, *18*, e02145. [CrossRef]
18. Viscione, N.; Veropalumbo, R.; Oreto, C.; Biancardo, S.A.; Abbondati, F.; Russo, F. Additional procedures for characterizing the performance of recycled polymer modified asphalt mixtures. *Measurement* **2022**, *187*, 110238. [CrossRef]
19. Li, Y.; Ge, D.; Ju, Z.; Lv, S.; Xue, Y.; Xue, Y.; Peng, L. Study on Performance and Mechanism of SBR and Bio-Oil Recycled SBS Modified Asphalt. *Polymers* **2022**, *14*, 5096. [CrossRef]
20. *AASHTO T 49-15*; Standard Method of Test for Penetration of Bituminous Materials. American Association of State Highway and Transportation Officials: Washington, DC, USA, 2015.
21. *AASHTO T 53-09*; Standard Method of Test for Softening Point of Bitumen (Ring-and-Ball Apparatus). American Association of State Highway and Transportation Officials: Washington, DC, USA, 2013.
22. *AASHTO T 313-19*; Standard Method of Test for Determining the Flexural Creep Stiffness of Asphalt Binder Using the Bending Beam Rheometer (BBR). American Association of State Highway and Transportation Officials: Washington, DC, USA, 2019.
23. Eisa, M.S.; Mohamady, A.; Basiouny, M.E.; Abdulhamid, A.; Kim, J.R. Mechanical properties of asphalt concrete modified with carbon nanotubes (CNTs). *Case Stud. Constr. Mater.* **2022**, *16*, e00930. [CrossRef]
24. *JTG E20—2011*; Standard Test Methods of Bitumen and Bituminous Mixture. China Communications Press: Beijing, China, 2011.
25. Zhang, R.; You, Z.; Wang, H.; Chen, X.; Si, C.; Peng, C. Using bio-based rejuvenator derived from waste wood to recycle old asphalt. *Constr. Build. Mater.* **2018**, *189*, 568–575. [CrossRef]
26. Wang, L.; Si, B.; Han, X.; Yi, W.; Li, Z.; Zhang, A. Study on the effect of red mud and its component oxides on the composition of bio-oil derived from corn stover catalytic pyrolysis. *Ind. Crops Prod.* **2022**, *184*, 114973. [CrossRef]
27. Hu, M.; Ling, S.; Sun, D.; Lu, T.; Ma, J.; Sun, Y. A sustainable high-viscosity modified asphalt modified with multiple anti-aging agents: Micro-chemical analysis and macro-rheological characterization. *Constr. Build. Mater.* **2022**, *339*, 127701. [CrossRef]
28. Lu, C.; Zheng, M.; Gao, Y.; Ding, X.; Zhang, W.; Chen, W. Study of long-term skid and wear resistance of waterborne epoxy resin-SBR compound modified emulsified asphalt microsurfacing. *J. Appl. Polym. Sci.* **2023**, *141*, e54937. [CrossRef]
29. Wang, Q.; Li, S.; Wu, X.; Wang, S.; Ouyang, C. Weather aging resistance of different rubber modified asphalts. *Constr. Build. Mater.* **2016**, *106*, 443–448. [CrossRef]
30. Sun, G.; Li, B.; Sun, D.; Zhang, J.; Wang, C.; Zhu, X. Roles of aging and bio-oil regeneration on self-healing evolution behavior of asphalts within wide temperature range. *J. Clean. Prod.* **2021**, *329*, 129712. [CrossRef]
31. Kong, P.; Xu, G.; Fu, L.; Feng, H.; Chen, X. Chemical structure of rubber powder on the compatibility of rubber powder asphalt. *Constr. Build. Mater.* **2023**, *392*, 131769. [CrossRef]
32. Dong, D.; Huang, X.; Li, X.; Zhang, L. Swelling process of rubber in asphalt and its effect on the structure and properties of rubber and asphalt. *Constr. Build. Mater.* **2012**, *29*, 316–322. [CrossRef]

33. Xiao, M.M.; Fan, L. Ultraviolet aging mechanism of asphalt molecular based on microscopic simulation. *Constr. Build. Mater.* **2022**, *319*, 126157. [CrossRef]
34. Li, S.; Xu, W.; Zhang, F.; Wu, H. A study on the rheological properties and modification mechanism of graphene oxide/polyurethane/SBS-modified asphalt. *PLoS ONE* **2022**, *17*, e0262467. [CrossRef]
35. Chaihad, N.; Karnjanakom, S.; Abudula, A.; Guan, G. Zeolite-based cracking catalysts for bio-oil upgrading: A critical review. *Resour. Chem. Mater.* **2022**, *1*, 167–183. [CrossRef]
36. Daryaee, D.; Ameri, M.; Mansourkhaki, A. Utilizing of waste polymer modified bitumen in combination with rejuvenator in high reclaimed asphalt pavement mixtures. *Constr. Build. Mater.* **2020**, *235*, 117516. [CrossRef]
37. Xu, J.; Luo, X.; Qiu, X.; Hu, G. Wavelet and fractal analysis of acoustic emission characteristic of fatigue damage of asphalt mixtures. *Constr. Build. Mater.* **2022**, *349*, 128643. [CrossRef]
38. Zhang, C.; Yu, H.; Zhu, X.; Yao, D.; Peng, X.; Fan, X. Unified Characterization of Rubber Asphalt Mixture Strength under Different Stress Loading Paths. *J. Mater. Civ. Eng.* **2024**, *36*, 04023498. [CrossRef]
39. Vamegh, M.; Ameri, M.; Chavoshian Naeni, S.F. Experimental investigation of effect of PP/SBR polymer blends on the moisture resistance and rutting performance of asphalt mixtures. *Constr. Build. Mater.* **2020**, *253*, 119197. [CrossRef]
40. Zeng, M.; Tian, W.; Zhu, Y. Study on performance of castor oil bio asphalt blended asphalt mixture. *J. Hunan Univ.* **2017**, *44*, 177–182.
41. Asadi Azadgoleh, M.; Modarres, A.; Ayar, P. Effect of polymer modified bitumen emulsion production method on the durability of recycled asphalt mixture in the presence of deicing agents. *Constr. Build. Mater.* **2021**, *307*, 124958. [CrossRef]
42. Chen, S.; Gong, F.; Ge, D.; You, Z.; Sousa, J.B. Use of reacted and activated rubber in ultra-thin hot mixture asphalt overlay for wet-freeze climates. *J. Clean. Prod.* **2019**, *232*, 369–378. [CrossRef]
43. Kabir, S.F.; Mousavi, M.; Fini, E.H. Selective adsorption of bio-oils' molecules onto rubber surface and its effects on stability of rubberized asphalt. *J. Clean. Prod.* **2020**, *252*, 119856. [CrossRef]
44. Liao, M.; Liu, Z.; Gao, Y.; Liu, L.; Xiang, S. Study on UV aging resistance of nano-TiO_2/montmorillonite/styrene-butadiene rubber composite modified asphalt based on rheological and microscopic properties. *Constr. Build. Mater.* **2021**, *301*, 124108. [CrossRef]

buildings

MDPI

Article

Research on the Properties of Nano-Al_2O_3- and Nano-SiO_2-Modified Bio-Asphalt and Mixtures

Wei Qiao [1], Feilong Yuan [1,*], Xiaojian Xu [1] and Qingzhou Wang [2]

[1] CHECSC Highway Maintenance and Test Technology Co., Ltd., Beijing 102200, China; xjqw530@126.com (W.Q.); xxj123xxj014@126.com (X.X.)
[2] School of Civil and Transportation Engineering, Hebei University of Technology, Xiping Road, Beichen District, Hebei University of Technology Beichen Campus, Tianjin 300401, China; 2006126@hebut.edu.cn
* Correspondence: yflong0417@126.com

Abstract: Bio-oil as a substitute or modifier for petroleum asphalt has significant environmental, economic, and sustainable development implications. However, the insufficient high-temperature and anti-aging characteristics of bio-asphalt lead to the limited application of bio-oil in road engineering. In this study, the characteristics of bio-asphalt were reinforced by nano-Al_2O_3 and nano-SiO_2. The nanomaterial-modified bio-asphalt was evaluated using basic characteristics, viscosity, and high- and low-temperature rheological characteristic tests. Meanwhile, the road behavior of nanomaterial-modified bio-asphalt mixtures was assessed. The test results demonstrated that incorporating bio-oil improved the low-temperature performance of PA. Compared with PA, BA's ductility was increased by 83.87%. Nano-Al_2O_3 and nano-SiO_2 positively affected the bio-asphalt's temperature sensitivity and high-temperature and anti-aging characteristics but inhibited the low-temperature characteristics. Combining the results of the high- and low-temperature characteristics of nanomaterial-modified bio-asphalt, a nano-Al_2O_3 dosage between 4 and 6% or a nano-SiO_2 dosage between 6 and 8% can realize the enhancement of high-temperature and anti-aging characteristics of bio-asphalt, and the low-temperature characteristics are better than those of petroleum asphalt. Meanwhile, the dynamic stability, crack resistance, and water loss resistance of the nanomaterial-modified bio-asphalt mixtures meet the specification requirements. The adjustable and designable high- and low-temperature characteristics of bio-asphalt can be realized by adjusting the dosage of nanomaterials. The application of bio-oil and nanomaterials promotes the resourceful utilization of waste and provides new possibilities for road engineering materials.

Keywords: bio-asphalt; nano-Al_2O_3; nano-SiO_2; rheological characteristics; road performance

Citation: Qiao, W.; Yuan, F.; Xu, X.; Wang, Q. Research on the Properties of Nano-Al_2O_3- and Nano-SiO_2-Modified Bio-Asphalt and Mixtures. *Buildings* **2024**, *14*, 3366. https://doi.org/10.3390/buildings14113366

Academic Editor: Pengfei Liu

Received: 1 October 2024
Revised: 21 October 2024
Accepted: 22 October 2024
Published: 24 October 2024

1. Introduction

With quick economic development, the construction of highways is growing. Asphalt pavement has a pivotal position in highway construction [1]. As a non-renewable resource, petroleum asphalt (PA) is an essential raw material for asphalt pavement construction. However, with the continuous growth in energy costs and the gradual depletion of petroleum resources [2,3], the search for a renewable resource that can replace the PA material or a resource that can be reused repeatedly is imminent [4,5].

Recently, biomass energy sources (BESs) have received significant attention and research from experts due to their benefits, such as broad availability, easy access to materials, large reserves, recyclability, eco-friendliness, and low cost [6–8]. BESs mainly comprise straw, woodchips, grass clippings, discarded edible fats and grease, and gutter oils, which are renewable materials [9,10]. If BESs are substituted or blended into PA, the use of PA can be effectively reduced, and the dependence on petroleum resources and costs can be reduced [1,11]. Therefore, domestic and foreign research scholars have conducted a series of studies.

Wen et al. [12] used waste food oils to replace PA and blended them into PA at 0%, 10%, 30%, and 60%. The results demonstrated that incorporating bio-oil diminished the fatigue, anti-rutting, and stiffness of the asphalt mixture but boosted the resistance to thermal cracking. Williams et al. [13] chose oak and corn stover as biomass sources, obtained bio-oil through pyrolysis and fractionation, and mixed it with PA. The results demonstrated that incorporating bio-oil enhanced the anti-rutting behavior of PA but impaired the low-temperature behavior. You et al. [14] extracted bio-oil from swine manure, blended it with petroleum asphalt at 2%, 5%, and 10% and tested the properties. The results demonstrated that raising the dosage of bio-oil enhanced the low-temperature behavior of asphalt pavements. The 10% bio-oil decreased the cracking temperature by 4.6 to 4.9 °C. The greater the dosage of bio-oil, the more dramatic the reduction in the low-temperature cracking. Fini et al. [15,16] obtained bio-oil from swine manure with PA for property studies. The results demonstrated that bio-oil extracted from swine manure enhanced PA's low-temperature characteristics and viscosity but impaired the high-temperature anti-rutting characteristics. Yang et al. [17] examined the characteristics of PA modified with bio-oil generated from waste wood resources. The results demonstrated that the incorporation of bio-oil degraded the mixing temperature and enhanced the high-temperature characteristics of PA but negatively impacted the low-temperature characteristics. Silvia et al. [18] extracted bio-oils from bagasse, corn cobs, and rice husks to investigate the effect on PA characteristics. Sun [19] et al. incorporated gutter oil into PA. It was discovered that there was a good agreement between the doping amount of gutter oil and the low-temperature flexibility of PA. Numerous studies have demonstrated that the use of waste food oil, swine manure, waste wood resources, and other products to extract bio-oil can dramatically increase the low-temperature characteristics of PA, decrease the compaction temperature of asphalt mixtures, and effectively diminish the accumulation of waste and the emission of hazardous gases. However, it will weaken PA's high-temperature rutting and aging resistance characteristics. Therefore, the efficient and large-scale application of bio-oil is an effective way to minimize environmental pollution and economic costs.

To compensate for the high-temperature and anti-aging characteristics of bio-asphalt (BA), Yan et al. [20] utilized rock asphalt for modification. The results demonstrated that incorporating rock asphalt can boost BA's high-temperature characteristics and anti-shear strength. Zhang et al. [21] utilized a styrene–butadiene–styrene block copolymer (SBS) to improve the high-temperature characteristics of BA, and the results demonstrated that SBS-modified bio-asphalt not only has higher high-temperature characteristics, but its temperature sensitivity will also be degraded. Hou et al. [22] utilized silica fume for the performance enhancement of BA. The results also proved that silica fume can boost the high-temperature characteristics of BA. Although it diminishes the low-temperature characteristics, the characteristics are all better than those of PA. At the same time, scholars also used a variety of modifiers such as liquefied 4,4′-diphenylmethane diisocyanate (MDI), waste rubber powder, waste polyethylene, mineral powder filler, polyethylene (PE), and ethylene vinyl acetate (EVA) to enhance the high-temperature and aging-resistant characteristics of the BA, and all obtained better results [23–27].

Recently, nanomaterials have shown potential as asphalt modifiers with many great benefits in size, specific surface area, and strain resistance [28]. Nanomaterials can dramatically enhance asphalt's rutting, cracking, and aging resistance and improve the compatibility between asphalt and aggregates [29–32]. Fini et al. [33] found that nano-silica as an asphalt modifier can improve its rheological properties and oxidative aging resistance. Xu et al. [34] found that nano-zinc oxide was significant in enhancing the aging resistance of asphalt, while the rheological and fatigue properties were also enhanced. Bach Le, M.S. Eisa, and Xue et al. [35–37] modified the characteristics of asphalt and asphalt mixtures using carbon nanotubes (CNTs), which demonstrated that the CNTs exhibited a significant benefit in enhancing the anti-rutting characteristics of asphalt and asphalt mixtures. Golestani, Miglietta, and Ameri et al. [38–40] utilized nano-clay to boost the high-temperature behavior and fatigue resistance of asphalt mixtures. Crucho et al. [41] found that nano-SiO_2-modified asphalt mixtures have higher self-healing potential, resistance to permanent deformation, and fatigue durability. Faheem Sadiq Bhat et al. [28,42] added nano-Al_2O_3 to asphalt and asphalt mixtures, dramatically boosting the medium- and high-temperature characteristics. The fatigue resistance was increased by about 130%. CNTs, nano-clay, nano-SiO_2, and nano-Al_2O_3 have displayed great potential in improving the high-temperature behavior and fatigue durability of asphalt and asphalt mixtures. Abdulnaser et al. [43] modified palm oil asphalt with nano-SiO_2 and conducted rheological characterization. The results demonstrated that incorporating nanoparticles enhances the high-temperature behavior of BA. Liu [44] et al. investigated the characteristics of composite-modified asphalt with nano-SiO_2 and castor oil, and the results demonstrated that the composite modification of nano-SiO_2 with bio-oil has better overall characteristics. Bie et al. [45] used nano-SiO_2, nano-ZnO, and nano-TiO_2 to strengthen the high-temperature characteristics and ultraviolet (UV) aging resistance of BA, and the results demonstrated that the nanomaterials could effectively boost the UV aging resistance of BA to prolong the service life of asphalt pavements. Yadykova, AY et al. [46] demonstrated that the synergistic effect of montmorillonite clay and castor oil can improve the low-temperature cracking resistance and high-temperature permanent deformation resistance of asphalt. Nanomaterials perform well in enhancing the characteristics of BA. However, there is less research on nanomaterials for bio-oil-modified asphalt mixtures.

Therefore, in this study, for the poor high-temperature anti-rutting and anti-aging characteristics of BA and the unknown effect of nanomaterials on the behavior of bio-oil-modified asphalt mixtures, nano-Al_2O_3 and nano-SiO_2 were selected to compensate for the characteristics of bio-oil-modified asphalt and mixtures. The high- and low-temperature properties, temperature sensitivity, and anti-aging characteristics of the composite-modified asphalt were detected. Meanwhile, the high-temperature rutting, low-temperature cracking resistance, and water damage characteristics of nano-Al_2O_3 and nano-SiO_2 incorporated into bio-oil-modified asphalt mixtures were detected to assess the potential of nanomaterials in improving the road behavior of bio-oil-modified asphalt pavements.

2. Test Materials and Methods

2.1. Original Materials

2.1.1. Asphalt

The asphalt used in this study was 70# petroleum asphalt (PA) produced by Fuwei Chemical Co., Ltd. in Maoming City, China. The performance was estimated according to the JTG E20-2011 specification [47]. The detailed parameters are displayed in Table 1.

Table 1. Performance parameters for PA.

Parameters	Results	Requirement
Penetration (25 °C)/0.1 mm	69	60~80
Ductility (10 °C)/cm	31	≥15
Softening point/°C	58	≥45
60 °C dynamic viscosity/Pa·s	199	≥160
Residue after RTFOT		
Quality changes	−0.083	≤±0.8
Residual penetration ratio (25 °C)	81	≥61

2.1.2. Bio-Oil

The bio-oil used in this study was the bio-oil produced by Shandong Yi Neng Bioenergy Co., Ltd. (Binzhou, China), which was obtained from wood chips, straw, grain husk, and other plant-based waste biomass by a rapid thermal cracking process. The key parameters are illustrated in Table 2.

Table 2. Key technical indicators of bio-oil.

Parameters	Results
Density (15 °C)/(g/mL)	0.95
Viscosity (60 °C)/(Pa·s)	127
Acid value/(mgKOH/g)	51
Moisture content/%	≤0.3
Ignition point/°C	244
Flash point/°C	210
Fatty acid content/%	65

2.1.3. Nanomaterial

The nano-SiO_2 (NS) was from Shanghai Xiaohuang Nano Technology Co. (Shanghai, China). The nano-Al_2O_3 (NA) was from Anhui Kerun Nano Technology Co. (Lu'an, China). The basic characteristics of NS and NA are displayed in Table 3.

Table 3. Key indicators of nanomaterials.

Property	Test Results	
	NS	NA
Appearance	White powder	White powder
Purity	≥99.9	≥99.9
Average particle size/nm	30–50	30–50
Specific surface area/m^2/g	200–250	100–160

2.1.4. Aggregate

Limestone was used as the coarse and fine aggregates in this study. The JTG E42-2005 [48] specification was used for testing, and the technical parameters are presented in Tables 4 and 5, which meet the technical requirements of the specification.

Table 4. Technical parameters of coarse aggregates.

Property	Results	Requirement
Los Angeles abrasion test/%	25	≤28
Crushing value/%	11.3	≤28
Acicular and flaky grain in aggregate/%	10	≤15
Mud content/%	≤0.4	≤1

Table 5. Technical parameters of fine aggregates.

Property	Results	Requirement
Water absorption/%	0.6	≤2.0
Mud content/%	1.8	≤3
Sand equivalent/%	70	≥60

2.1.5. Gradation

This study used the mineral gradation of AC-13 for asphalt mixtures, and Table 6 presents the target gradation.

Table 6. Asphalt mixture gradation curve.

Sieve Size (mm)	16	13.2	9.5	4.75	2.36	1.18	0.6	0.3	0.15	0.075
Gradation upper limit	100	100	85	68	50	38	28	20	15	8
Lower gradation limit	100	90	68	38	24	15	10	7	5	4
Target grade	100	95	74	48.5	34	23.5	15	11	8.5	6

2.2. Preparation Process of Modified Asphalt

The dispersion homogeneity of nanomaterials in asphalt has a significant influence on the characteristics of the asphalt and asphalt mixtures. High-speed shear mixing technology has been shown to effectively achieve nanomaterial dispersion [40]. In this study, an FM300 high-speed shear mixer was used to prepare the bio-oil/nanomaterial composite-modified asphalt. The preparation procedure was as follows: 400 g of PA was heated in an oven at 135 °C for 1 h. Once it reached a flowable state, the PA was heated to 150 ± 5 °C, and 5% bio-oil (by mass of PA) was added. The bio-asphalt (BA) was obtained by mixing at 150 °C for 30 min using a high-speed shear mixer at 3000 r/min. Nanomaterials were then immediately added to the BA. Nanomaterial-modified bio-asphalt (NMBA) with different dosages was prepared by shearing at a rotational speed of 5000 r/min for 1 h at 150 ± 5 °C. The dosages of NS and NA were 2%, 4%, 6%, and 8% (by mass of PA). The abbreviations for NMBA are displayed in Table 7.

Table 7. The abbreviations for NMBA.

Nanomaterial	NMBA				
	Dosage of Nanomaterials/%				
	0	2	4	6	8
NA	BA	Bio + 2%NA	Bio + 4%NA	Bio + 6%NA	Bio + 8%NA
NS		Bio + 2%NS	Bio + 4%NS	Bio + 6%NS	Bio + 8%NS

2.3. Test Methods

2.3.1. Basic Characteristics Test

According to the specifications of AASHTO T 49 [49], AASHTO T 53 [50], and AASHTO T 51 [51], NMBA with different dosages was tested for penetration, softening point, and ductility. The test temperature for penetration was 25 °C, and the test temperature for ductility was 15 °C.

2.3.2. Aging Test

The Rotary Film Oven Test (RTFOT) and the Pressure Aging Test (PAV) were performed on PA, BA, and NMBA according to the specifications of ASTM D 2872 [52] and ASTM D 6521 [53].

2.3.3. Brookfield Rotational Viscosity Test (BRV)

Viscosity tests were performed on unaged PA, BA, and NMBA using a Brookfield viscometer according to AASHTO T 316 [54]. The test was conducted using rotor No. 27. The test temperatures were 135 °C, 165 °C, and 175 °C. The shear rate at 135 °C was 6.8 s^{-1}. At 165 °C and 175 °C, the shear rate was 17.0 s^{-1}. Torque readings were required to be in the range of 10–98%. After the viscosity value stabilized, measurements were taken three times at 60 s intervals, and the average was recorded as the viscosity value. Two parallel tests were conducted for each asphalt sample.

2.3.4. Temperature Sweep (TS) Tests

In accordance with the AASHTO T 315-20 specification [55], a TS test was conducted using a dynamic shear rheometer. The temperature range was from 46 °C to 82 °C with a temperature step of 6 °C. The angular frequency of the test was 10 rad/s. The strain value was 12% for unaged asphalt and 10% for short-term-aged asphalt. Two parallel tests were conducted for each type of asphalt.

2.3.5. Multiple Stress Creep Recovery (MSCR) Tests

In accordance with the AASHTO T 315-20 specification [55], MSCR tests were conducted on short-term-aged PA, BA, and NMBA using a dynamic shear rheometer. The test temperature was 58 °C. Two parallel tests were conducted for each asphalt.

2.3.6. Bending Beam Rheology (BBR) Test

The low-temperature performances of long-term-aged PA, BA, and NMBA were measured according to the AASHTO T 313-19 specification [56]. The dimensions of the test specimen were 127 mm × 6.35 mm × 12.7 mm. The test temperature was −12 °C. The applied contact load was 35 ± 10 mN, and the test load was maintained at 980 ± 50 mN. The stiffness modulus (S) and creep rate (m) were read separately at a time of 60 s. Two parallel tests were conducted for each asphalt.

2.3.7. Asphalt Mixture Road Performance Test

According to the JTG E20-2011 specification [47], the appropriate mixing and compaction temperatures were selected based on the BRV test. The Marshall test method determined that the optimum oil–rock ratio was 5.0%. The high-temperature rutting, low-temperature bending, and freeze–thaw splitting tests were conducted on the nanomaterial-modified bio-asphalt mixtures (NMBAMs), respectively. The tests obtained three indicators of dynamic stability (DS), failure strain (FS), and freeze–thaw split strength ratio (TSR), respectively. Three parallel tests were conducted for each asphalt mixture.

3. Test Results and Discussion

3.1. Basic Characteristic Analysis

Figure 1 exhibits the results of the penetration, ductility, and softening point tests of NMBA.

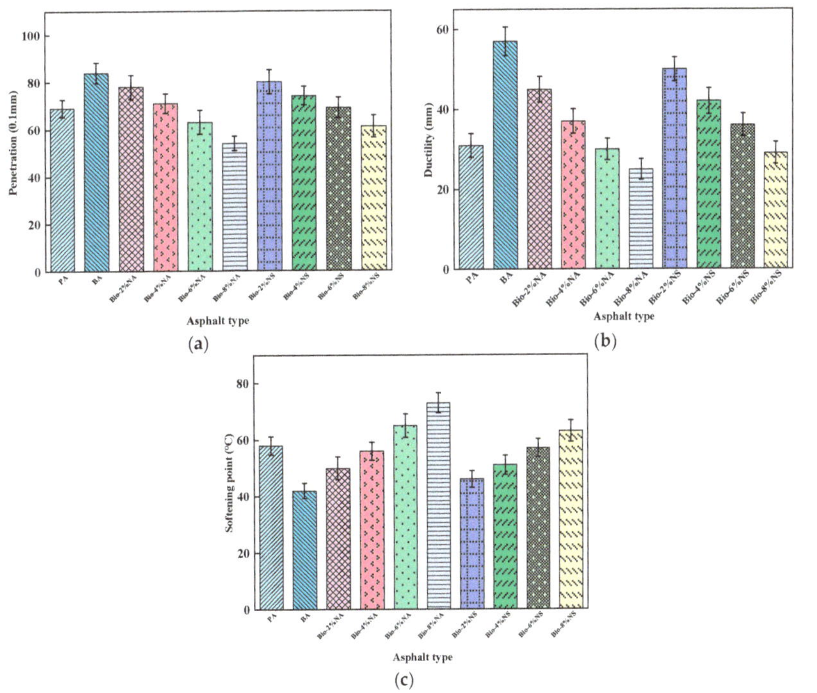

Figure 1. Basic characteristic results of NMBA. (**a**) Penetration results; (**b**) ductility results; (**c**) softening point results.

The incorporation of bio-oil increased the penetration and ductility and decreased the softening point of PA, as presented in Figure 1. Compared with PA, the penetration and ductility of BA increased by 21.74% and 83.87%, respectively, and the softening point decreased by 27.59%. This may be because bio-oil supplements the light components of the asphalt, reduces the viscosity, and makes asphalt prone to flow and deformation, which can better generate deformation to resist the damage caused by stresses. After the incorporation of NA and NS into BA, the penetration and ductility of NMBA decreased continuously with the growth in nanomaterial content, and the softening point exhibited an opposite trend. When the dosage of NA was 2–4% or that of NS was 2–6%, the ductility of NMBA was superior to that of PA.

The incorporated nanomaterials may alter the intermolecular forces of BA, leading to an enhanced brittleness of BA at low temperatures, causing the ductility of BA to diminish with increasing nanomaterial dosage. Compared with NS, NA-modified bio-asphalt had a higher softening point. And the penetration and ductility were less. Compared with Bio + 6% NS, the softening point of Bio + 6% NA increased by 12.31%. The penetration and ductility declined by 9.52% and 20.00%. It demonstrated that NA can remarkably promote the high-temperature characteristics of BA.

3.2. BRV Analysis

The viscosity of asphalt can remarkably assess the construction and workability of asphalt pavement [22]. Figure 2 illustrates the BRV values of NMBA at different temperatures.

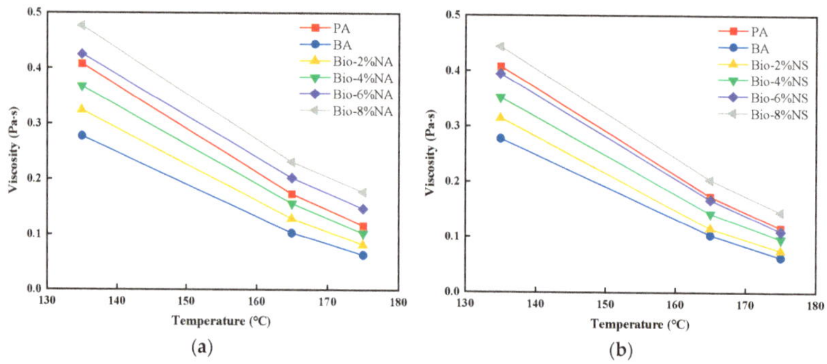

Figure 2. The BRV values of NMBA. (**a**) NA; (**b**) NS.

As depicted in Figure 2, the BRV values of BA, BA, and NMBA gradually diminished with the temperature rise. The rise in temperature leads to the intensification of thermal movement within the asphalt molecules; the relative motion between asphalt molecules is accelerated, and the intermolecular interaction force is diminished, leading to the enhancement of asphalt mobility and a decrease in viscosity. The incorporation of bio-oil decreased the BRV values of PA. When the temperature was 135 °C, the BRV of BA decreased by 31.94% compared with that of PA. This is mainly because the bio-oil supplemented the light components in the asphalt and added fluidity to the asphalt, resulting in the reduced viscosity of PA. The BRV value of BA gradually increased after incorporating nanomaterials into BA. At 135 °C, compared with BA, the BRV values of Bio-8%NA, Bio-6%NA, Bio-4%NA, and Bio-2%NA were boosted by 16.61%, 32.49%, 53.43%, and 71.84%, respectively. The BRV values of Bio-8%NS, Bio-6%NS, Bio-4%NS, and Bio-2%NS were boosted by 13.36%, 26.71%, 42.24%, and 59.93%. The nanomaterials gradually increased BA's BRV values, and BA's deformation resistance was continuously improved. The viscosity of NMBA was less than that of PA when NA was 2–4% or NS was 2–8%.

As illustrated in Figure 2, asphalt is susceptible to temperature. The viscosity–temperature curve is fitted to determine the sensitivity of asphalt to temperature, which is calculated as presented in Formula (1) [57].

$$log(log(\eta \times 10^3)) = n - mlg(T + 273.13) \quad (1)$$

where m is called the viscosity–temperature index; the larger the value, the more the viscosity of the asphalt is affected by temperature and the poorer the temperature stability. Figure 3 exhibits the results of viscosity–temperature curve fitting for NMBA. The fitted parameters are displayed in Table 8.

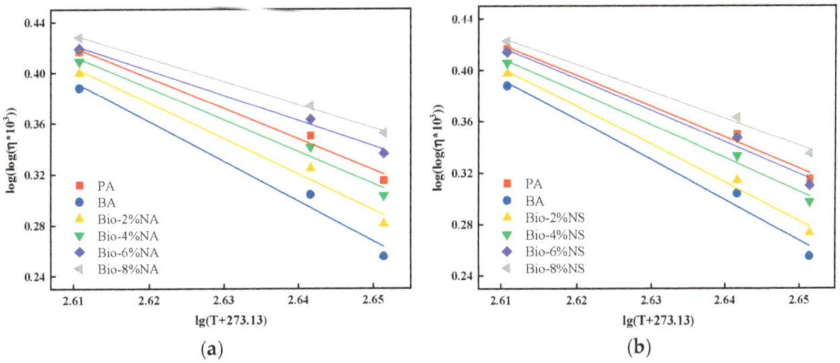

Figure 3. The results of viscosity–temperature curve fitting for NMBA. (**a**) NA; (**b**) NS.

Table 8. Viscosity–temperature curve fitting parameters.

Types of Asphalt	m	n	R^2
PA	2.424	6.748	0.987
BA	3.138	8.584	0.981
Bio + 2%NA	2.810	7.737	0.981
Bio + 4%NA	2.516	6.980	0.983
Bio + 6%NA	1.993	5.625	0.992
Bio + 8%NA	1.844	5.242	0.998
Bio + 2%NS	2.978	8.173	0.997
Bio + 4%NS	2.600	7.194	0.991
Bio + 6%NS	2.476	6.882	0.985
Bio + 8%NS	2.109	5.931	0.992

As indicated in Figure 3 and Table 8, the m-value of BA increased by 29.45% compared with that of PA. The addition of bio-oil degraded the temperature stability of PA. The incorporation of nanomaterials significantly degraded the m-value of BA. The greater the nanomaterial incorporation, the lesser the viscosity of NMBA was influenced by temperature. Compared with BA, the m-values of Bio-8%NA, Bio-6%NA, Bio-4%NA, and Bio-2%NA decreased by 10.45%, 19.82%, 36.49%, and 41.24%, respectively. The m-values of Bio-8%NS, Bio-6%NS, Bio-4%NS, and Bio-2%NS decreased by 5.10%, 17.14%, 21.10%, and 32.79%, respectively. The viscosity–temperature index of NMBA was less than that of PA when the dosage of NA was 6–8% or when the dosage of NS was more than 6%. The temperature stability of BA was effectively enhanced.

3.3. TS Result Analysis

The rutting factor (G*/sinδ) is an evaluation indicator of asphalt's high-temperature characteristics. The larger its value, the stronger the asphalt's ability to resist permanent deformation at high temperature [58]. Figures 4 and 5 present the G*/sinδ test results of NMBA in its original state and after short-term aging, respectively.

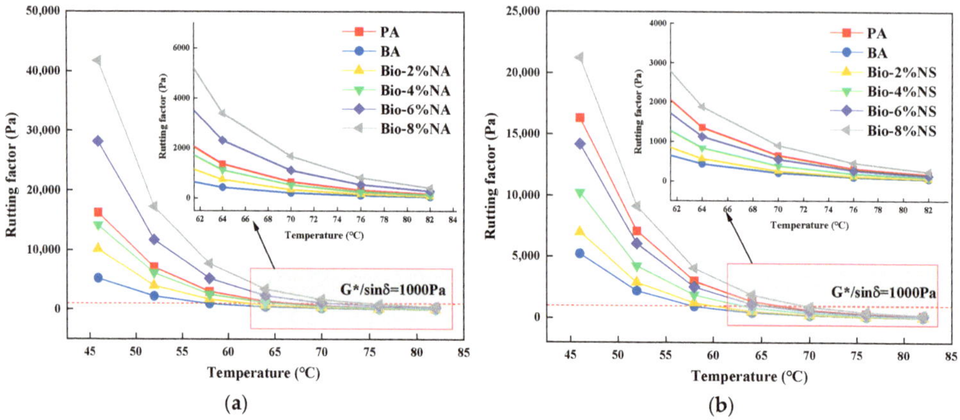

Figure 4. Rutting factor of unaged NMBA. (**a**) NA; (**b**) NS.

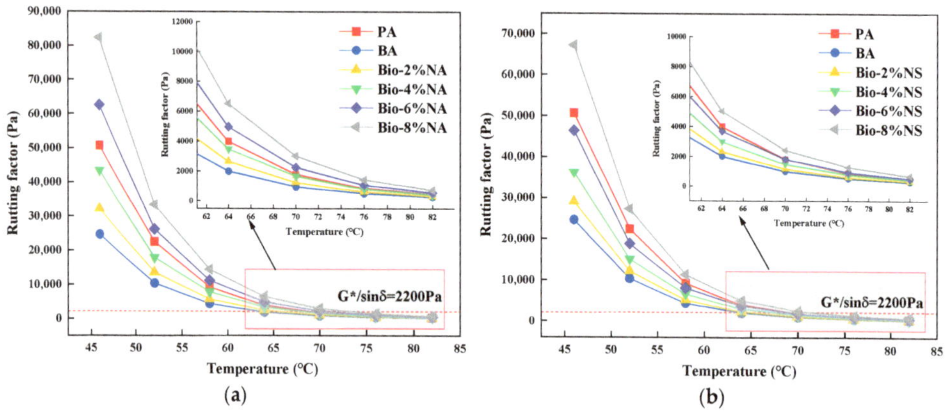

Figure 5. Rutting factor of short-term-aging NMBA. (**a**) NA; (**b**) NS.

As illustrated in Figure 4, with the rise in temperature, the G*/sinδ of asphalt diminished gradually. The incorporation of bio-oil markedly degraded the G*/sinδ of petroleum asphalt. At 64 °C, the G*/sinδ of PA decreased by 68.47%. The incorporation of nanomaterials remarkably enhanced the G*/sinδ value of BA. At 64 °C, compared with the G*/sinδ values of BA, the G*/sinδ of Bio-8%NA, Bio-6%NA, Bio-4%NA, and Bio-2%NA were boosted by 74.46%, 160.56%, 435.09%, and 689.45%, respectively. The G*/sinδ of Bio-8%NS, Bio-6%NS, Bio-4%NS, and Bio-2%NS were boosted by 32.58%, 94.79%, 160.99%, and 335.82%, respectively. At constant temperature, the G*/sinδ of NMBA was ranked as Bio-8%NA > Bio-8%NS > Bio-6%NA > PA > Bio-6%NS > Bio-4%NA > Bio-4%NS > Bio-2%NA > Bio-2%NS > BA. When the NA content exceeded 4% or the NS exceeded 6%, the high-temperature characteristics of NMBA were superior to those of PA. As presented in Figure 5, the G*/sinδ value of NMBA increased remarkably after experiencing short-term aging, which was mainly due to the loss of the light components of asphalt during aging, resulting in an increase in the heavy components of asphalt, which enhanced the deformation resistance of NMBA under high temperature.

The G*/sinδ value is a parameter for the high-temperature PG grading of asphalt. The G*/sinδ value of unaged asphalt should be greater than 1.0 kPa, and the G*/sinδ value of short-term-aged asphalt should be greater than 2.2 kPa. The high-temperature PG grading was carried out based on the G*/sinδ value of NMBA in Figures 4 and 5, and the results are displayed in Table 9.

Table 9. The high-temperature PG grading of asphalt.

Types of Asphalt	PG Grading
PA	PG64
BA	PG52
Bio-2%NA	PG58
Bio-4%NA	PG64
Bio-6%NA	PG70
Bio-8%NA	PG70
Bio-2%NS	PG58
Bio-4%NS	PG58
Bio-6%NS	PG64
Bio-8%NS	PG64

As illustrated in Table 9, nanomaterials can effectively promote the high-temperature capability of BA and make up for the shortcomings of the lack of high-temperature capability of BA. Among them, NA has a remarkably reinforcing effect on the high-temperature characteristics of BA.

3.4. Anti-Aging Characteristic Analysis

The rutting factor aging index (RFAI) is a crucial index for assessing the anti-aging capability of asphalt, which is calculated as in Formula (2). The smaller the value of RFAI is, the better the aging resistance of asphalt is [22]. Figure 6 displays the RFAI values of NMBA.

$$RFAI = \frac{G^*/\sin\delta_{Short-term\ aging}}{G^*/\sin\delta_{unaged}} \tag{2}$$

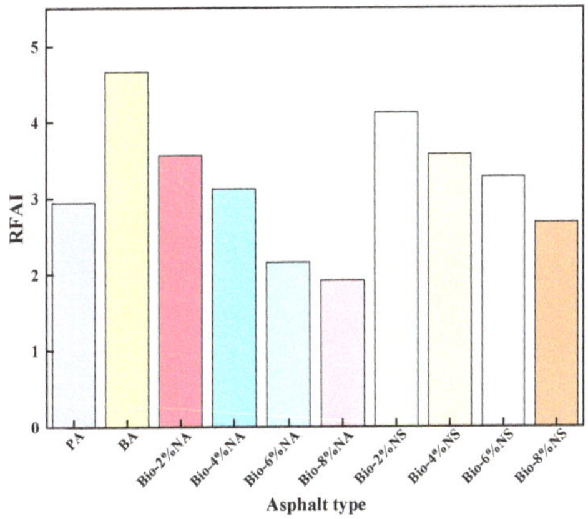

Figure 6. RFAI value of NMBA.

As depicted in Figure 6, the RFAI value of BA rose by 58.51% compared with that of PA, and the incorporation of bio-oil diminished the aging resistance of the asphalt. The RFAI value of BA declined continuously after the blending of nanomaterials. Compared with BA, the RFAI values of Bio-8%NA, Bio-6%NA, Bio-4%NA, and Bio-2%NA declined by 23.57%, 33.07%, 53.67%, and 58.82%, respectively. The RFAI values of Bio-8%NS, Bio-6%NS, Bio-4%NS, and Bio-2%NS declined by 11.58%, 23.35%, 29.71%, and 42.59%, respectively. Nanomaterials can improve the anti-aging properties of asphalt. The aging resistance of NMBA was superior to that of PA at a dosage of greater than 4% of NA or greater than 6% of NS.

3.5. MSCR Result Analysis

The recovery rate (R) and the non-recoverable creep compliance (Jnr) are critical parameters in MSCR testing. The R-value assesses the elastic recovery ability of asphalt. The bigger the R-value, the faster the asphalt can recover its characteristics after stress unloading. The Jnr value characterizes the non-recoverable strain of asphalt. The lesser the Jnr value is, the stronger the asphalt's high-temperature stability is [21,26]. Figures 7 and 8 present the R- and Jnr values of NMBA under different stress conditions.

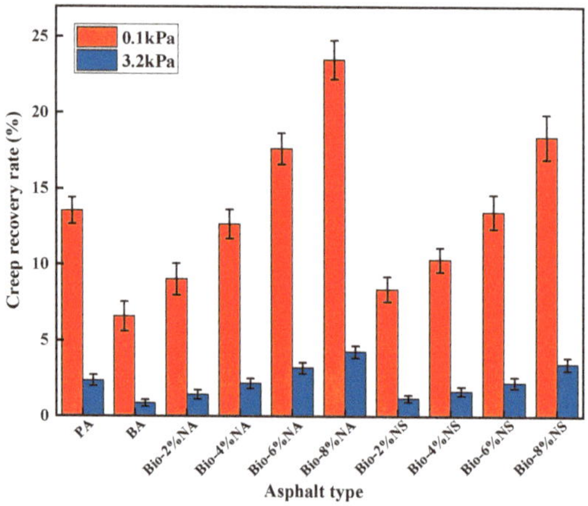

Figure 7. R-value of NMBA.

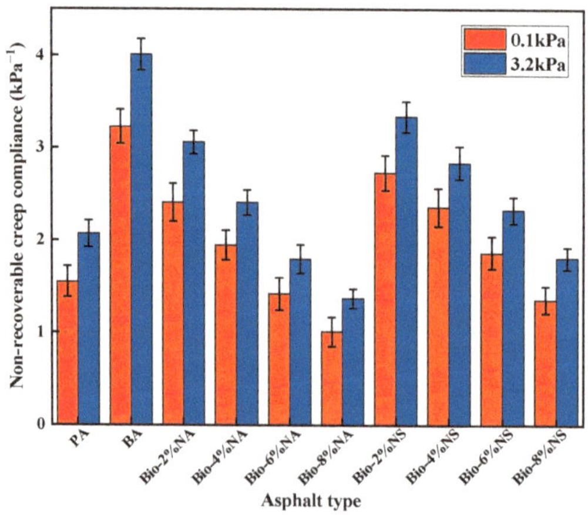

Figure 8. Jnr value of NMBA.

As depicted in Figure 7, the greater the stress, the lesser the deformation recovery ability of the asphalt. The incorporation of bio-oil dramatically decimated the R-value of PA at the same stress level. At the 0.1 kPa level, the R-value of BA diminished by 51.40% compared with that of PA, demonstrating that bio-oil incorporation remarkably diminished the elastic recovery ability of asphalt under high temperatures. This is mainly due to bio-oil supplementing the light components of asphalt, resulting in a diminishing elastic component of the asphalt. Incorporating nanomaterials into the BA resulted in a gradual rise in the R-value of the BA. At 0.1 kPa, the R-values of Bio-8%NA, Bio-6%NA, Bio-4%NA, and Bio-2%NA were promoted by 37.39%, 92.31%, 168.08%, and 256.84%, respectively. The R-values of Bio-8%NS, Bio-6%NS, Bio-4% NS, and Bio-2%NS were boosted by 27.51%, 56.90%, 104.71%, and 179.88%, respectively. The nanomaterials remarkably improved the deformation recovery ability of BA under high temperatures. The R-value of NMBA was greater than PA's when the NA content exceeded 4% or the NS exceeded 6%, which was mainly attributed to the nanomaterials' surface and volume effects.

As depicted in Figure 8, the Jnr value of NMBA rose with the stress level. The bigger the stress, the worse the rutting resistance characteristic of asphalt under a high-temperature environment. At 0.1 kPa, the Jnr value of BA was raised by 108.39% compared with that of PA, and the non-resistant deformation of BA was dramatically cut down. The introduction of nanomaterials in BA can remarkably diminish the Jnr value. When compared with BA, the Jnr values of Bio-8%NA, Bio-6%NA, Bio-4%NA, and Bio-2%NA declined by 25.39%, 39.63%, 56.04%, and 68.73%, respectively. The Jnr values of Bio-8%NS, Bio-6%NS, Bio-4%NS, and Bio-2%NS were raised by 15.48%, 26.93%, 42.11%, and 57.89%. The Jnr value of NMBA was greater than that of PA when the NA content was greater than 4% or the NS was greater than 6%.

The comprehensive G*/sinδ, R-, and Jnr values revealed that incorporating bio-oil degraded the high-temperature characteristics of PA. Still, incorporating nanomaterials can compensate for the shortcomings of BA's lack of high-temperature capability and aging endurance to achieve the characteristics of BA reinforcement. When the NA content was more than 4% or the NS was more than 6%, the high-temperature capability and aging resistance of NMBA were better than PA's.

3.6. BBR Result Analysis

S characterizes the capacity of asphalt to withstand loading at low temperatures. m represents the rate of change of asphalt stiffness over time. The specification specifies that the value of S should be less than 300 MPa, and the value of m should be more than 0.3. The lesser the value of S and the larger the value of m, the stronger the low-temperature characteristics of the asphalt [26,57]. Figures 9 and 10 display the results of the S- and m-values for NMBA.

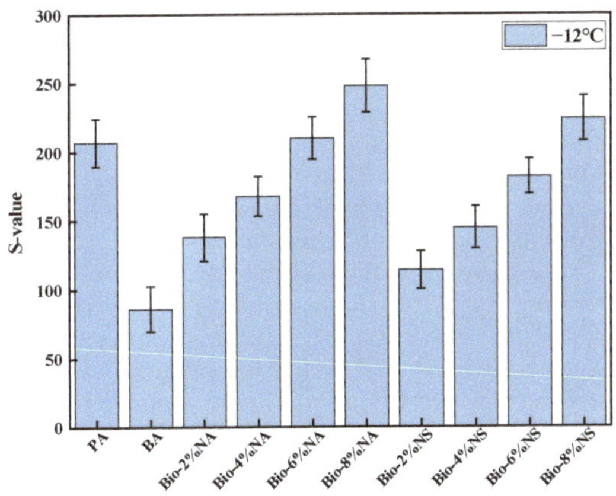

Figure 9. S-value of NMBA.

As depicted in Figure 9, after the incorporation of bio-oil into PA, the S-value of BA decreased by 58.45%. The anti-loading capacity of BA was dramatically strengthened at low temperatures. The incorporation of bio-oil supplements the lightweight component, which makes the asphalt more susceptible to deformation when subjected to external stresses, thus diminishing the stiffness of the asphalt. The S-value of BA gradually rose after the nanomaterials were incorporated into BA. At −12 °C, the S-values of NMBA all met the specification requirements. Compared with BA, the S-values of Bio-8%NA, Bio-6%NA, Bio-4%NA, and Bio-2%NA were boosted by 60.46%, 95.35%, 144.19%, and 188.37%, respectively. The S-values of Bio-8%NS, Bio-6%NS, Bio-4%NS, and Bio-2%NS were boosted by 32.56%,

68.60%, 111.63%, and 160.46%, respectively. The S-value of NMBA was less than that of PA when the content of NA was less than 6% or when NS was less than 8%. Nanomaterials with a large specific surface area were prone to absorbing the lightweight components of the BA, thereby increasing the hardness of the BA, which led to an increase in the brittleness of the BA at low temperatures and poorer resistance to loading.

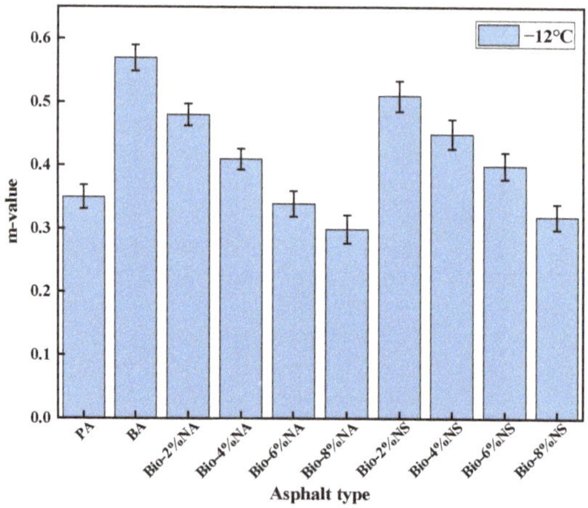

Figure 10. m-value of NMBA.

As indicated in Figure 10, the incorporation of 5% bio-oil in PA resulted in a significant rise in the m-value of PA, which increased by 62.86%, indicating that bio-oil accelerated the rate of change of PA stiffness with time. The incorporation of nanomaterials diminished the m-value of BA. At −12 °C, the m-values of NMBA met the specification requirements. Compared with BA, the m-values of Bio-8%NA, Bio-6%NA, Bio-4%NA, and Bio-2%NA decreased by 15.79%, 28.07%, 40.35%, and 47.37%, respectively. The m-values of Bio-8%NS, Bio-6%NS, Bio-4%NS, and Bio-2%NS declined by 10.53%, 21.05%, 29.82%, and 43.86%, respectively. When the content of NA was less than 6% or NS was less than 8%, the m-value of NMBA was greater than that of PA. The combined S-value and m-value results indicated that the low-temperature capability of NMBA was superior to that of PA when the dosage of NA was less than 6% or NS was less than 8%.

Comprehensive analysis of the basic characteristics, viscosity, and high- and low-temperature capability of NMBA showed that an NA dosage between 4 and 6% or an NS dosage between 6 and 8% can be realized to promote the high-temperature capability and aging resistance of BA, and the low-temperature characteristics are better than those of PA. At the same time, according to the actual requirements of pavement materials, the dosage of nanomaterials can be adjusted to achieve adjustable and designable characteristics of asphalt materials.

3.7. Road Behavior Analysis

The results of the dynamic stability, crack resistance, and water stability tests of the NMBAMs are displayed in Figures 11–13.

As illustrated in Figure 11, the DS of the petroleum asphalt mixture (PAM) was 2346 times/mm. The result of the DS of bio-asphalt mixture (BAM) was only 1637 times/mm. The incorporation of bio-oil degraded the high-temperature capability of the PAM. After the incorporation of nanomaterials, the DS of the NMBAMs improved continuously with the increase in the dosage so that the BAM possessed better high-temperature characteristics. The rutting test was consistent with the asphalt softening point and high-temperature rheology test results. When the content of NA was more than 4% or NS was more than 6%, the high-temperature capability of the NMBAMs was better than that of the PAM.

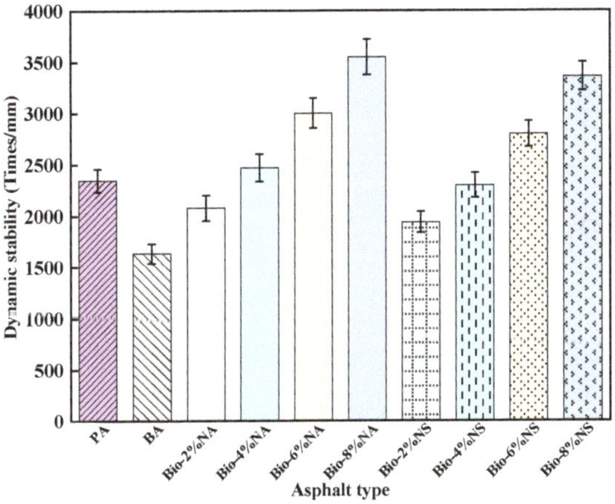

Figure 11. DS results of NMBAMs.

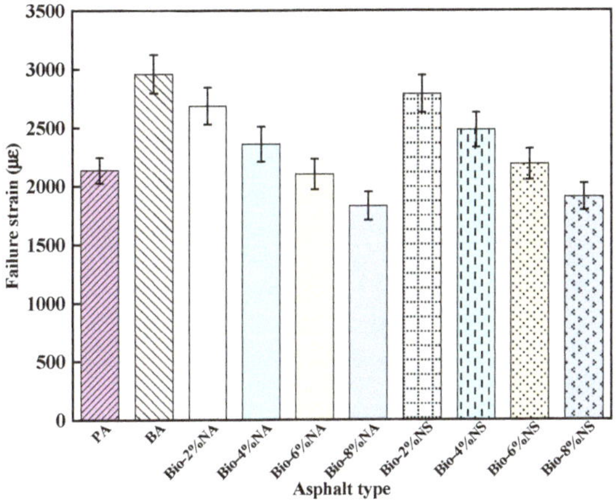

Figure 12. FS results of NMBAMs.

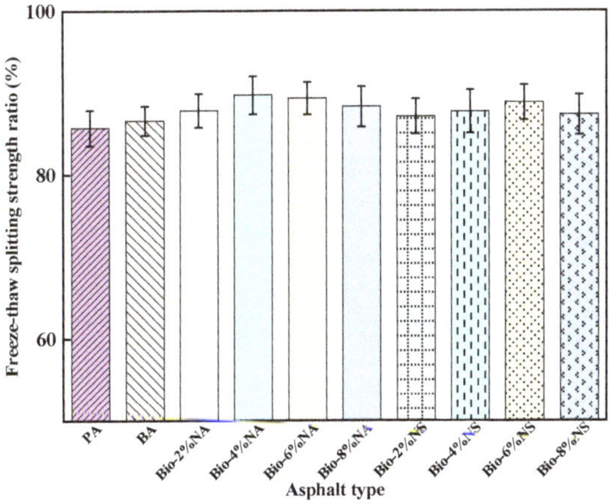

Figure 13. TSR results of NMBAM.

As indicated in Figure 12, the FS of PAM was 2134 με. Compared with PAM, the FS of BAM was boosted by 38.52%. The low-temperature characteristics of BAM were remarkably strengthened. The FS of NMBAM exhibited a decreasing trend with the rise in the dosage after the nanomaterials were incorporated. When the NA content was less than 6% or NS was less than 8%, the FS of the NMBAMs was better than the PAM's. This indicates that, although the nanomaterials will cut down the low-temperature cracking resistance of the BAM, the low-temperature capability of the NMBAMs is superior to that of the PAM in the reasonable dosage range.

As illustrated in Figure 13, the TSR of PAM was 85.7%. Compared with PAM, the TSR value of BAM increased by 1.05%. After incorporating nanomaterials, the TSR of the NMBAMs tended to increase and then decrease. The small dosage of nanomaterials incorporated into the BA will raise the adhesion and bonding between bio-oil and asphalt so that the NMBA forms a stable network structure, which boosts the water damage resistance of the NMBAMs. However, when the dosage of nanomaterials is too large, the nanomaterials are prone to aggregation, resulting in poor dispersion in asphalt, which diminishes the water loss resistance of the NMBAMs.

4. Conclusions

A partial or total replacement of PA by bio-oil is a practical approach to save non-renewable resources. Still, the insufficient high-temperature and anti-aging characteristics of BA lead to limitations in the broad application of bio-oil. In this study, the characteristics of BA were reinforced by nano-SiO_2 and nano-Al_2O_3, respectively. By evaluating the performance of nanomaterial-modified bio-asphalt and mixtures under different nanomaterial dosages, the following conclusions are drawn:

(1) Incorporating bio-oil reduced the temperature sensitivity, high-temperature characteristics, and anti-aging characteristics of PA, but the low-temperature cracking resistance was dramatically strengthened.
(2) The incorporation of NA and NS improved the temperature sensitivity, high-temperature characteristics, and anti-aging capability of BA. At −12 °C, the S and m indexes display that incorporating NA and NS will negatively affect the low-temperature capability of BA.
(3) Combining the results of the high- and low-temperature capability of NMBA, an NA content between 4 and 6% or an NS content between 6 and 8% can realize the enhancement of BA's high-temperature and anti-aging capability, and the low-temperature capability is better than that of PA. Through the adjustment of the content of nanomaterials, the performance of BA can be adjustable and designable.
(4) When the NA dosage was 4–6% or the NS dosage was 6–8%, the dynamic stability, cracking resistance, and water loss resistance of the NMBAMs met the specification requirements.

The experimental results of nanomaterial-modified bio-asphalt and mixtures provide a theoretical basis for applying biomass resources in road engineering. However, only one type of bio-oil was considered in this study. Future research will explore the effects of nanomaterials on different types and dosages of bio-oil, asphalt components, and micro-mechanisms.

Author Contributions: W.Q.: Conceptualization, Resources, Writing—original draft, Data curation, Writing—reviewing and editing. F.Y.: Software, Validation, Writing—Reviewing and Editing. X.X.: Visualization, Investigation, Formal analysis, Writing—reviewing and editing. Q.W.: Methodology, Conceptualization, Validation, Formal analysis. All authors have read and agreed to the published version of the manuscript.

Funding: This research received no external funding.

Data Availability Statement: Some or all data, models, or codes generated or used during this study are available from the corresponding author by request.

Conflicts of Interest: Wei Qiao, Feilong Yuan and Xiaojian Xu were employed by the CHECSC Highway Maintenance and Test Technology Co., Ltd. The remaining authors declare that the research was conducted in the absence of any commercial or financial relationships that could be construed as a potential conflict of interest.

References

1. Zhang, Z.; Fang, Y.; Yang, J.; Li, X. A comprehensive review of bio-oil, bio-binder and bio-asphalt materials: Their source, composition, preparation and performance. *J. Traffic Transp. Eng. (Engl. Ed.)* **2022**, *9*, 151–166. [CrossRef]
2. Al-Sabaeei, A.M.; Napiah, M.B.; Sutanto, M.H.; Alaloul, W.S.; Usman, A. A systematic review of bio-asphalt for flexible pavement applications: Coherent taxonomy, motivations, challenges and future directions. *J. Clean. Prod.* **2020**, *249*, 119357. [CrossRef]
3. Dong, R.; Zhao, M.; Tang, N. Characterization of crumb tire rubber lightly pyrolyzed in waste cooking oil and the properties of its modified bitumen. *Constr. Build. Mater.* **2019**, *195*, 10–18. [CrossRef]
4. Mills-Beale, J.; You, Z.; Fini, E.; Zada, B.; Lee, C.H.; Yap, Y.K. Aging Influence on Rheology Properties of Petroleum-Based Asphalt Modified with Biobinder. *J. Mater. Civ. Eng.* **2014**, *26*, 358–366. [CrossRef]
5. Wang, H.; Ma, Z.; Chen, X.; Hasan, M.R.M. Preparation process of bio-oil and bio-asphalt, their performance, and the application of bio-asphalt: A comprehensive review. *J. Traffic Transp. Eng. (Engl. Ed.)* **2020**, *7*, 137–151.
6. Bridgwater, A.V. Review of fast pyrolysis of biomass and product upgrading. *Biomass Bioenergy* **2012**, *38*, 68–94. [CrossRef]
7. Yadykova, A.Y.; Ilyin, S.O. Bitumen improvement with bio-oil and natural or organomodified montmorillonite: Structure, rheology, and adhesion of composite asphalt binders. *Constr. Build. Mater.* **2023**, *364*, 129919. [CrossRef]
8. Zhang, R.; Shi, Q.; Hu, P.; Ji, J.; Suo, Z. Influence of castor oil-based bio-oil on the properties and microstructure of asphalt binder. *Constr. Build. Mater.* **2023**, *408*, 133564. [CrossRef]
9. Raouf, M.A.; Williams, C.R. General Rheological Properties of Fractionated Switchgrass Bio-Oil as a Pavement Material. *Road Mater. Pavement Des.* **2010**, *11* (Suppl. S1), 325–353. [CrossRef]
10. Alessio, M.; Giulia, T.; Piergiorgio, T.; Ernesto, S.; Cesare, S. Use of bio-based products towards more sustainable road paving binders: A state-of-the-art review. *J. Road Eng.* **2024**, *4*, 151.
11. Xue, Z.; Hui, L.; Harvey, J.T.; Butt, A.A.; Ming, J.; Jiawen, L. A review of converting woody biomass waste into useful and eco-friendly road materials. *Transp. Saf. Environ.* **2022**, *4*, tdab031.
12. Wen, H.; Bhusal, S.; Wen, B. Laboratory Evaluation of Waste Cooking Oil-Based Bioasphalt as an Alternative Binder for Hot Mix Asphalt. *J. Mater. Civ. Eng.* **2013**, *25*, 1432–1437. [CrossRef]
13. Williams, R.C.; Satrio, J.; Rover, M.; Brown, R.C.; Teng, S. Utilization of Fractionated Bio-Oil in Asphalt. In Proceedings of the 88th Annual Meeting of the Transportation Research Board, Washington, DC, USA, 11–15 January 2009.
14. You, Z.; Mills-Beale, J.; Fini, E.; Goh, S.W.; Colbert, B. Evaluation of Low-Temperature Binder Properties of Warm-Mix Asphalt, Extracted and Recovered RAP and RAS, and Bioasphalt. *J. Mater. Civ. Eng.* **2011**, *23*, 1569–1574. [CrossRef]
15. Fini, E.; Yang, S.H.; Xiu, S. Characterization and application of manure-based bio-binder in asphalt industry. In Proceedings of the Transportation Research Board 89th Annual Meeting, Washington DC, USA, 10–14 January 2010.
16. Fini, E.H.; Kalberer, E.W.; Shahbazi, A.; Basti, M.; You, Z.; Ozer, H.; Aurangzeb, Q. Chemical characterization of biobinder from swine manure: Sustainable modifier for asphalt binder. *J. Mater. Civ. Eng.* **2011**, *23*, 1506–1513. [CrossRef]
17. Yang, X.; You, Z.-P.; Dai, Q.-L. Performance Evaluation of Asphalt Binder Modified by Bio-oil Generated from Waste Wood Resources. *Int. J. Pavement Res. Technol.* **2013**, *6*, 431–439.
18. Caro, S.; Vega, N.; Husserl, J.; Alvarez, A.E. Studying the impact of biomodifiers produced from agroindustrial wastes on asphalt binders. *Constr. Build. Mater.* **2016**, *126*, 369–380. [CrossRef]
19. Sun, Z.; Yi, J.; Huang, Y.; Feng, D.; Guo, C. Properties of asphalt binder modified by bio-oil derived from waste cooking oil. *Constr. Build. Mater.* **2016**, *102*, 496–504. [CrossRef]
20. Yan, K.; Zhang, M.; You, L.; Wu, S.; Ji, H. Performance and optimization of castor beans-based bio-asphalt and European rock-asphalt modified asphalt binder. *Constr. Build. Mater.* **2020**, *240*, 117951. [CrossRef]
21. Zhang, R.; Wang, H.; Gao, J.; You, Z.; Yang, X. High temperature performance of SBS modified bio-asphalt. *Constr. Build. Mater.* **2017**, *144*, 99–105. [CrossRef]
22. Hou, G.; Xue, Y.; Li, Z.; Lu, W. Rheological Properties of Silica-Fume-Modified Bioasphalt and Road Performance of Mixtures. *Materials* **2024**, *17*, 2090. [CrossRef]
23. Zheng, W.; Wang, H.; You, Z.; Shao, L.; Golroo, A.; Chen, Y. Mechanism and rheological characterization of MDI modified wood-based bio-oil asphalt. *Constr. Build. Mater.* **2021**, *309*, 125113. [CrossRef]
24. Chen, Y.; Ji, C.; Wang, H.; Su, Y. Evaluation of crumb rubber modification and short-term aging on the rutting performance of bioasphalt. *Constr. Build. Mater.* **2018**, *193*, 467–473. [CrossRef]
25. Fernandes, S.R.M.; Silva, H.M.R.D.; Oliveira, J.R.M. Developing enhanced modified bitumens with waste engine oil products combined with polymers. *Constr. Build. Mater.* **2018**, *160*, 714–724. [CrossRef]
26. Lei, Y.; Wang, H.; Chen, X.; Yang, X.; You, Z.; Dong, S.; Gao, J. Shear property, high-temperature rheological performance and low-temperature flexibility of asphalt mastics modified with bio-oil. *Constr. Build. Mater.* **2018**, *174*, 30–37. [CrossRef]
27. Xiaocui, Z.; Guangyuan, Z.; Jiasheng, G.; Jiaolong, R. Study on preparation and pavement properties of high performance modified bio-asphalt binder. *J. Shandong Univ. Technol. (Nat. Sci. Ed.)* **2020**, *34*, 28–32.

28. Bhat, F.S.; Gilani, T.A.; Din, I.M.U.; Aziz, G.; Mir, M.S.; Shah, A.H.; Sheikh, I.R.; Mudasir, P. Integration of nano Al_2O_3 and nano SiO2 in asphalt mixes: A comprehensive performance and durability evaluation. *Constr. Build. Mater.* **2024**, *412*, 134687. [CrossRef]
29. Goh, S.W.; Akin, M.; You, Z.; Shi, X. Effect of deicing solutions on the tensile strength of micro- or nano-modified asphalt mixture. *Constr. Build. Mater.* **2011**, *25*, 195–200. [CrossRef]
30. Amirkhanian, A.N.; Xiao, F.; Amirkhanian, S.N. Evaluation of High Temperature Rheological Characteristics of Asphalt Binder with Carbon Nano Particles. *J. Test. Eval.* **2010**, *39*, 583–591. [CrossRef]
31. Calandra, P.; Loise, V.; Porto, M.; Oliviero Rossi, C.; Lombardo, D.; Caputo, P. Exploiting Nanoparticles to Improve the Properties of Bitumens and Asphalts: At What Extent Is It Really Worth It? *Appl. Sci.* **2020**, *10*, 5230. [CrossRef]
32. Ashish, P.K.; Singh, D. Use of nanomaterial for asphalt binder and mixtures: A comprehensive review on development, prospect, and challenges. *Road Mater. Pavement Des.* **2021**, *22*, 492–538. [CrossRef]
33. Fini, E.H.; Hajikarimi, P.; Rahi, M.; Moghadas Nejad, F. Physiochemical, Rheological, and Oxidative Aging Characteristics of Asphalt Binder in the Presence of Mesoporous Silica Nanoparticles. *J. Mater. Civ. Eng.* **2016**, *28*, 04015133. [CrossRef]
34. Xu, X.; Guo, H.; Wang, X.; Zhang, M.; Wang, Z.; Yang, B. Physical properties and anti-aging characteristics of asphalt modified with nano-zinc oxide powder. *Constr. Build. Mater.* **2019**, *224*, 732–742. [CrossRef]
35. Bach Le, V.; Phuc Le, V. Performance evaluation of carbon nanotubes as a binder modifier for asphalt mixtures. *Int. J. Civ. Eng.* **2021**, *19*, 1143–1153. [CrossRef]
36. Eisa, M.S.; Mohamady, A.; Basiouny, M.E.; Abdulhamid, A.; Kim, J.R. Mechanical properties of asphalt concrete modified with carbon nanotubes (CNTs). *Case Stud. Constr. Mater.* **2022**, *16*, e00930. [CrossRef]
37. Xue, Y.; Liu, C.; Lv, S.; Ge, D.; Ju, Z.; Fan, G. Research on rheological properties of CNT-SBR modified asphalt. *Constr. Build. Mater.* **2022**, *361*, 129587. [CrossRef]
38. Golestani, B.; Nam, B.H.; Moghadas Nejad, F.; Fallah, S. Nanoclay application to asphalt concrete: Characterization of polymer and linear nanocomposite-modified asphalt binder and mixture. *Constr. Build. Mater.* **2015**, *91*, 32–38. [CrossRef]
39. Miglietta, F.; Underwood, B.S.; Tsantilis, L.; Baglieri, O.; Kaloush, K.E.; Santagata, E. Fatigue properties of nano-reinforced bituminous mixtures: A viscoelastic continuum damage approach. *Int. J. Pavement Res. Technol.* **2018**, *11*, 766–773. [CrossRef]
40. Ameri, M.; Mohammadi, R.; Vamegh, M.; Molayem, M. Evaluation the effects of nanoclay on permanent deformation behavior of stone mastic asphalt mixtures. *Constr. Build. Mater.* **2017**, *156*, 107–113. [CrossRef]
41. Lopes Crucho, J.M.; Coelho das Neves, J.M.; Capitao, S.D.; de Picado-Santos, L.G. Mechanical performance of asphalt concrete modified with nanoparticles: Nanosilica, zero-valent iron and nanoclay. *Constr. Build. Mater.* **2018**, *181*, 309–318. [CrossRef]
42. Hamedi, G.H. Evaluating the effect of asphalt binder modification using nanomaterials on the moisture damage of hot mix asphalt. *Road Mater. Pavement Des.* **2017**, *18*, 1375–1394. [CrossRef]
43. Al-Sabaeei, A.M.; Napiah, M.; Sutanto, M.; Alaloul, W.; Md Yusoff, N.I.; Imran Khan, M.; Modibbo Saeed, S. Physicochemical, rheological and microstructural properties of nano-silica modified bio-asphalt. *Constr. Build. Mater.* **2021**, *297*, 123772. [CrossRef]
44. Xu, L.; Ling, W. Study on rheological properties of nano-SiO_2/castor oil composite modified asphalt. *West. China Commun. Sci. Technol.* **2023**, *3*, 12–16.
45. Changjun, B. Study on the influence of nanomaterials on the rheological properties and UV aging resistance of bio asphalt. *Highw. Automot. Appl.* **2023**, *3*, 51–56+60.
46. Yadykova, A.Y.; Shariati, S.; Fini, E.H.; Ilyin, S.O. Novel Bio-Nanocomposite for Asphalt Pavements: Montmorillonite Intercalated with Castor oil. *ACS Sustain. Chem. Eng.* **2024**, *12*, 13062–13079. [CrossRef]
47. *JTG E20-2011*; Standard Test Methods of Bitumen and Bituminous Mixtures for Highway Engineering—Industry Standard—Transportation. Renmin Communication Press, B.C.: Beijing, China, 2011.
48. *JTG E42-2005*; Test Methods of Aggregate for Highway Engineering—Industry Standard-Traffic. Institute of Highway Science, M. o. C.: Beijing, China, 2005.
49. *AASHTO T 49-15*; Standard Method of Test for Penetration of Bituminous Materials. American Association of State and Highway Transportation Officials: Washington, DC, USA, 2015.
50. *AASHTO T 53-09*; Standard Method of Test for Softening Point of Bitumen (Ring-and-Ball Apparatus). American Association of State and Highway Transportation Officials: Washington, DC, USA, 2006.
51. *AASHTO T 51-09*; Standard Method of Test for Ductility of Asphalt Materials. American Association of State Highway and Transportation Officials: Washington, DC, USA, 2018.
52. *ASTM D 2872*; Test Method for Effect of Heat and Air on a Moving Film of Asphalt (Rolling Thin-Film Oven Test). American Society for Testing Materials: West Conshohocken, PA, USA, 2012.
53. *ASTM D 6521*; Accelerated Aging of Asphalt Binder Using a Pressurized Aging Vessel (PAV). American Society for Testing Materials: West Conshohocken, PA, USA, 2012.
54. *AASHTO T 316*; Viscosity Determination of Asphalt Binder Using Rotational Viscometer. American Association of State and Highway Transportation Officials: Washington, DC, USA, 2017.
55. *AASHTO T 315-20*; Standard Method of Test for Determining the Rheological Properties of Asphalt Binder Using a Dynamic Shear Rheometer (DSR). American Association of State Highway and Transportation Officials: Washington, DC, USA, 2010.
56. *AASHTO T 313-19*; Standard Method of Test for Determining the Flexural Creep Stiffness of Asphalt Binder Using the Bending Beam Rheometer (BBR). American Association of State Highway and Transportation Officials: Washington, DC, USA, 2006.

57. Xue, Y.; Liu, C.; Qu, J.; Lv, S.; Ju, Z.; Ding, S.; An, H.; Ren, K. Research on pavement performance of recycled asphalt mixture based on separation technology of asphalt and aggregate in RAP. *Constr. Build. Mater.* **2023**, *393*, 132103. [CrossRef]
58. Xue, Y.; Liu, C.; Shi, Q.; Ju, Z.; Fan, G.; Zhang, C.; Lv, S. Road performance and mechanism of Hot in-place recycling asphalt mixture modified by direct-to-plant SBS. *Constr. Build. Mater.* **2024**, *416*, 135122. [CrossRef]

buildings

MDPI

Article

Analysis of Equivalence between Loading Rate and Stress Level of Fatigue Characteristics of Asphalt Mixture

Liang Tao [1,2,*], Zhengnan Liu [1] and Xinglan Ding [1]

1 National Engineering Laboratory of Highway Maintenance Technology, Changsha University of Science & Technology, Changsha 410004, China; lzn@csust.edu.cn (Z.L.); 19101030033@stu.csust.edu.cn (X.D.)
2 Guangxi Beitou Highway Construction Investment Group Co., Ltd., Nanning 530029, China
* Correspondence: jc@stu.csust.edu.cn

Abstract: The accurate characterization of fatigue life affects the durability and reliability of asphalt pavement for the asphalt mixtures. This paper proposed a new fatigue test method and analysis method which, in addition to increasing the accuracy of fatigue characterization, reduces the time and cost consumed in laboratory tests of asphalt mixtures. According to the loading speed corresponding to different loading frequency and stress levels, the corresponding strength value is selected to determine the stress ratio in the fatigue equation. Therefore, the stress ratio can truly reflect the fatigue loading stress conditions and correspond to the fatigue loading speed one by one, avoiding the unscientific fatigue resistance design problem caused by the traditional S-N fatigue equation stress ratio, which takes a single strength value at a constant loading speed as the standard. Since the stress ratio takes into account the influence of loading speed under different loading frequencies and stress levels, a unified representation model of the fatigue equation of the asphalt mixture under different loading frequencies can be established; that is, the fatigue performance of the asphalt mixture under different loading frequencies can be normalized. The results showed that the loading rate in the strength test differed from the loading rate in the traditional fatigue test, which means that the fatigue curve could not be extended to point (1,1) when the fatigue equation was extended to a fatigue life $N_f = 1$. The fatigue test related to the velocity-dependent stress ratio in multiple experimental works, and data discreteness had a remarkable influence on fatigue characterization. The fatigue curve obtained by the fatigue equation under a constant loading rate was consistent with the fatigue curve under different loading rates. This method can reduce the duration of strength tests and the impact of the strength results of discreteness on fatigue characterization. The fatigue characterization model under constant loading rates considered the viscoelasticity of asphalt mixtures and guaranteed the correspondence of test conditions among stress level t, standard strength S_t, and fatigue life N_f. The stress levels and loading rates of the asphalt mixtures were confirmed to be equivalent. Furthermore, this paper established the relationship between strength and fatigue and obtained the fatigue life curve of the strength values. The fatigue performance of the asphalt mixtures is evaluated comprehensively.

Keywords: asphalt mixtures; fatigue property; S-N fatigue equation; loading rates; stress ratio

Citation: Tao, L.; Liu, Z.; Ding, X. Analysis of Equivalence between Loading Rate and Stress Level of Fatigue Characteristics of Asphalt Mixture. *Buildings* **2024**, *14*, 3102. https://doi.org/10.3390/buildings14103102

Academic Editor: Pengfei Liu

Received: 9 August 2024
Revised: 19 September 2024
Accepted: 22 September 2024
Published: 27 September 2024

1. Introduction

With the rapid development of national economies, lots of traffic problems are emerging, such as large-scale vehicles, overloading vehicles, and the enhancement of traffic volume. In this case, different forms of early damage to asphalt pavement occurred in service. Fatigue damage is one of the main failure modes of asphalt concrete pavement. Cracking is the most common fatigue failure mode. Fatigue cracking is generally divided into top-down cracks and down-top cracks. It significantly effects the road performance and the service life of asphalt pavements. In recent decades, the fatigue life of asphalt pavement has been a major concern in the world [1–5].

The asphalt mixtures are typical viscoelastic materials, and their fatigue performances are closely related to temperature, load, and other factors. The effect of viscoelasticity on fatigue characterization has been heeded in existing theoretical research. Zhang et al. [6] built a damage evolution model considering the effect of loading frequency, based on the traditional Chaboche damage evolution model. The model explained the basic regularity of damage evolution and accumulation. It also reflected the influence of loading frequency on the damage evolution rates and fatigue life of the asphalt mixtures. Kuai et al. [7] established the fatigue cracking growth model of asphalt mixtures based on viscoelasticity and fracture mechanics, reasonably predicting the fatigue cracking growth of asphalt mixtures under different loads and temperatures. H.Sharma [8], H.Sadek [9], and P.Singh [10] proposed a probabilistic fatigue life prediction method including viscoelastic continuum damage theory, considering the data discreteness of fatigue tests. The result showed that the probabilistic fatigue curve precisely described the fatigue damage of the mixture. De Mello [11] analyzed the impact of temperature on the results of the four-point beams fatigue test as used on hot-mixed asphalt by employing three methods. The observed results showed that there was a clear correlation between fatigue parameters and temperature. Based on the simplified viscoelastic continuum damage mechanics (SCECD) of the Schapery criterion [12], Kim Y.R. [13] and E. Masad [14] introduced pseudo strain and dissipative pseudo strain energy to evaluate the fatigue cracking of asphalt mixtures.

However, in the structure design of asphalt pavements, the viscoelasticity is not fully considered. The construction of asphalt pavements is usually designed by the mechanistic–empirical method. The design parameters of fatigue life are determined by the S-N fatigue equation based on phenomenology. Using the remaining loading frequencies, a variety of fatigue tests under four or five stress levels are carried out, and S–N curves are obtained to reveal the fatigue performance. Each stress level has three to four parallel specimens. In brief, there are approximately 15 specimens [15,16]. In this design, the loading rates have remarkable influence on the fatigue properties, which are simply to be supplemented subsequently with reduction factors considering the modulus and with voids filled with asphalt. This results in unreasonably designing and inaccurately revealing the fatigue characterization of the materials. The fatigue life of asphalt pavements was obviously reduced.

Songtao Lv [17,18] revealed the nonlinear characteristics of the strength of asphalt mixtures through a variety of strength tests under different loading rates. The S-N fatigue equation modified with the velocity-dependent stress ratio was proposed, and a normalized model under different stress states was established to characterize the fatigue performance of asphalt mixtures based on the yield criterion idea. The structural coefficient of tensile strength recalculated by the model considered the viscoelasticity of the materials, eliminated the influence of the test methods and conditions on the test results, and improved the accuracy of fatigue characterization to a degree. The research results show that a unified fatigue model was obtained under different stress states, stress levels, loading frequencies, and other test conditions based on the normalized model characterization. This meant that the test conditions and other external factors do not affect the results of the fatigue characterization, which means that test results are more reliable depending on which test methods and conditions are used in the engineering, which could also improve the scientific reliability of pavement design. But, determining the parameters in the fatigue equation requires performing multiple strength tests, which may exceed the time and budget constraints in the tests. At the same time, the discreteness of fatigue data produces some errors in the fatigue performance characterization because of the strength obtained by the loading rates on the S-N curve.

The fatigue performance of asphalt mixtures is usually revealed by fatigue tests under the frequency of 10 Hz. Existing tests have shown that the frequency does not affect the fatigue performance of asphalt mixtures, while also considering the loading rates of fatigue tests for asphalt mixtures. Thus, in this paper, corresponding with the test conditions of t, S_t, and N_f, the loading rate was constant under different stress levels while adjusting the loading frequency in the fatigue tests. Combining the time–temperature equivalence principle with the WLF equation, the fatigue life master curve and viscoelastic parameter values under different loading rates for asphalt mixtures were obtained. And the equivalence of the stress levels and loading rate was verified. The connection between strength and fatigue was established directly, and the accuracy and efficiency of fatigue characterization for asphalt mixtures were enhanced.

2. Methodology

In the technical specifications for asphalt pavement construction in China [19], the fatigue tests are conducted under the constant loading frequency of 10 Hz, and the stress levels were confirmed using the strength value. The following S-N fatigue equation is obtained:

$$N_f = k \cdot \left(\frac{1}{t}\right)^n \quad (1)$$

where, N_f is the fatigue life; t is the stress ratio; and k, n are the regression parameters of the S-N fatigue equation.

The fatigue curves confirmed by the S-N fatigue equation are extended to $N_f = 1$. It means that the asphalt mixtures fail after being loaded for the first time. In this case, the tensile stress obtained by the fatigue equation is considered the fatigue dynamic strength. Therefore, the structural coefficient of the tensile strength is calculated. In the fatigue equation, the n parameter is usually used to evaluate the stress sensitivity of the fatigue performance. The fatigue curves acquired by this method reveal several problems:

1. Non-correspondence of test conditions between strength tests and fatigue tests:

While determining the fatigue equation, the stress ratio t is calculated by the results of the strength tests at a constant loading rate. But the loading rate changes by changing the stress levels and loading frequencies. This could cause inconsistencies in the loading rates between the strength tests and fatigue tests. The effect of the loading rates on strength and fatigue is not considered while analyzing test results through the S-N fatigue equation.

2. The loading rates of strength tests varied from the actual loading rates

The loading frequency is approximately 10 Hz when the vehicle drives at 60–80 km/h. Laboratory fatigue tests usually adopted this frequency. However, precious studies have shown that the loading frequency has a weak influence on the fatigue characterization of asphalt mixtures.

Therefore, when the S-N fatigue equation and the current fatigue test methods are applied to analyze the fatigue performance of asphalt mixtures, some of the following phenomena arise:

1. The stress ratio is not 1, and the fatigue curves cannot pass through the point (1,1) after the fatigue equation is extended to N_f = 1. Actually, the fatigue life is far greater than 1 and is not regular. It shows the arbitrary K value, which results in the inaccuracy of fatigue dynamic load strengths. Moreover, the calculation of the structural coefficient of tensile strength would cause great errors.
2. The fatigue properties obtained by different test methods are different, which are showed as the differences between the n values in the S-N fatigue equation. The stress sensitivity of the material is diverse under different test methods.

Based on these problems, researchers proposed a fatigue equation which considered the velocity dependence of a material. For example, Songtao Lv defined the real stress ratio and put forward the fatigue equation related to the velocity-dependent stress ratio. The stress ratio was corrected by establishing the relationship between the strengths and the loading rates. This guaranteed the correspondence of the test conditions among t, S_t, and N_f. The fatigue curve obtained by the optimized S-N fatigue equation passed through point (1,1). The n values of the same material under different test methods were similar, uniformly characterizing the stress sensitivity of the material. Also, the relevance of the fatigue dynamic load strengths and loading rates was determined [17]. The S-N fatigue equation is

$$N_f = k' \cdot \left(\frac{1}{t_s}\right)^{n'} \tag{2}$$

where, $k' = 1$ (because the fatigue curve passed through the point of (1,1) after using the true stress ratio), and t_s is the real stress ratio which is defined as the ratio of the stress levels of fatigue dynamic load strengths at different loading rates:

$$t_s = \frac{\sigma}{S_{dz}} \tag{3}$$

The stress versus the fatigue life relationship appears to be linear on the dual logarithmic coordinate diagram. Two parameters of fatigue equation of k', n' are the intercept and slope of the straight line, respectively. The fatigue performance of the asphalt mixture is reflected by the following two parameters of the fatigue equation: the greater the value of n', the steeper the fatigue curve and the more sensitive the fatigue life with the changes in the stress levels. The value of k indicates the level of the fatigue curve line. The larger the value of k', the higher the level of the fatigue curve line and the better the fatigue durability.

However, multiple strength tests at different loading rates were employed to determine the model and the relationship between strengths and loading rates. And the strength value calculated by the model was deviated from the actual measured value. Even slight errors will cause large deviations in the fatigue equation. The comparison of fatigue life is not sharp for different materials because the stress ratio is selected by the strengths of the material at corresponding loading rates.

Considering the above problems, a research method was proposed to eliminate the effect of loading rates on the fatigue performance of asphalt mixtures in this paper. The loading frequency was adjusted according to the stress levels to maintain the loading rates and finish the fatigue tests. The strength tests of the asphalt mixtures were performed at the loading rate v to obtain the strength value S_t. Then, six stress levels of $0.2S_t$, $0.3S_t$, $0.4S_t$, $0.5S_t$, $0.6S_t$, and $0.7S_t$ were taken, and the corresponding loading frequencies according to Equation (4) were calculated. While the loading rates during the fatigue tests remain, the loading frequency was f under the stress level of $0.2S_t$, and the corresponding loading frequencies of other stress levels are listed in Table 1.

$$\mathrm{v} = \frac{2\sigma}{\mathrm{T}} = 2\sigma \cdot f \tag{4}$$

where v is the loading rate; σ is the stress level; t is the loading cycle; and f is the loading frequency.

Table 1. The corresponding relationship between the stress levels and the loading frequency under constant loading rates.

Number	The Stress Levels	The Loading Frequency
1	$0.2S_t$	f
2	$0.3S_t$	$\frac{2}{3}f$
3	$0.4S_t$	$\frac{1}{2}f$
4	$0.5S_t$	$\frac{2}{5}f$
5	$0.6S_t$	$\frac{1}{3}f$
6	$0.7S_t$	$\frac{2}{7}f$

The set test conditions guaranteed the constant loading rates of the fatigue tests under different stress levels and the correspondence of the test conditions, including the stress level t, standard strength S_t, and fatigue life N_f. Also, lots of laboratory tests were conducted while determining the relationship between the strengths and the loading rates.

3. Materials and Design

In this paper, the dense-graded asphalt mixture (AC-16C) was selected to investigate its strengths and fatigue performances, which comprised styrene–butadiene–styrene (SBS)-modified asphalt and limestone. Along with the technique standards already referred to, the test results of the performance of SBS-modified asphalt and aggregate are shown in Tables 2 and 3, respectively. The test results in the two tables show that SBS-modified asphalt and the aggregate met the requirements of JTG F40-2004 [20], which were the construction technical specifications of asphalt pavement in China.

Table 2. Test results of SBS (I-D)-modified asphalt in this research.

Test Properties		Test Standard: JTG F40-2004 (China)		
		Test Results	Sprcification	Test Methods
Density (15 °C)		1.07	/	T 0603-1993
Flash point (°C)		265	≥230	T 0611-1993
Solubility (%)		100.1	≥99	T 0607-1993
135 °C dynamic viscosity (Pa·s)		2.49	≤3	T 0615-2000
Softening point (Ring ball) (°C)		76.0	≥60	T 0606-2000
Penetration (25 °C, 100 g, 5 s) (0.1 mm)		54	30–60	T 0604-200
Ductility (5 cm/min, 5 °C) (cm)		31	≥20	T 0605-1993
Penetration index PI		0.54 (R^2 = 0.997)	≥0	T 0604-2000
Roll Thin Film Oven Test (RTFOT) (163 °C, 85 min)	Mass loss (%)	0.2	<±1.0	T 0609-1993
	Residual ductility (5 °C) (cm)	68	≥65	T 0605-1993
	Residual penetration ratio (25 °C) (%)	17	≥15	T 0604-2000

Table 3. Properties of the coarse aggregate in this research.

Test	Values	Specification Limit	Standard Method
Apparent relative density (g/cm^3)	2.9	≥2.6	T 0321-2005
Firmness (%)	5	≤12	T 0314-2000
Los Angeles wearied stone value (%)	19.7	≤28	T 0317-2005
Crushed stone value (%)	18.4	≤26	T 0316-2005
Water absorption (%)	0.4	≤2.0	T 0304-2005
<0.075 mm particle content (Washing methods) (%)	0.5	≤1	T 0310-2000
Percentage of flat and elongated particles (%)	8	≤15	T 0312-2005
Content of soft stone (%)	2.3	≤3	T 0320-2000
Asphalt adhesion/grade	6	≥4	T 0616-1993

Figure 1 shows the designed gradation curve range and the design gradation of dense-graded asphalt mixtures (AC-16C). The optimum asphalt content was determined by the Marshall test. The pieces, sized φ101.6 mm × 63.5 mm, was compacted 75 times on two sides in accordance with the specification. After 24 h, the physical index of the specimen at 25 °C water temperature was measured, and then the Marshall stability and flow values at 60 °C were obtained [21]. Test results and technique standards [20] are shown in Table 4.

Table 4. Results of the Marshall test at the optimum asphalt content in this research.

Test Projects	Asphalt Aggregate Ratio (%)	Volume Filled with Asphalt VFA (%)	Volume of Air Voids VV (%)	Bulk Specific Gravity (g·cm^{-3})	Marshall Stability (KN)	Flow Value (0.1 mm)
Test results	5.2	68.9	4.4	2.51	14.23	31
Technical requirements	/	65–75	3–5	/	>8	20–40

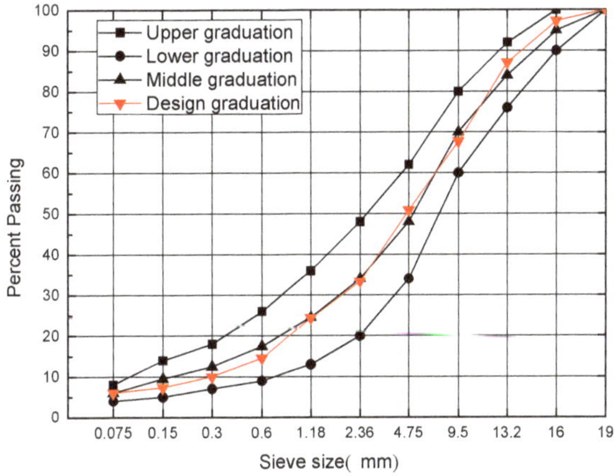

Figure 1. Aggregate grading curve of the dense-graded asphalt mixtures (AC-16C).

4. Test

4.1. Traditional Fatigue Test and the Fatigue Test Modified with the Stress Rates

The standard strength value, S_t, was measured at a loading rate of 50 mm/min and a temperature of 15 °C. Then, direct tensile fatigue tests were performed at stress levels of $0.2S_t$, $0.3S_t$, $0.4S_t$, $0.5S_t$, $0.6S_t$, and $0.7S_t$, a temperature of 15 °C, and a half-sine wave frequency of 10 Hz (as shown in Figure 2).

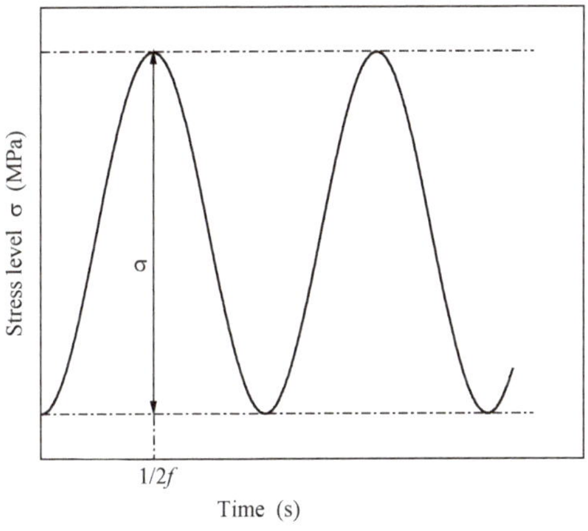

Figure 2. Loading wave of fatigue test [22].

4.2. Modified Fatigue Test in This Research

The mineral material utilized in this paper undergoes a meticulous layer-by-layer sifting process, followed by a four-hour preheating in an oven to ensure complete dryness. Upon its introduction into the mixing pot, either the mineral material or asphalt is stirred vigorously for 90 s to achieve the uniform blending of the aggregate. The roller molding test plate adheres to dimensions of 400 mm × 300 mm × 50 mm. Notably, within this study, a 40 × 30 × 5 cm rutting plate us initially prepared to be subsequently meticulously sliced into 25 × 5 × 5 cm trabecular specimens for specific experimental use. Acknowledging potential cutting inaccuracies, the specimen's area is determined through the average measurement of three distinct cross-sectional areas.

A total of eight rutting plates were formed, yielding 50 trabecular specimens that were divided for direct tensile strength and direct tensile fatigue tests. For these experiments, the MTS-Landmark material test system was employed. Prior to testing, the direct tensile specimens were conditioned in an incubator at 15 °C for 4–5 h to stabilize, and the entire testing process was conducted within this controlled environment.

The loading program was carefully crafted and implemented, with the assistance of the built-in MPT system, which facilitated the setting of the relevant load and data acquisition parameters. During the strength test, the maximum load and corresponding displacement were automatically captured, while the fatigue test leveraged the data acquisition system to seamlessly record the force applied during each loading cycle. In the first step, the direct tensile strength tests were conducted to determine the strength values S_{dz} of the asphalt mixtures at the five calculated loading rates. In the second step, a modified fatigue test was used with five stress levels, namely $0.2S_t$, $0.3S_t$, $0.4S_t$, $0.6S_t$, and $0.7S_{t,}$, which were accomplished at loading frequencies calculated by the corresponding stress levels according to Equation (3) at the five loading rates.

5. Test Results and Analysis

5.1. Traditional Fatigue Tests and the Fatigue Test Modified with the Velocity-Dependent Stress Ratio

Under the loading rate referred to in the specifications, the direct tensile strength of asphalt mixtures is 0.76 MPa. But the asphalt mixtures show viscoelasticity, and their strength values increase with the increase in loading rates. Figure 3 shows the strength test results under six loading rates.

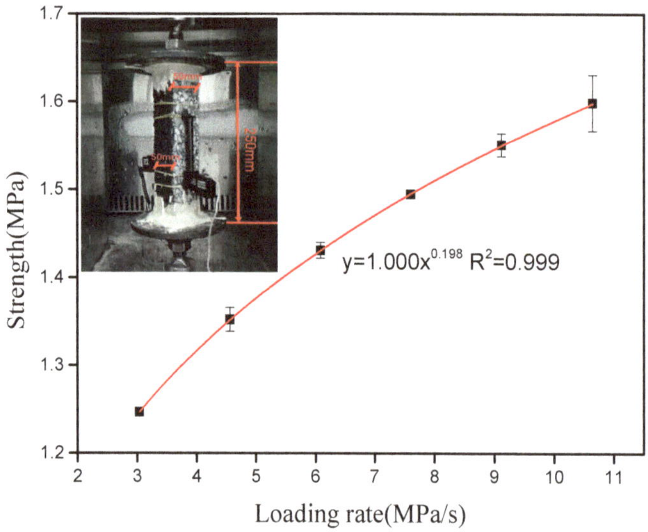

Figure 3. Nonlinear characteristic of direct tensile strength tests for asphalt mixtures.

The loading rates had a significant impact on the strength of the asphalt mixtures, and these followed a power function that described in Figure 3. However, there is a remarkable difference between the strength value under the standard test conditions and the corresponding loading rate under various stress levels in the fatigue test. And the standard strength value is not representative. In this context, the deviation of fatigue performance characterization exists, as shown in Figure 4.

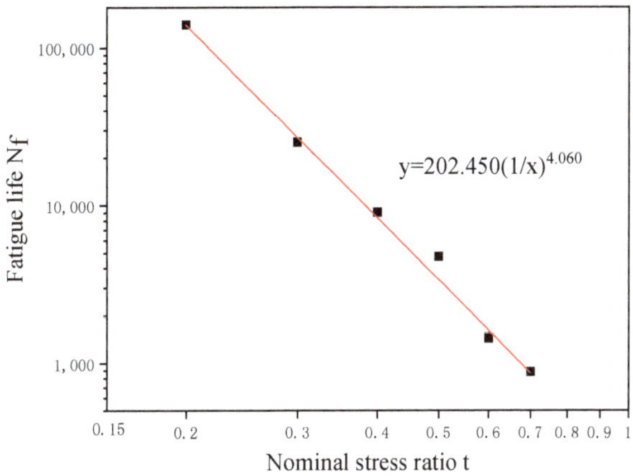

Figure 4. Direct tensile fatigue curve of asphalt mixtures under traditional S-N fatigue equation.

As shown in the above figure, although the fatigue equation expressed by the nominal stress ratio was an equation of the straight line under double logarithmic coordinates, the fatigue curve did not pass through the point (1,1) and the intersection point with abscissa was far larger than the stress ratio of 1. Artificial extension leads to the arbitrariness and unscientific value of anti-fatigue parameters.

Therefore, Lv, et al. [17] proposed the fatigue equation related to the velocity-dependent stress ratio, by which the fatigue test results were fitted. And the results are summarized in Figure 5.

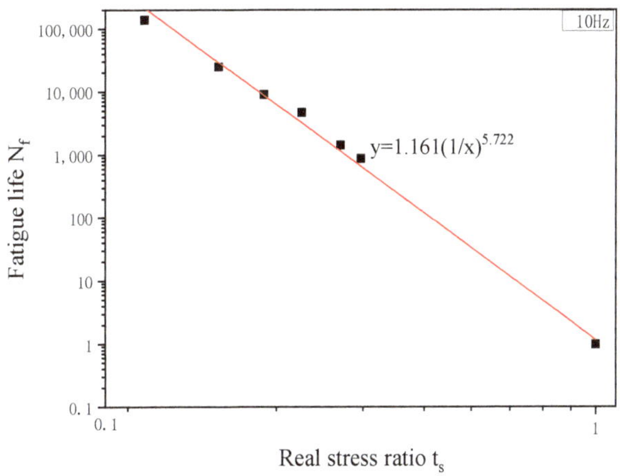

Figure 5. Direct tensile fatigue curve of asphalt mixtures under the S-N fatigue equation modified with the velocity-dependent stress ratio.

Obviously, the fatigue curve passed through the point (1,1), And the relationship between fatigue failure and strength failure was established. However, the method needs to establish the relationship between strengths and loading rates firstly, so as to determine the strength values corresponding to the loading rates under different stress levels in fatigue tests and to calculate the stress ratios related to the loading rates. The strength tests under different loading rates increased the amount of tests and the errors of the stress ratio of the calculation results due to test errors.

5.2. Results and Analysis of Modified Direct Tensile Fatigue Tests

According to the strength value S_t under different loading rates in Figure 3, six stress levels of $0.2S_t$, $0.3S_t$, $0.4S_t$, $0.6S_t$, and $0.7S_t$ were taken. And the loading rate reached a constant value by adjusting the loading frequency according to the stress level. The corresponding fatigue test conditions are listed in Table 5.

Table 5. Fatigue life under constant loading rates.

Number	Loading Rate (MPa/s)	Strength (MPa)	Stress Ratio (MPa/MPa)	Stress Level (MPa)	Loading Frequency (Hz)
1	3.04	1.25	0.2	0.25	6.08
			0.3	0.38	4.05
			0.4	0.50	3.04
			0.5	0.63	2.43
			0.6	0.75	2.03
			0.7	0.88	1.74
2	4.56	1.35	0.2	0.27	8.44
			0.3	0.41	5.63
			0.4	0.54	4.22
			0.5	0.68	3.38
			0.6	0.81	2.81
			0.7	0.95	2.41
3	6.08	1.43	0.2	0.29	10.63
			0.3	0.43	7.09
			0.4	0.57	5.31
			0.5	0.72	4.25
			0.6	0.86	3.54
			0.7	1.00	3.04
4	9.12	1.55	0.2	0.31	14.71
			0.3	0.47	9.81
			0.4	0.62	7.35
			0.5	0.78	5.88
			0.6	0.93	4.90
			0.7	1.09	4.20
5	10.64	1.60	0.2	0.32	16.63
			0.3	0.48	11.08
			0.4	0.64	8.31
			0.5	0.80	6.65
			0.6	0.96	5.54
			0.7	1.12	4.75

The fatigue test results obtained by the modified fatigue test method and stress levels are presented in Figure 6 on the dual logarithmic coordinate diagram.

Figure 6 shows the fatigue life curves of six stress levels under different loading rates. It can be seen from the figure that the fatigue life curves under five different loading rates are parallel to each other, with little difference in slope. And the slight difference may be caused by personal error. There may be an equivalent relationship between stress levels and loading rates in direct tensile fatigue tests of asphalt mixtures.

The solution of the shift factor Φ_T is the key to solving the main curve of the dynamic modulus by using the time–temperature equivalence principle. As a horizontal displacement, the shift factor represents the distance from the main curve at different test temperatures to the main curve at the reference temperature. The displacement factor can be determined by the WLF equation. The WLF equation reflects the relationship between the viscosity and temperature of polymer derived from the free volume theory, which was first summarized through a large number of experiments while checking time temperature conversion. M. L. Willianms, R. F. Lanbel, and J. D. Ferry found that the modulus–temperature curves of several times at different temperatures were horizontally shifted and superimposed into a main curve for all amorphous polymers. And the relationship between the horizontal displacement $lg\Phi_T$ and the temperature T on the time axis basically conformed to an empirical equation, which is the WLF equation [23].

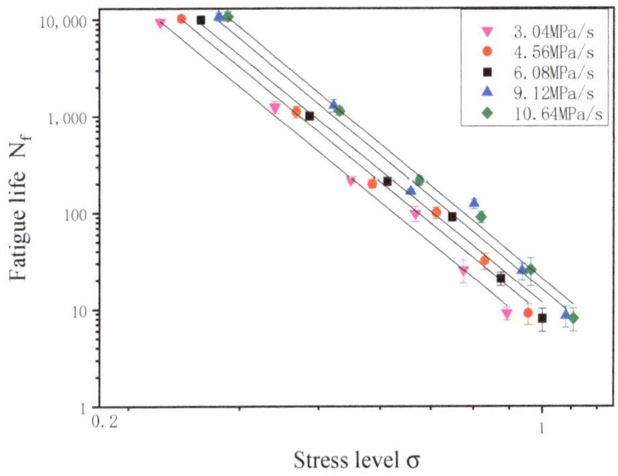

Figure 6. Fatigue curves obtained by the modified fatigue method.

According to the free volume theory, the relationship between the viscosity and free volume fraction for the materials satisfies the Doolittle equation [24]:

$$lg\eta = lnA + B\left(\frac{1}{f} - 1\right) \tag{5}$$

in which A and B are material parameters.

According to the time–temperature equivalent principle, the mechanical behavior of viscoelastic materials on different time scales can be realized by changing the temperature, and the essence lies in the temperature dependence on the viscoelastic relaxation spectrum of materials. If the free volume fraction of material is linear with the change in temperature,

$$f = f_0 + \alpha_T(T - T_0) \tag{6}$$

where α_T is the thermal expansion coefficient of the free volume fraction and f_0 is the free volume fraction of the material at the reference temperature T_0. Φ_T is the time temperature shift factor, $\Phi_T = \tau/\tau_0 = \eta/\eta_0$, where τ_0 and η_0 are the relaxation time and viscosity of the material at a certain temperature T_0 of the set temperatures, respectively. And τ and η are the relaxation time and viscosity of the material at the test temperature T, respectively. $lg\Phi_T = B(f^{-1} - f_0^{-1})/2.303$. Then,

$$lg\Phi_T = -\frac{B}{2.303f_0}\left(\frac{T - T_0}{\frac{f_0}{\alpha(T)} + T - T_0}\right) = \frac{-C_1(T - T_0)}{C_2 + T - T_0} \tag{7}$$

where $C_1 = \frac{B}{2.303f_0}$, $C_2 = \frac{f_0}{\alpha(T)}$ are material parameters.

In this study, the shift factor is a horizontal displacement of the stress level and represents the distance from the fatigue life curve under different stress levels to the fatigue life curve under the reference stress level. Compared with the main modulus curve, this was obtained by translating the modulus curve at different temperatures. The shift factor is a function of temperature. This paper aims to obtain the main fatigue life curve at different loading rates; here, the shift factor is a function of loading rate, as shown in Equation (8):

$$lg\Phi_v = \frac{-C_1(v - v_0)}{C_2 + v - v_0} \tag{8}$$

The fatigue life under different loading rates v (v = 3.04, 4.56, 6.08, 9.12, 10.64 MPa/s) was obtained through experiments. The test curve with v = 3.04 MPa/s was taken as the reference line, and these test points were horizontally shifted to the reference line until they coincided with the corresponding fatigue life points on the reference test curve. The distance of movement was calculated as the displacement factor Φ_v.

A group of data concerning the shift factor Φ_v and the loading rate v were obtained through experiments at different loading rates. The parameters C_1 and C_2 in the formula can be obtained by fitting these data to a nonlinear curve according to Equation (8), and thus the expression of the shift factor Φ_v can be obtained.

The relative shift factor $lg\Phi_v{}'$ and the shift factor $lg\Phi_v$ relative to the reference loading rate v = 3.04 MPa/s are shown in Table 6.

Table 6. Shift factor $lg\Phi_v$ under different loading rates v.

Difference in Loading Rate Δv	Relative Shift Factor $lg\Phi_v{}'$	Shift Factor $lg\Phi_v$
0	0	0
1.52	0.024	0.024
3.04	0.05	0.074
6.08	0.021	0.095
7.6	0.036	0.131

The difference value Δv and $lg\Phi_v$ of the loading rate in Table 6 are plotted as shown in Figure 7, and the parameters C_1 and C_2 are obtained through nonlinear curve fitting according to Equation (8), as shown in Figure 8.

As Figure 8 shows, the fitting results are $C_1 = -0.413$, $C_2 = 17.476$ and the fitting degree of curve R^2 = 0.951 for dense-graded asphalt mixtures (AC-16C), resulting in the following:

$$lg\Phi_v = \frac{0.413(v - v_0)}{17.476 + v - v_0} \tag{9}$$

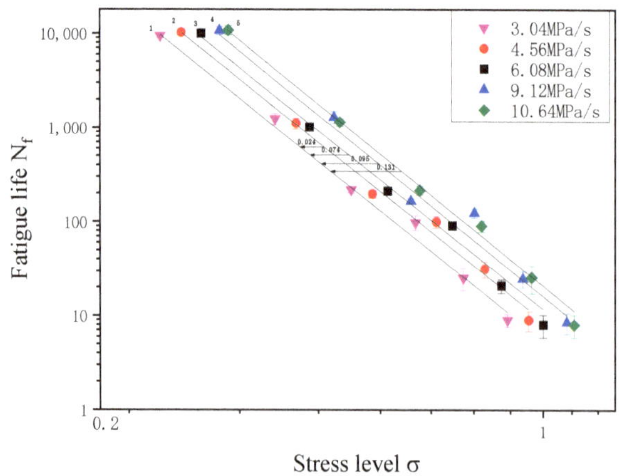

Figure 7. Shift factor corresponding to the change in loading rate.

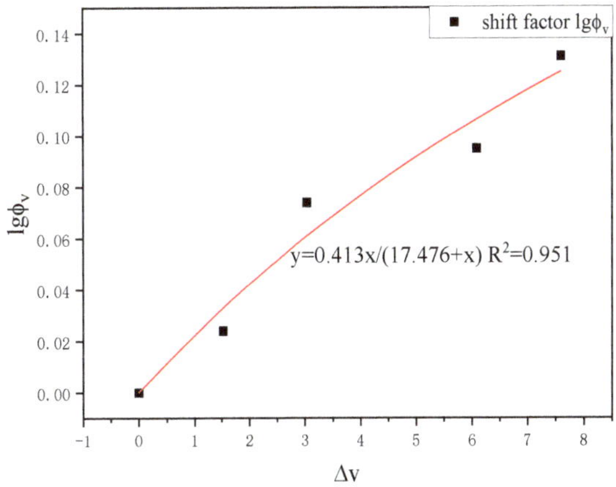

Figure 8. Fitting parameter diagram of WLF equation.

The fatigue life curve under different loading rates shown in Figure 6 is horizontally shifted to the fatigue life curve with reference loading rate v = 3.04 MPa/s by the shift factor, and the main curve of the fatigue life of asphalt mixtures is obtained as shown in Figure 9.

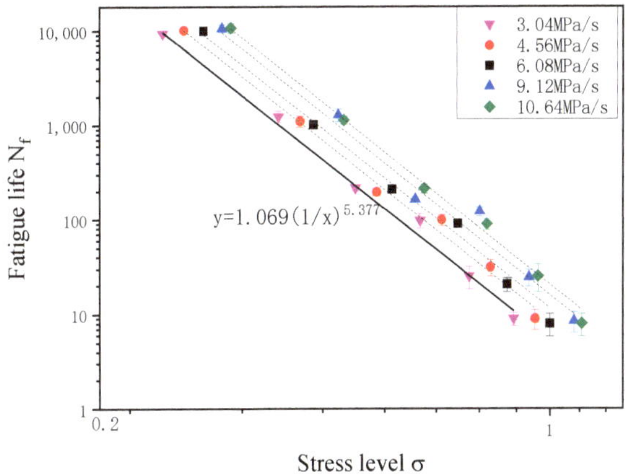

Figure 9. Main curve of fatigue life for asphalt mixtures under different loading rates.

Obviously, there is an equivalent relationship between the stress level and loading rate of fatigue characteristics for the asphalt mixtures. The fatigue life curve under different loading rates and stress levels can be shifted to the main curve under the reference loading rate. The test, taking a long time originally, can be completed in a short time by the method of shifting. The accelerated test method has important theoretical significance and application value in loading tests of the long-term mechanical behavior of materials and life prediction.

5.3. Comparative Analysis of Fatigue Curves Related to the Velocity-Dependent Stress Ratio

Figure 10 compares the fatigue curve obtained after shifting with the fatigue curve relating to the velocity-dependent stress ratio.

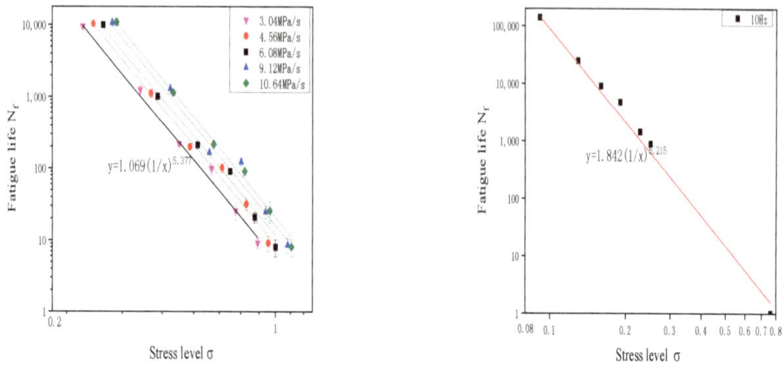

(**a**) (**b**)

Figure 10. σ-N curves (**a**) Main curve of fatigue life after shifting; (**b**) fatigue curve related to the velocity-dependent stress ratio.

Figure 10 shows that the equation of the fatigue curve obtained after shifting is very close to the equation of the fatigue curve relating to the velocity-dependent stress ratio. The equivalent relationship between the stress level and loading rate of fatigue characteristics for the asphalt mixtures has been further verified. In addition, it shows that the fatigue analysis method based on the WLF equation and the time–temperature equivalent principle is effective and feasible.

1. The data of the new method are expressed by the stress ratio, which is the fatigue equation relating to the velocity-dependent stress ratio.

Figure 11a shows that the S-N fatigue life curves under different loading rates can be normalized into one curve. The fatigue life decreases gradually with the increase in the stress ratio. The fatigue curve pass through the point (1,1), and the fatigue life of the strength failure point is one. The fatigue curve reflects the strength failure characteristics of the asphalt mixtures, which is consistent with the fatigue curve related to the velocity-dependent stress ratio. The effectiveness and accuracy of the method was verified. Compared with Figure 11b, the parameters of the fatigue curve under a constant loading rate and constant loading frequency are very similar. It verifies the correctness of the fatigue test method under a constant loading rate to a degree.

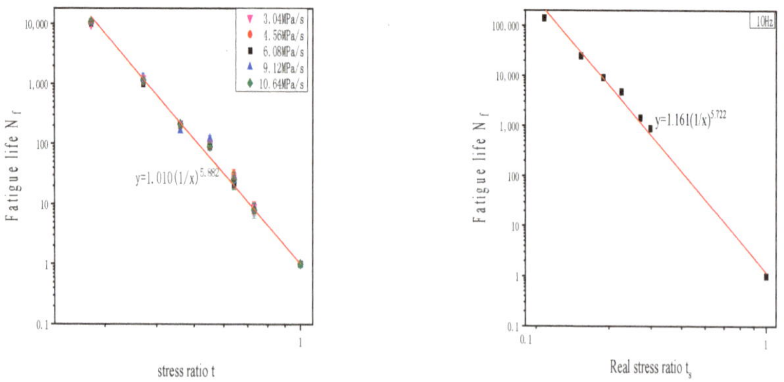

(**a**) (**b**)

Figure 11. t-N curves (**a**) Normalization of fatigue curve under difference loading rates; (**b**) fatigue curve related to the velocity-dependent stress ratio.

6. Conclusions

The traditional fatigue tests and the fatigue tests and analyses under the constant loading rate mentioned in the paper for the asphalt mixtures were carried out. To summarize, the following conclusions can be drawn:

1. Compared with the traditional fatigue test method, the fatigue test method under a constant loading rate considers the viscoelastic properties of asphalt mixtures and ensures the correspondence of test conditions among the stress level t, standard strength S_t, and fatigue life N_f. When compared with the fatigue test equation relating to the velocity-dependent stress ratio, it saves a lot of strength tests and reduces the discreteness of data to a certain extent.
2. According to the time–temperature equivalent principle and WLF equation, the main fatigue life curve of the asphalt mixture with a reference loading rate of 3.04 MPa/s and its viscoelastic parameters are obtained by the fatigue life of difference stress levels under difference loading rates for the asphalt mixtures. The fatigue life curves with other loading rates can be obtained by shifting the fatigue life curves with loading rates of 3.04 MPa/s. The shifting distance is the shift factor.
3. The equivalent relationship between the stress level and the loading rate of the asphalt mixture is verified, and the fatigue characteristics of the asphalt mixture under different stress levels are normalized. The relationship between strength and fatigue life can be directly established, and the fatigue characteristics of the asphalt mixture can be comprehensively evaluated by strength.

Author Contributions: Conceptualization, L.T.; Methodology, L.T. and Z.L.; Investigation, L.T. and X.D.; Data curation, Z.L.; Writing—original draft, L.T., Z.L. and X.D. All authors have read and agreed to the published version of the manuscript.

Funding: This research was funded by key technology projects in the transportation industry of Guangxi Zhuang Autonomous Region: "Field scientific observation and research on the long-term performance of major and medium maintenance structures of national and provincial trunk highways".

Data Availability Statement: Some or all data, models, or code generated or used during the study are available from the corresponding author by request.

Conflicts of Interest: Author Liang Tao was employed by the company Guangxi Beitou Highway Construction Investment Group Co., Ltd. The remaining authors declare that the research was conducted in the absence of any commercial or financial relationships that could be construed as a potential conflict of interest.

References

1. Kocak, S.; Kutay, M.E. Fatigue performance assessment of recycled tire rubber modified asphalt mixtures using viscoelastic continuum damage analysis and AASHTOW are pavement ME design. *Constr. Build. Mater.* **2020**, *248*, 118658. [CrossRef]
2. Wang, H.; Dang, Z.; Li, L.; You, Z. Analysis on fatigue crack growth laws for crumb rubber modified (CRM) asphalt mixture. *Constr. Build. Mater.* **2013**, *47*, 1342–1349. [CrossRef]
3. Ziari, H.; Amini, A.; Goli, A. The effect of different aging conditions and strain levels on relationship between fatigue life of asphalt binders and mixtures. *Constr. Build. Mater.* **2020**, *244*, 118345. [CrossRef]
4. Jia, Y.; Yang, Y.; Liu, G.; Gao, Y.; Yang, T.; Hu, D. Reliability assessment of flexural fatigue failure of asphalt mixture: A new perspective. *Constr. Build. Mater.* **2020**, *257*, 119553. [CrossRef]
5. Reddy, K.S.; Yong-Rak, K.; Ribeiro, R.L.d. Fatigue characterization of binder with aging in two length scales: Sand asphalt mortar and parallel plate binder film. *Constr. Build. Mater.* **2020**, *237*, 117588.
6. Zhang, J.; Wang, Y.D.; Su, Y. Fatigue damage evolution model of asphalt mixture considering influence of loading frequency. *Constr. Build. Mater.* **2019**, *218*, 712–720. [CrossRef]
7. Kuai, H.D.; Lee, H.J.; Lee, J.H.; Mun, S. Fatigue crack propagation model of asphalt concrete based on viscoelastic fracture mechanics. *Transp. Res. Rec.* **2010**, *2181*, 11–18. [CrossRef]
8. Sharma, H.; Swamy, A.K. Development of probabilistic fatigue curve for asphalt concrete based on viscoelastic continuum damage mechanics. *Int. J. Pavement Res. Technol.* **2016**, *9*, 270–279. [CrossRef]
9. Sadek, H.; Sadeq, M.; Masad, E.; Al-Khalid, H.; Sirin, O. Probabilistic viscoelastic continuum damage analysis of fatigue life of warm-mix asphalt. *J. Transp. Eng. Part B Pavements* **2019**, *145*, 0000128. [CrossRef]

10. Singh, P.; Swamy, A.K. Probabilistic characterisation of damage characteristic curve of asphalt concrete mixtures. *Int. J. Pavement Eng.* **2019**, *20*, 659–668. [CrossRef]
11. De Mello, L.G.R.; de Farias, M.M.; ZKaloush, K.E. Effect of temperature on fatigue tests parameters for conventional and asphalt rubber mixes. *Road Mater. Pavement Des.* **2018**, *19*, 417–430. [CrossRef]
12. Micaelo, R.; Pereira, A.; Quaresma, L.; Cidade, M.T. Fatigue resistance of asphalt binders: Assessment of the analysis methods in strain-controlled tests. *Constr. Build. Mater.* **2015**, *98*, 703–712. [CrossRef]
13. Kim, Y.R.; Little, D.N.; Lytton, R.L. Fatigue and healing characterization of asphalt mixtures. *J. Mater. Civ. Eng.* **2003**, *15*, 75–83. [CrossRef]
14. Masad, E.; Branco, V.T.F.C.; Little, D.N.; Lytton, R.L. A unified method for analysis of controlled-strain and controlled-stress fatigue testing. *Int. J. Pavement Eng.* **2008**, *9*, 233–246. [CrossRef]
15. Collins, J.A.; Smith, C.O. Failure of materials in mechanical design. *J. Mech. Des.* **1984**, *106*, 107–108. [CrossRef]
16. Xie, L.; Liu, J.; Wu, N.; Qian, W. Backwards statistical inference method for P–S–N curve fitting with small-sample experiment data. *Int. J. Fatigue* **2014**, *63*, 62–67. [CrossRef]
17. Lv, S.; Liu, C.; Chen, D.; Zheng, J.; You, Z.; You, L. Normalization of fatigue characteristics for asphalt mixtures under different stress states. *Constr. Build. Mater.* **2018**, *177*, 33–42. [CrossRef]
18. Liu, C.; Lv, S.; Peng, X.; Zheng, J.; Yu, M. Analysis and Comparison of Different Impacts of Aging and Loading Frequency on Fatigue Characterization of Asphalt Concrete. *J. Mater. Civ. Eng.* **2020**, *32*, 0003317. [CrossRef]
19. *JTG D50-2017;* Specifications for Design of Highway Asphalt Pavement. Renmin Communication Press: Beijing, China, 2017.
20. *JTG F40-2004;* Technical Specifications for Construction of Highway Asphalt Pavement. Renmin Communication Press: Beijing, China, 2004.
21. *JTG E20-2011;* Specifications and Test Methods of Bitumen and Biminous Mixtures for Highway Engineering. Renmin Communication Press: Beijing, China, 2011.
22. Fan, X.; Lv, S.; Ge, D.; Liu, C.; Peng, X.; Ju, Z. Time-temperature equivalence and unified characterization of asphalt mixture fatigue properties. *Constr. Build. Mater.* **2022**, *359*, 129118. [CrossRef]
23. Wu, W.; Tian, X.; Lv, S.; Xu, Z. Time-temperature-aging equivalent relationship of relaxation characteristics of asphalt mixture. *J. Build. Mater.* **2011**, *14*, 340–344.
24. Liu, C.; Qin, X.; Yang, Z.; Zhang, H.; Lv, S. Assessing bio-oil cutback asphalt for prime coat on semi-rigid base asphalt pavement: Lab experiment & indoor test section application. *Constr. Build. Mater.* **2024**, *422*, 135814.

buildings

MDPI

Article

Dynamic Viscoelastic Behavior of Epoxy Asphalt Mixture under Four-Point Bending

Baiqing He [1,2], Zhiyong Huang [3,4,*], Jingsong Chen [1,2], Mu He [1,2], Yan Wang [1,2], Jian Li [4] and Shaohuai Wang [3]

[1] Guangzhou North Second Ring Road Traffic Technology Co., Ltd., Guangzhou 510540, China; he.boqing@yuexiutransport.com (B.H.); chen.jingsong@yuexiutransport.com (J.C.); he.mu@yuexiutransport.com (M.H.); wang yan4@yuexiutransport.com (Y.W.)
[2] Yuexiu (China) Transportation Infrastructure Investment Co., Ltd., Hong Kong SAR, China
[3] School of Civil Engineering and Transportation, South China University of Technology, Guangzhou 510641, China; 15391104229@163.com
[4] Xiaoning Institute of Roadway Engineering, Guangzhou 510000, China; m15738523861@163.com
* Correspondence: 201910101471@mail.scut.edu.cn; Tel.: +86-18825181976

Abstract: Given the dominant failure mode of steel bridge deck pavement layers, which is flexural–tensile damage, the dynamic modulus parameters conventionally determined through uniaxial compression testing are found to be inadequate for the design or performance analysis of these layers. In order to simulate the actual stress of a pavement structure under wheel load, the four-point bending fatigue test method and uniaxial compression test method are used to measure the dynamic modulus of an epoxy asphalt mixture, and the differences between the two test methods are analyzed. Furthermore, the four-point bending fatigue test is employed to investigate the dynamic modulus and phase angle properties across varying temperatures and frequencies, facilitating the creation of master curves for these properties and utilizing Sigmoidal models to correlate dynamic modulus data at diverse temperature conditions. This study delves into the influence of epoxy resin content, mixture composition, and aging on the dynamic modulus. The experimental results show that the dynamic modulus measured by uniaxial compression exceeds that obtained from bending fatigue tests, with the difference initially increasing and then decreasing as temperature rises. This discrepancy significantly impacts the mechanical calculations of pavement layers, underscoring the importance of selecting the appropriate testing method. Temperature, frequency, and epoxy resin content have pronounced effects on the viscoelastic properties of the mixtures. Specifically, as temperature increases, the dynamic modulus undergoes a decrease, whereas the phase angle exhibits an increase. Additionally, the dynamic modulus augments with an increase in loading frequency, while the phase angle exhibits varied trends with frequency shifts across different temperatures. Both the WLF and Sigmoidal models are effective in constructing master curve representations for the dynamic flexural modulus and phase angle. The incorporation of epoxy resin transforms asphalt from a primarily viscous to a more elastic material, significantly enhancing the viscoelastic properties of the mixture. Notably, mixtures with 50% and 60% epoxy resin content exhibit comparable dynamic moduli and phase angles, while displaying notably superior performance compared to those with 40% epoxy resin content. For large-scale steel bridge deck pavement, 50% epoxy resin content is recommended. Moreover, epoxy asphalt mixtures demonstrate robust aging resistance, with minimal variations in the dynamic modulus and phase angle before and after aging. The research results can enable the acquisition of dynamic modulus and phase angle data in the whole temperature domain and the whole frequency domain, and provide reliable mixed performance parameters for the study of different application environmental performance of steel bridge deck pavement.

Keywords: dynamic modulus; phase angle; bending test; epoxy asphalt mixture

Citation: He, B.; Huang, Z.; Chen, J.; He, M.; Wang, Y.; Li, J.; Wang, S. Dynamic Viscoelastic Behavior of Epoxy Asphalt Mixture under Four-Point Bending. *Buildings* **2024**, *14*, 3061. https://doi.org/10.3390/buildings14103061

Academic Editors: Songtao Lv, Xinghai Peng and Boumediene Nedjar

Received: 8 August 2024
Revised: 20 September 2024
Accepted: 23 September 2024
Published: 25 September 2024

1. Introduction

The determination of material design parameters is one of the main pieces of structural design work, and the study of material design parameters is indispensable in the design of a steel bridge pavement layer. Among the material design parameters, the dynamic modulus parameter is one of the most important of all design parameters [1–5]. The dynamic modulus can more reasonably reflect the actual stress condition of the road surface under wheel load. The dynamic modulus of the mixture is studied to make the modulus value more reasonable. It is of great significance for pavement material design, and performance evaluation [6–8]. The indoor dynamic modulus evaluation primarily encompasses uniaxial compression testing, indirect tensile assessment, and four-point bending examination [9–11]. A uniaxial compression test serves as a means to evaluate the dynamic compressive modulus of various materials. Specifically, for asphalt mixtures, this technique is among the frequently employed methods for assessing their dynamic modulus [12,13].

A limitation lies in the fact that the cylinder specimen's central point experiences biaxial stress during loading, which significantly deviates from the authentic stress conditions encountered in pavement structures. Furthermore, the specimen exhibits noticeable localized distortions at its extremities, thereby influencing its stress pattern directly. A uniaxial compression test and indirect tensile test adopt the form of compression load, and the obtained modulus parameters are the dynamic compression modulus and tensile modulus. A steel box girder is mainly composed of orthotropic plates. Under the action of temperature load, the whole bridge has huge deflection changes [14]. Relevant research shows that steel deck pavement mainly undergoes bending–tensile failure under the coupling action of traffic load, temperature stress, and steel plate deformation, while a uniaxial compression test and indirect tensile test cannot simulate bending–tensile load, and the obtained modulus parameters are not suitable for epoxy asphalt pavement [15–19].

The four-point bending test primarily focuses on the fatigue characteristics of test materials [20,21]. The specimens were produced by compacting asphalt concrete into slabs through wheel rolling, subsequently cutting the slabs into standard-sized specimens for testing [22,23]. The loading modes include stress control and strain control, and the failure judgment varies according to different control methods [24]. When stress control is adopted, the failure condition is defined by either material fracturing or an increase in tensile strain to twice the initial value. The modulus is deemed to have failed when it drops to half of its initial stiffness value, under strain control conditions [25–27]. The repeated bending–tension mode of the four-point bending test can better simulate the bending–tension stress state of epoxy asphalt pavement under the coupling action of traffic load, temperature stress, and steel plate deformation, and the obtained modulus parameters belong to the dynamic bending–tension modulus, which can better reflect the performance level of steel deck pavement materials.

Researchers often use shear rheometers to analyze the dynamic modulus of epoxy asphalt blends, and sometimes opt for epoxy asphalt mortar as an alternative material for such investigations [28–33]. Xue et al. introduced the viscoelastic properties of asphalt using DSR tests [34]. Xue and colleagues characterized the viscoelastic attributes of epoxy asphalt through dynamic shear rheological examinations, encompassing frequency sweep and creep assessments [35]. Apostolidis examined the compositional and rheological alterations in epoxy asphalt, both with and without filler, utilizing infrared spectroscopy and a dynamic shear rheometer, along with conducting dynamic modulus assessments [36,37]. Guo used four epoxy resin asphalt materials with different crosslinking degrees to carry out dynamic rheological tests [38]. Zhang et al. carried out dynamic frequency scanning for an epoxy asphalt mortar by a dynamic shear rheometer, and obtained dynamic shear modulus parameters [39]. Extensive research has been conducted on the dynamic modulus parameters of epoxy asphalt and epoxy asphalt mastic and many studies have been carried out on the factors affecting the dynamic modulus parameters. However, these studies have been carried out on mastic or mortar specimens and have not considered the influence of the aggregate gradation of the mixes. The related dynamic modulus test has been well studied in common asphalt mixtures and formed a standard test. However, the experimental investigation into the dynamic modulus of epoxy asphalt mixtures is relatively scarce. Luo et al. elucidated the dynamic modulus measurement through uniaxial compression testing and formulated a master curve model for this property [40]. Chen et al. established a dynamic modulus prediction model for an epoxy asphalt mixture through compression tests [41,42]. Yao and their team conducted compression tests to assess the dynamic modulus of epoxy asphalt mixtures [43,44], but they only simply analyzed the changing trend of the dynamic modulus, and did not comprehensively analyze the factors affecting the dynamic modulus.

Reflecting upon the aforementioned analysis, it is evident that significant research on the dynamic modulus of epoxy asphalt mixtures has been conducted using uniaxial compression tests. However, these tests do not accurately replicate the bending and tensile loading conditions typical of steel bridge pavements. Furthermore, the literature lacks studies comparing the dynamic modulus results from four-point bending fatigue tests to those from uniaxial compression tests. There is a notable scarcity of research examining the factors that influence the dynamic modulus parameters of epoxy asphalt mixes, particularly those derived from flexural fatigue tests. In addition, there are few studies on the change in viscoelastic properties of epoxy asphalt mixtures with different grades and different aging degrees, and there are few studies on this part at present.

The four-point bending test is employed to measure the dynamic modulus and phase angle of epoxy asphalt mixtures, examining their trends across varying temperatures and frequencies. Leveraging the time–temperature equivalence principle, master curves for both the dynamic modulus and phase angle are constructed, facilitating data acquisition across the full temperature and frequency spectra. This provides crucial performance parameters for studying steel bridge deck pavement in diverse environments. Furthermore, the experiment serves to contrast the viscoelastic behaviors between epoxy asphalt blends with and without the inclusion of epoxy resin, as well as their changes due to aging, comprehensively elucidating the factors influencing these properties.

2. Methods

The four-point bending test and uniaxial compression test were used to test the difference in the dynamic modulus of the same grade epoxy asphalt mixture at the same frequency and temperature. At the same time, the four-point bending fatigue test method was used to test dynamic modulus and phase angle parameters at different temperatures and frequencies based on the viscosity–temperature equivalent principle, and the main curve of the dynamic modulus and phase angle was constructed to fit the Sigmoidal model of the dynamic modulus at different temperatures. Finally, the effects of epoxy resin parameters, mixture composition, and mixture aging on the dynamic modulus were analyzed. The flow chart is shown in Figure 1:

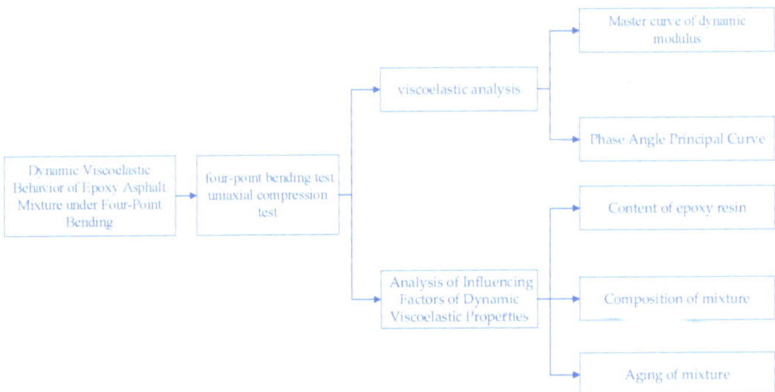

Figure 1. Test flow chart.

3. Experimental Plan

3.1. Materials

The composition of epoxy asphalt comprises TAF epoxy resin sourced from Japan and Shell matrix asphalt. Epoxy resin can enhance the crosslinked network structure of the mixture, thereby improving its strength, stiffness, durability, stability, and adhesion and other key performance indicators. Asphalt gradually changes from the original viscous material to elastic material, and the viscoelastic property of the mixture has been greatly improved, while the Japanese version of TAF epoxy resin has more excellent mechanical properties and thermal stability.

The epoxy resin and curing agent were preheated at 60 °C for about 1 h, and then manually stirred at the ratio of the main agent to curing agent of 56:44 for about 3 min. The mixed epoxy resin is mixed with 165 °C base asphalt in a certain proportion, and then sheared and dispersed by a high-speed shear instrument at 5000 r/min and 5 min rotational speed, mixing into epoxy asphalt with different epoxy ratios. The key technical specifications for the main and curing agents are detailed in Table 1. Diabase is used as an aggregate, and the specific mixture gradation is shown in Table 2. In Gradation 1, the mixture's asphalt-to-aggregate ratio stands at 6.5% with an air void content of 1.4%, while for Gradation 2, they are 6.2% and 2.0%, respectively. The technical specifications of both mixtures are presented in Table 3.

Table 1. Technical Index of TAF Epoxy Resin.

Physical Properties	Main Agent	Curing Agent
Specific gravity	1.15	0.862
Epoxy equivalent	204	-
Acid number (mg, KOH/g)	-	168
Flash point (°C)	239	169

Table 2. Gradation of Epoxy Asphalt Mixture.

Gradation	Percentage of Mass through the Sieve (Square Hole Sieve, mm)								
	13.2	9.5	4.75	2.36	1.18	0.6	0.3	0.15	0.075
Gradation 1	100	99.7	75	58	45.3	36	24.4	18.1	11.1
Gradation 2	100	97.4	44.2	31.1	24.2	19.7	16.3	13.2	9.8

Table 3. Technical Index of Epoxy Asphalt Mixture [45].

Technical Index	Gradation 1	Gradation 2
Apparent density (g/cm^3)	2.586	2.594
Air voids (%)	1.4	2.0
Marshall stability (kN)	68.3	46.3
Flow value (mm)	5.0	4.0
Residual stability (%)	92.5	93.6
Freeze–thaw splitting strength ratio TSR (%)	92.9	92.4

3.2. Experimental Method

The Cooper NU-14 (An independent four-point bending trabecular test system produced by Cooper Technologies, UK) four-point bending testing apparatus is utilized for conducting the four-point bending test, depicted in Figure 2. The size of the test specimen is 385 mm × 65 mm × 50 mm. Firstly, the epoxy asphalt mixture should be mixed, and the specimen with the size of 400 mm × 300 mm × 75 mm should be formed by vibrating wheel rolling equipment, and then the standard specimen with the size of 385 mm × 65 mm × 50 mm should be cut by double-sided dramatic cutting. Relevant research shows that epoxy asphalt pavement is in the linear elastic range during use, and the deformation of epoxy asphalt pavement can be quickly restored after stress is revoked [46]. Existing research indicates that the linear viscoelastic region of asphalt mixtures is typically confined to less than 100 macro-strains [47,48], so the test load strain is controlled to 100 micro-strains, and the load mode adopts a sine wave as the standard loading waveform. Relevant research shows that when the load frequency is 10 Hz, it corresponds to the vehicle load with a road speed of 60~65 km/h [49]. According to the requirements of the dynamic modulus test, the frequencies are set to 15 Hz, 10 Hz, 5 Hz, 1 Hz, 0.5 Hz, and 0.1 Hz, respectively. Considering the actual working temperature range of steel deck pavement, the test temperatures are set at −10 °C, 5 °C, 20 °C, 35 °C, and 50 °C, respectively. In order to compare the difference in the dynamic modulus between the flexural fatigue test and uniaxial compression test, MTS810 (The MTS 810 material test system is manufactured by MTS Systems) is used for the uniaxial compression test, the test frequency is 10 Hz, and the test temperature is consistent with the flexural fatigue test.

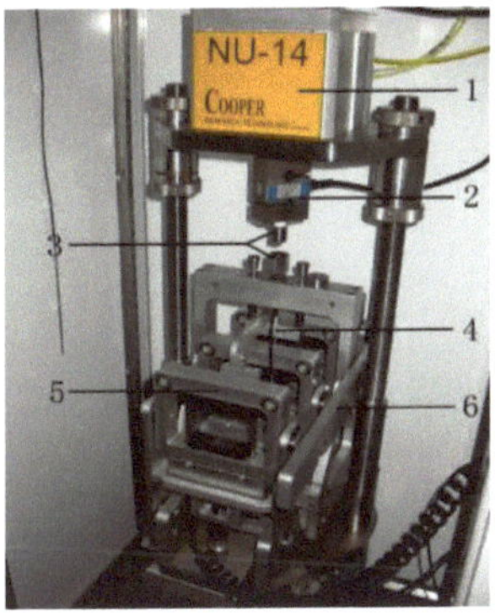

Figure 2. Cooper NU-14 test system (1. Pneumatic service loading device; 2. Force sensor; 3. Clamp joint; 4. Displacement sensor; 5. Four-point chuck; 6. Positioning plate).

4. Test Results

Four-point bending specimens and uniaxial compression specimens are made by Gradation 1, and corresponding tests are carried out. Figures 3 and 4, along with Table 4, present the results. Notably, Figure 3 reveals that at a test frequency of 10 Hz, the dynamic modulus obtained from the four-point bending test is lower compared to that from the uniaxial compression test. As can be seen from Figure 4, with the change in temperature, the dynamic modulus difference between the bending test and uniaxial compression test is constantly changing. The difference between them first increases with temperature. The difference between the dynamic modulus of asphalt mixtures measured by the uniaxial compression test and four-point bending test is the largest at 20 °C, mainly because the viscosity of asphalt at this time is at a medium level and the internal microstructure of the mixture is highly sensitive to the loading method. With the increase in temperature, the viscosity of asphalt decreases gradually and the fluidity of mixtures increases. This makes the stress distribution and transfer mechanism inside the mixture gradually converge in the uniaxial compression test and four-point bending test. At the same time, the elastic recovery ability of asphalt at a high temperature is weakened, and the viscosity is prominent, resulting in a decrease in the dynamic modulus measured by the two test methods. Since this downward trend has a similar rule in both test methods, the difference between them gradually decreases. The research shows that the test result of the bending test and uniaxial compression test is similar at low-temperature and high-temperature stages, but there is a big difference between the two tests' results at the medium-temperature stage. It can be seen from Table 4 that at 20 °C, the dynamic modulus of the bending test is 4944 MPa lower than that of the uniaxial compression test, and the deviation ratio is as high as 58%. The design parameters usually take the dynamic modulus at 20 °C as the calculation standard.

It can be seen that the uniaxial compression test will have a great impact on the parameters of steel deck pavement.

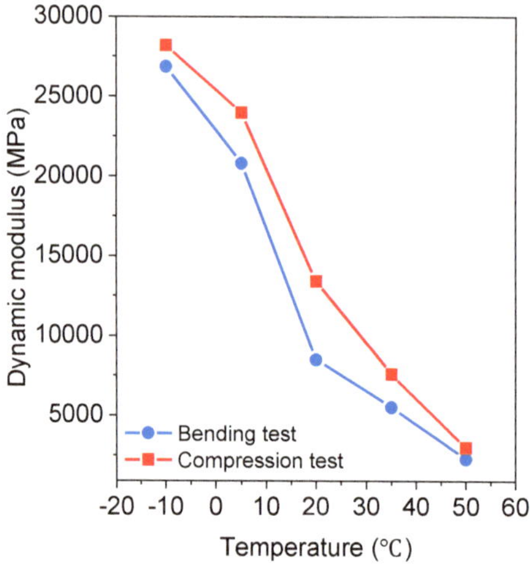

Figure 3. Comparison of dynamic modulus from four-point bending and uniaxial compression tests across various temperatures.

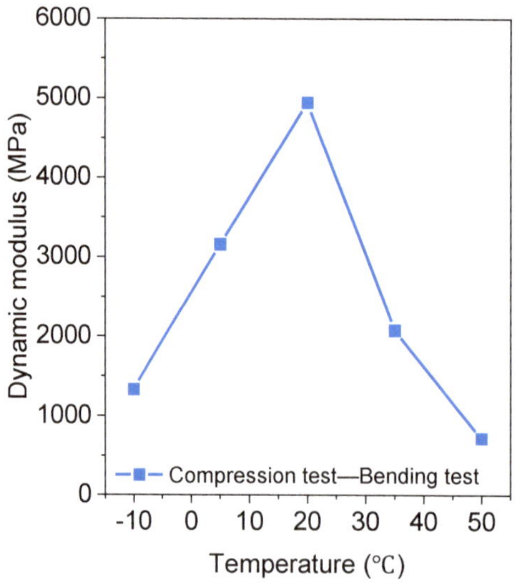

Figure 4. Difference between bending test and compression test at different temperatures.

Table 4. Dynamic Modulus Test Results of Different Tests (10 Hz).

Method	Dynamic Modulus (MPa)				
	−10 °C	5 °C	20 °C	35 °C	50 °C
Bending test	26,848	20,810	8470	5505	2266
Compression test	28,177	23,969	13,414	7581	2982
Difference	1329	3159	4944	2076	716
Proportion	5%	15%	58%	38%	32%

The mixture specimen was made with Gradation 1, and the four-point bending test results are shown in Figures 5 and 6. As depicted in Figure 4, the dynamic modulus of the epoxy asphalt blend diminishes with rising temperatures at a constant load frequency, highlighting a significant influence of temperature on the blend's dynamic modulus. As temperature rises, the epoxy asphalt binder softens, weakening the adhesion between it and the aggregate, resulting in a gradual shift from elasticity to plasticity in the mixture and a consequent decrease in the dynamic modulus. Conversely, at a constant temperature, an increase in load frequency leads to an enhancement in the dynamic modulus of the epoxy asphalt mixture. The epoxy asphalt mixture is a kind of viscoelastic material, which does not deform under the action of external force. After the external force is revoked, the deformation does not recover instantaneously and completely, which shows the hysteresis of load response. With the increase in load frequency, the load time is shortened, and the deformation of the mixture is also small. Therefore, an increase in load frequency leads to a corresponding increase in the dynamic modulus value.

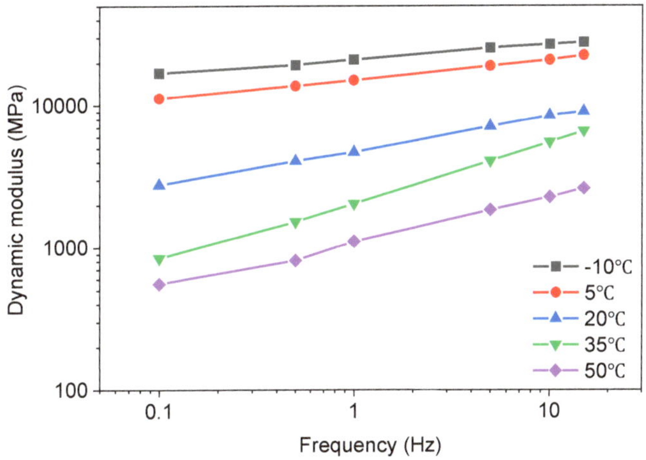

Figure 5. Comparison of dynamic modulus measurement results at different temperatures and frequencies.

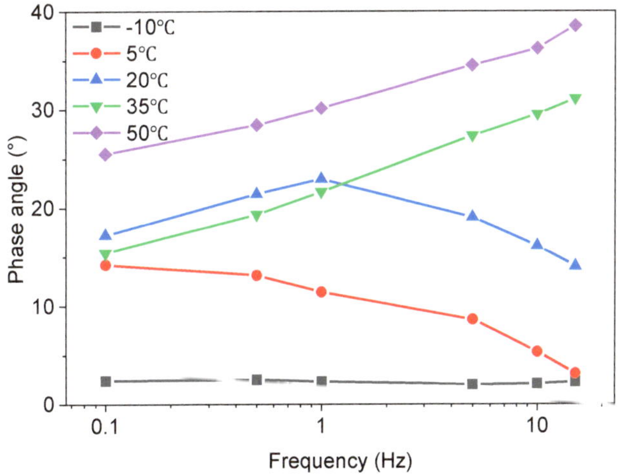

Figure 6. Comparison of phase angle measurement results at different temperatures and frequencies.

As depicted in Figure 6, the epoxy asphalt mixture exhibits a small and stable phase angle at −10 °C, indicative of robust elastic characteristics with minimal dependence on frequency. As temperature rises, the impact of frequency on the mixture's phase angle intensifies. Specifically, at 5 °C, the phase angle diminishes with increasing load frequency, whereas, at 20 °C, it initially rises and subsequently declines. At higher temperatures of 35 °C and 50 °C, the phase angle consistently increases with frequency. In low-temperature settings, the asphalt binder significantly influences the mixture's modulus. When the loading frequency is low (i.e., longer loading durations), the mixture displays pronounced viscosity, leading to an increase in the phase angle as frequency decreases. As temperature increases, the epoxy asphalt binder softens, diminishing its bonding strength with aggregates, thereby transitioning the mixture from elastic to plastic behavior. Under low load frequencies (longer action times), the rigidity of the mixture is largely governed by the embedded mineral aggregate skeleton, contributing to a smaller phase angle. Conversely, with higher load frequencies (shorter action times), the mixture exhibits enhanced hysteretic load response, equivalent to an enlarged phase angle, highlighting its more evident external delay in responding to load changes.

5. Viscoelastic Analysis

5.1. Master Curve of Dynamic Modulus

The research indicates that the mechanical state of asphalt mixtures remains consistent under varying load frequencies and temperatures. It implies that the mechanical behavior of viscoelastic materials during their action process is influenced by time and temperature in a manner that exhibits certain comparable processes [50]. This equivalence can be achieved by a shift factor. This suggests that the mechanical data measured at one temperature can be converted into mechanical data at another temperature by the shift factor, which is called the time–temperature equivalence principle. Based on this principle, the main curve at a certain temperature is formed. This displacement, $\lg\alpha_T$, is called the shift factor, and the shift factor $\lg\alpha_T$ is only related to temperature. The WLF equation is used to calculate the shift factor [51], and the calculation equation is shown in Equation (1).

$$\lg\alpha_T = \frac{-8.86(T - T_s)}{101.6 + (T - T_s)} \tag{1}$$

where T is any other temperature, and T_s is the fiducial temperature.

In practice, the fiducial temperature T_s is unknown and is not included in the measured temperature conditions achieved. Therefore, it is necessary to calculate the fiducial temperature T_s by derivation. A certain temperature T_0 is selected as the reference temperature in the measured temperature conditions, and the shift factor $\lg\alpha_T$ of other temperatures T_i to the reference temperature T_0 is as follows: assuming T is any temperature, Equation (1) is applied for T_0 and T.

$$\lg\alpha_{T_0} = \frac{-8.86(T_0 - T_s)}{101.6 + (T_0 - T_s)} \tag{2}$$

$$\lg\alpha_T = \frac{-8.86(T - T_s)}{101.6 + (T - T_s)} \tag{3}$$

The viscoelastic characteristic function curve is moved at temperature T to T_0, and the shift factor is recorded at this time as $\lg\alpha'_{T_0}$. Therefore, from Equations (2) and (3), we can obtain

$$\lg\alpha'_{T_0} = -8.86 \times 101.6 \frac{(T - T_0)}{(101.6 + T - T_s)(101.6 + T_0 - T_s)} \tag{4}$$

In the above Equation (4), there is only one unknown number T_s. By solving the quadratic equation about T_s, the fiducial temperature T_s corresponding to $T \sim T_0$ is obtained. After the fiducial temperature T_s is obtained, Equation (4) is substituted to obtain the shift factor when the required temperature translates to the selected reference temperature T_0. Meanwhile, the master curve can be obtained under the reference temperature condition. According to Equations (1) and (4), the reference temperature is selected as 35 °C, and the value of the fiducial temperature T_s is calculated inversely. Table 5 presents the detailed computation outcomes:

Table 5. The reference temperature is the base temperature T_s at 35 °C.

Temperature/°C	Displacement Factor $\lg\alpha'_{T_0}$	T_{s1}/°C	T_{s2}/°C
20	2.50	55.23	202.97
35	0	-	-
50	−1.8	57.16	231.04
Average	0.233	56.2	217

It can be seen from Table 5 that two different analytical solutions can be obtained by solving the quadratic equation about T_s, which are denoted as T_{s1} and T_{s2}, respectively. Zhang verified that the fiducial temperature range of asphalt materials is generally between 30 °C and 60 °C [44]. Therefore, T_{s1} is the fiducial temperature value when the reference temperature is 35 °C, and its average value is 56.2 °C. By substituting the fiducial temperature into Equation (4), the translation factors of dynamic modulus viscoelastic curves of the epoxy asphalt mixture at −10 °C, 5 °C, 35 °C, and 50 °C relative to those at 20 °C are calculated, respectively. The outcomes are presented in Table 6 for specific details.

Table 6. Displacement factors of different temperature conditions relative to 20 °C temperatures.

Temperature/°C	−10	5	35	50
$\lg\alpha_T$	11.66	4.10	−2.57	−4.33

Figure 7 portrays the primary curve showcasing the dynamic modulus of the epoxy–asphalt blend at 20 °C, utilizing the shift factors derived from Table 6 for various temperatures in relation to 20 °C. The Sigmoidal modulus approach is adopted to align with the experimental outcomes of the dynamic modulus tests conducted on the blend. As a result, a Sigmoidal representation of the mixture's dynamic modulus primary curve is obtained at the reference temperature of 20 °C. This fitted curve is illustrated in Figure 7, while the corresponding fitting model is presented in Equation (5).

$$E^* = 3600 - \frac{35{,}560}{1 + e^{0.3806 \ln f - 2.02}} \tag{5}$$

where f is the load frequency, Hz. E^* is the dynamic modulus of the asphalt mixture.

The correlation coefficient R^2 is 0.99773, which indicates that the model fits the experimental results well. What is more, this paper evaluates the translocation factors under various alternative reference temperatures and obtains the master curve family of the dynamic modulus of the epoxy asphalt mixture through the time–temperature equivalence principle, as shown in Figure 8.

As evident from Figure 8, the primary curves for the dynamic modulus at varying temperatures exhibit comparable shapes, indicating a dependency of the mixture's dynamic modulus on load frequency across temperature ranges. The consistent pattern of morphological shifts with frequency changes further underscores this relationship. Distinct intervals exist between various master curves, reflecting varying temperature sensitivities of the mixtures under different thermal conditions. Notably, the extreme points of the viscoelastic curves for the same asphalt mixture remain consistent across different temperatures, suggesting that these extreme values are not constrained by temperature variations. The Sigmoidal function is applied to fit the modulus test outcomes of the epoxy asphalt mixture, yielding the Sigmoidal form of the dynamic modulus master curve for the mixture under various reference temperature conditions. Refer to Table 7 for specific modeling details. It can be seen from Table 7 that the correlation coefficient R2 is above 0.975, which indicates that the regression model has a good fitting effect on the test results. Using the model equation in Table 7, the dynamic modulus parameters under different temperatures and frequencies can be obtained, which is of great significance to the structural design of epoxy asphalt mixture pavement.

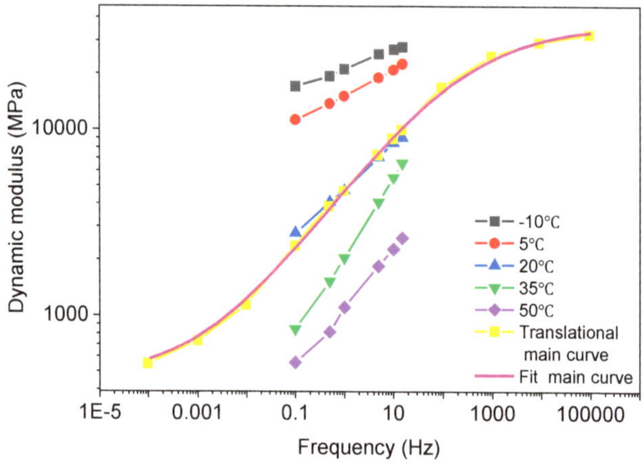

Figure 7. Master curve of dynamic modulus at 20 °C.

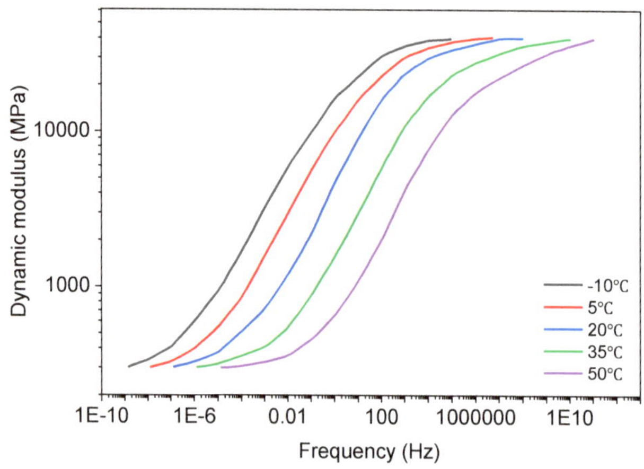

Figure 8. Main curve family of dynamic modulus of epoxy asphalt mixture.

Table 7. Sigmund model expressions at different reference temperatures.

Reference Temperature/°C	Model Expression	Correlation Coefficient R^2
−10	$E^* = 44{,}000 - \frac{43{,}800}{1+e^{0.2942\ln f-0.53}}$	0.986
5	$E^* = 42{,}000 - \frac{41{,}750}{1+e^{0.3086\ln f-1.24}}$	0.996
20	$E^* = 36{,}000 - \frac{355{,}600}{1+e^{0.3806\ln f-2.02}}$	0.994
35	$E^* = 33{,}000 - \frac{32{,}730}{1+e^{0.3463\ln f-3.12}}$	0.986
50	$E^* = 30{,}000 - \frac{29{,}710}{1+e^{0.3404\ln f-4.38}}$	0.976

Note: f in the table is the load frequency, Hz. E^* is the dynamic modulus.

As can be seen from Table 8, the difference between the dynamic modulus of the mixture obtained from the Sigmoidal model and the four-point bending test is small, which proves that the Sigmoidal model has strong reliability.

Table 8. Comparison of dynamic modulus between bending test and Sigmund model at 5 Hz.

Method	Dynamic Modulus (MPa)				
	−10 °C	5 °C	20 °C	35 °C	50 °C
Bending test	22,126	14,113	8241	3054	1105
Sigmund model	21,481	13,705	7400	2616	920
Difference	645	408	841	438	185
Proportion	3%	3%	11%	17%	20%

5.2. Phase Angle Principal Curve

The drawing method is the same as the main curve of the dynamic modulus of the epoxy asphalt mixture. According to Equations (1) and (4), the fiducial temperature T_s can be calculated inversely by selecting the reference temperature. According to the WLF equation, the shift factor of the phase angle curve under other temperature conditions relative to 20 °C temperature conditions is obtained, so as to make the phase angle master curve of the epoxy asphalt mixture under the 20 °C temperature conditions, as shown in Figure 9.

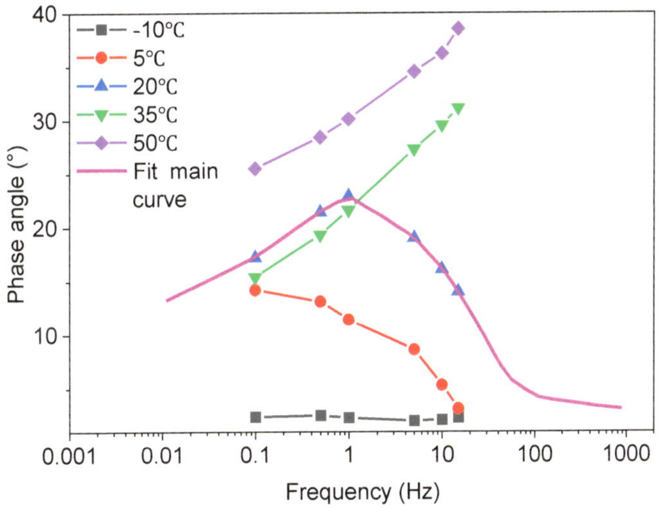

Figure 9. Principal curve of phase angle under condition of reference temperature of 20 °C.

Figure 9 reveals that the phase angle of the epoxy asphalt mixture is significantly influenced by load frequency, initially rising and subsequently declining as frequency increases. At low frequencies and longer durations, the stiffness of the mixture is primarily governed by the embedded mineral skeleton, resulting in a smaller phase angle. Conversely, at high frequencies with brief action times, the mixture's deformation is minimal, and external viscosity becomes less pronounced. In summary, the effect of load frequency on the phase angle of the epoxy asphalt mixture is different only in different frequency stages, which is obviously different from the dynamic modulus.

6. Analysis of Influencing Factors of Dynamic Viscoelastic Properties

6.1. Content of Epoxy Resin

Specimens of the epoxy asphalt mixture with epoxy resin contents of 30%, 40%, 50%, and 60% were prepared using Gradation 1. A dynamic frequency scanning test was conducted at various temperatures using a COOPER testing machine. The test results of the dynamic modulus and phase angle are shown in Figures 10 and 11.

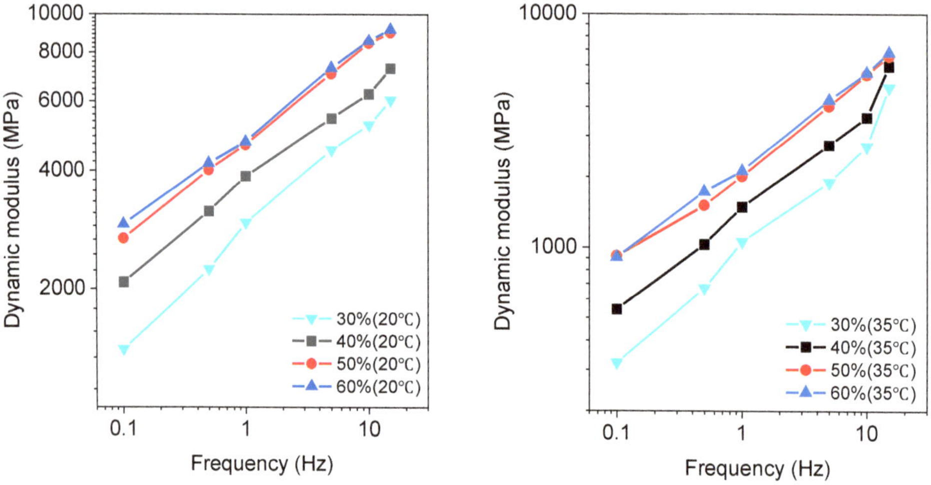

(**a**) 20 °C (**b**) 35 °C

Figure 10. The dynamic modulus of the mixture varies with differing epoxy resin content.

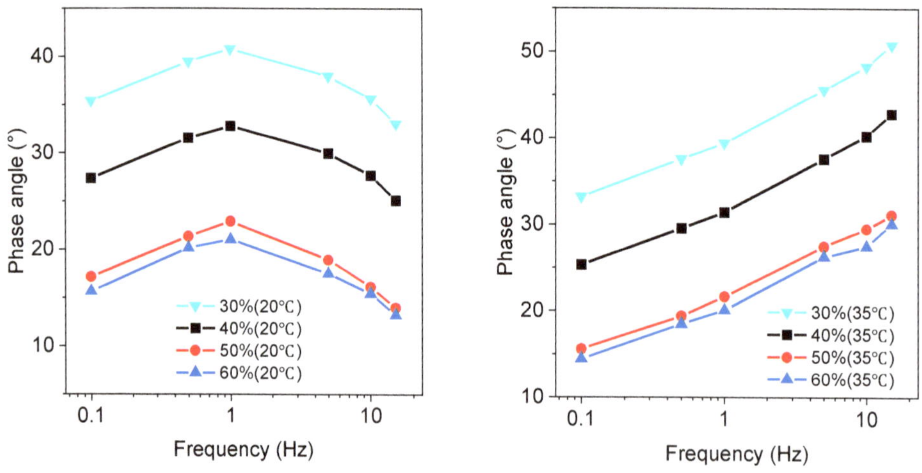

(**a**) 20 °C (**b**) 35 °C

Figure 11. Phase angle of mixture with different epoxy resin content.

According to Figures 10 and 11, under the conditions of 20 °C and 35 °C, the dynamic modulus of the epoxy asphalt mixture increases with the increase in epoxy resin content, and the phase angle decreases with the increase in epoxy resin content. This indicates that with the addition of epoxy resin, the crosslinked network structure of the mixture is enhanced, thus improving its strength, stiffness, durability, stability, adhesion, and other key performance indicators. The asphalt gradually changes from the original viscous material to the elastic material, and the viscoelastic property of the mixture is greatly improved. At the same time, when the epoxy resin content is 50% and 60%, the dynamic modulus and phase angle of the mixture have a small difference, but compared with the 40% content, the dynamic modulus of the mixture with 50% epoxy resin content is increased by 25.88% at 10 Hz and 20 °C, and by 34.48% at 10 Hz and 35 °C. The phase angle decreases by 58.04% at 10 Hz and 20 °C, and 36.23% at 10 Hz and 35 °C, showing better performance. Compared with the mixture with 60% epoxy resin content, the dynamic modulus only decreased by 1.45% at 10 Hz and 20 °C, and the phase angle only increased by 7.6%. Therefore, considering the influence of epoxy resin content on the performance of the mixture, combined with the actual situation such as the engineering cost, the epoxy resin content is recommended to be 50%.

At the same time, in order to analyze the changes in microscopic morphology of mixtures with different epoxy resin contents, SEM was used to conduct a microscopic analysis of mixtures with different epoxy resin contents, as shown in Figure 12.

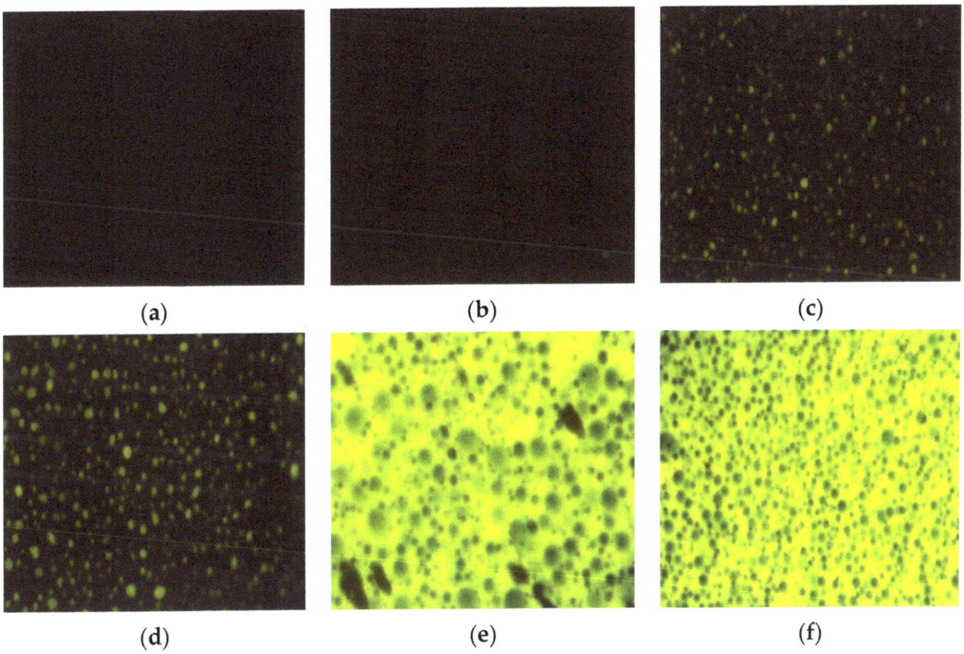

Figure 12. Fluorescence microscope photos: (**a**) 0%; (**b**) 10%; (**c**) 20%; (**d**) 30%; (**e**) 40%; (**f**) 50%.

The fluorescence microscope observation images are summarized in Figure 12. Yellow is for epoxy resin and black is for basic asphalt. From Figure 12, with the gradual increase in the epoxy resin content, the yellow particles in the picture gradually increased, and the size of the particles also enlarged. When the proportion of epoxy reaches 10%, the epoxy is just doped into the asphalt base in a tiny medium, which is only equivalent to the role of a modifier. When the proportion of epoxy resin reaches 20%, the number of yellow particles increases significantly. When the proportion of epoxy resin is increased to 30%, the epoxy resin particles aggregate with each other and form larger epoxy resin particles. The whole mixture is still mainly coated with asphalt epoxy resin. When the proportion of epoxy resin reaches 40%, epoxy asphalt has changed from bitumen-based to epoxy resin-based. The epoxy resin has formed a skeleton, and asphalt is distributed in the epoxy skeleton, but the particle size of the asphalt is not uniform. When the proportion of epoxy resin reaches 50%, asphalt particles are evenly distributed in epoxy. The performance of the epoxy asphalt mixture has been improved. The asphalt mixture with the epoxy resin content can be used for the pavement of a long-span steel bridge deck.

6.2. Composition of Mixture

The test specimens were made with Gradation 1 and Gradation 2, respectively, and the test frequency was 10 Hz. Figure 13 presents the outcomes of the dynamic modulus testing. Gradation 1 is typically used as the lower layer of steel deck pavement due to its fine gradation, high asphalt content, and good fatigue performance. On the other hand, Gradation 2 is usually used as the upper layer of steel deck pavement because of its coarse gradation, large surface structure, and good anti-skid performance. As can be seen from Figure 11, coarse gradation has a higher dynamic modulus. Due to the fact that a coarse aggregate accounts for a larger proportion and a fine aggregate is less evenly distributed in coarse gradation, it is possible to form a better skeleton structure, which increases the friction force within the material. Hence, the coarse gradation asphalt mixture exhibits a higher dynamic modulus than the fine gradation counterpart.

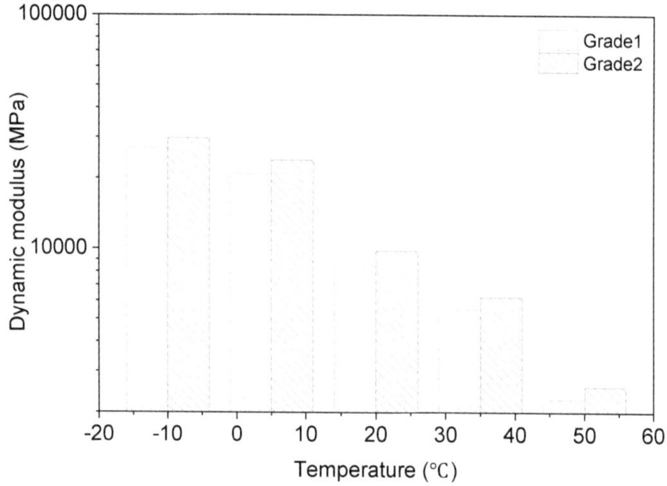

Figure 13. Dynamic modulus test results of different gradations.

6.3. Aging of Mixture

With the increase in service life of steel bridge deck pavement, pavement materials will continue to age, and its performance will also change accordingly. It can be seen that the anti-aging ability of pavement materials is an important factor affecting the service quality and life of pavement structures. The samples underwent a simulated long-term aging process in an oven at 85 °C for 5 days, mirroring the aging of real bridge deck pavement. Following this, a dynamic modulus scanning test was performed utilizing a COOPER material tester. Figures 14 and 15 present the outcomes, revealing minimal changes in the dynamic modulus and phase angle of the epoxy asphalt mixture before and after aging. This stability demonstrates the excellent aging resistance of the epoxy asphalt mixture.

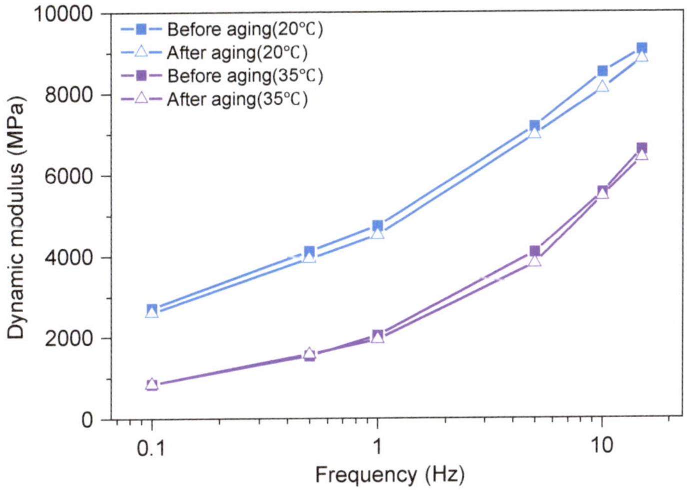

Figure 14. Dynamic modulus of mixture before and after aging.

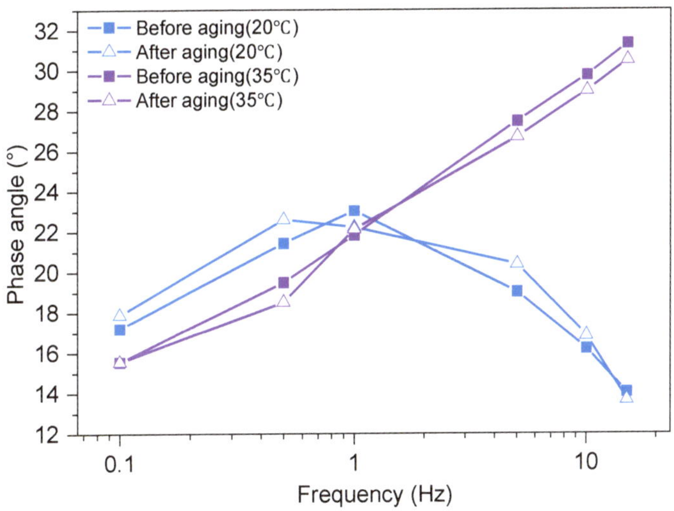

Figure 15. Phase angle of mixture before and after aging.

7. Conclusions

The viscoelastic parameters of the epoxy asphalt mixture are ascertained through four-point bending tests, upon which a master curve model of viscoelastic parameters is formulated; based on a laboratory test and microscopic analysis, the effects of epoxy resin content, mixture gradation, and mixture aging on the modulus of an epoxy asphalt mixture were investigated, leading to the derivation of research conclusions.

(1) The dynamic modulus ascertained via uniaxial compression tests exceeds that derived from bending tests; this disparity initially increases and then diminishes as temperature rises. At 20 °C, this discrepancy peaks, with a deviation of 58%.

(2) As temperature rises, the dynamic modulus diminishes while the phase angle augments. Conversely, the dynamic modulus intensifies with an increase in load frequency, but the phase angle's response to frequency alterations varies across temperatures.

(3) Utilizing the WLF and Sigmoidal models enables the construction of a master curve model for the dynamic bending–tensile modulus and phase angle, thereby facilitating the extraction of these parameters across the entire temperature and frequency spectra.
(4) When the epoxy resin content is 50% and 60%, the dynamic modulus and phase angle of the mixture have a small difference, but compared with the 40% content, the dynamic modulus of the mixture with 50% epoxy resin content is increased by 25.88% at 10 Hz and 20 °C, and by 34.48% at 10 Hz and 35 °C. The phase angle decreases by 58.04% at 10 Hz and 20 °C, and 36.23% at 10 Hz and 35 °C, showing better performance. Compared with the mixture with 60% epoxy resin content, the dynamic modulus only decreased by 1.45% at 10 Hz and 20 °C, and the phase angle only increased by 7.6%. Therefore, considering the influence of epoxy resin content on the performance of the mixture, combined with the actual situation such as the engineering cost, the epoxy resin content is recommended to be 50%.
(5) The epoxy asphalt blend demonstrates robust resistance to aging, exhibiting minimal variation in its dynamic modulus and phase angle before and after the aging process.

By comparing the dynamic modulus parameters obtained by the four-point bending test and uniaxial compression test, it is found that the results of the conventional uniaxial compression test are on the high side, which is not conducive to the mechanical calculation results of steel deck pavement. In the stage of pavement structure design, the dynamic modulus is one of the important parameters to calculate the stress and strain of the pavement structure. Through the dynamic modulus master curve proposed in this paper, the mechanical response of a pavement structure under vehicle load can be calculated more accurately, so as to optimize the thickness and combination of the pavement structure layer, and improve the overall bearing capacity and service life of pavement.

In order to ensure the repeatability of the test data, the standardized process, environmental conditions, and equipment calibration of the test should be strictly controlled. Since the asphalt mixture is a temperature-sensitive material, when measuring the dynamic modulus at different temperatures, the temperature of the test piece should be accurately controlled by using an incubator or heating/cooling device. The loading method should also be carried out in strict accordance with the test standards to ensure that the loading rate, loading direction, loading point position, and other parameters of each test are the same. In the four-point bending test, it is necessary to pay special attention to the position of the loading point and the distribution of the loading force to ensure the uniform force of the specimen in the bending process.

In this paper, the main curve equation of the dynamic modulus and phase angle is established by using the four-point bending fatigue test, and the factors affecting the dynamic modulus are studied. However, the influence of temperature and loading frequency on the dynamic modulus of an epoxy asphalt mixture is only carried out, and the influence of the oilstone ratio, voidage, and other factors on the dynamic modulus of an epoxy asphalt mixture needs to be further studied. In the next step, the microstructure of epoxy asphalt mixtures can also be studied in detail. Due to the limitation of space, this paper only studies the main curve of phase angle parameters at 20 °C, hoping to preliminarily analyze the trend of the phase angle through the main curve of a temperature, and then carry out systematic research on the phase angle.

Author Contributions: Conceptualization, B.H. and Z.H.; Data curation, Y.W.; Formal analysis, B.H., Z.H., J.C. and M.H.; Methodology, B.H., Z.H., J.C. and M.H.; Software, B.H., Z.H., J.C. and J.L.; Validation, B.H., M.H., J.L. and S.W.; Visualization, Z.H., J.C., Y.W. and J.L.; Writing—original draft, B.H. and Z.H.; Writing—review and editing, B.H. and S.W. All authors have read and agreed to the published version of the manuscript.

Funding: This research was funded by the Natural Science Fund of Guangdong Province, grant number 2019A1515011965, and Natural Science Foundation of China, grant number 51808228.

Data Availability Statement: The data presented in this study are available on request from the corresponding author.

Conflicts of Interest: Authors Baiqing He, Jingsong Chen, Mu He and Yan Wang were employed by the company Guangzhou North Second Ring Road Traffic Technology Co., Ltd. Authors Baiqing He, Jingsong Chen, Mu He and Yan Wang were employed by the company Yuexiu (China) Transportation Infrastructure Investment Co., Ltd. The remaining authors declare that the research was conducted in the absence of any commercial or financial relationships that could be construed as a potential conflict of interest.

References

1. Huang, W. Design of deck pavement for long-span steel bridges. *China Civ. Eng. J.* **2007**, *40*, 65–77.
2. Liu, Y.; Shen, Z.; Liu, J.; Chen, S.; Wang, J.; Wang, X. Advances in the application and research of steel bridge deck pavement. *Structures* **2022**, *45*, 1156–1174. [CrossRef]
3. Pan, Y.; Guo, Z. Design Research on Steel Deck Asphalt Pavement of Bridge with High Longitudinal Slope and Small Turning Radius. *J. Tongji Univ. (Nat. Sci.)* **2012**, *40*, 55–59.
4. Haibara, Y.; Ge, H.; Sun, J. Materials Optimization and Service Performance Evaluation of a Novel Steel Bridge Deck Pavement Structure: A Case Study. *Appl. Sci.* **2023**, *13*, 5930. [CrossRef]
5. Pan, Y.; Guo, Z. Design of steel deck pavement of long-span continuous steel box girder bridges. *J. Tongji Univ. (Nat. Sci.)* **2013**, *41*, 841–847.
6. Wang, B.; Tang, Y.; Liu, Y.; Wang, L. Research on multi-factor optimization of the orthotropic steel bridge deck pavement system based on orthogonal experimental design. *J. Xi'an Univ. Archit. Technol.* **2018**, *50*, 842–849+855.
7. Wu, C.; Liu, H.; Zhang, S.; Cui, B.; Ding, W. Influence of pavement on fatigue stress range of orthotropic steel deck. *Eng. Sci.* **2010**, *12*, 39–42.
8. Zheng, G.; Zhang, N.; Lv, S. Experimental Study on Dynamic Modulus of High Content Rubber Asphalt Mixture. *Buildings* **2024**, *14*, 434. [CrossRef]
9. Huang, Y.; Liu, Z.; Wang, X.; Li, S. Comparison of HMA dynamic modulus between trapezoid beam test and SPT. *J. Cent. South Univ. (Sci. Technol.)* **2017**, *48*, 3092–3099.
10. Liu, Y.; Zhang, X.-N.; Chen, S.-X. Method to Test Dynamic Modulus of Asphalt Mixture Based on Semi-Circular Specimens. *J. South China Univ. Technol.* **2010**, *38*, 41–46.
11. Yao, A.; Zheng, P. Test method and influencing factors for the dynamic compression elastic moduli of asphalt mixtures. *J. Univ. Sci. Technol. Beijing* **2011**, *33*, 1275–1279.
12. Gao, M.; Xue, Y.; Guan, P.; Yuan, F. Influence Mechanism of Epoxy Resin and Curing Agent on High-Temperature Performance of Asphalt. *Sains Malays.* **2020**, *49*, 661–669. [CrossRef]
13. Yan, X.; Zhou, Z.; Liu, Z.; Zhou, Y. Investigation of Dynamic Viscoelastic Characteristics of Permeable Asphalt. *Materials* **2024**, *17*, 2984. [CrossRef] [PubMed]
14. Weng, Y.-W.; Lin, D.; Huang, W.H.; Yeh, M.C. Research of Steel Deck Pavement Suitable Paving Materials. *Adv. Mater. Res.* **2011**, *243–249*, 4092–4096. [CrossRef]
15. Yao, B.; Cheng, G.; Wang, X. Dynamic modulus of epoxy asphalt mixture based on bending test mode. *J. Southeast Univ. (Nat. Sci. Ed.)* **2011**, *41*, 597–600.
16. Witczak, M.W.; Root, R.E. Summary of Complex Modulus Laboratory Test Procedures and Results. In *1973 Symposium on Fatigue and Dynamic Testing of Bituminous Mixtures*; ASTM International: Philadelphia, PA, USA, 1974.
17. Saito, M.; Kurata, K.; Arakawa, S. Applicability of indirect tensile test method with frame specimens to concrete. *J. Soc. Mater. Sci.* **1982**, *31*, 103–108. [CrossRef]
18. Liu, X.; Wu, Z.; Min, Z.; Zhang, L. Investigation on the Preparation and Performances of Epoxy-Modified Asphalt Binder and Its Mixtures. *Materials* **2024**, *17*, 2539. [CrossRef]
19. Xiao, M.; Chen, Y.; Feng, H.; Huang, T.; Xiong, K.; Zhu, Y. Evaluation of Fatigue Behavior of Asphalt Field Cores Using Discrete Element Modeling. *Materials* **2024**, *17*, 3108. [CrossRef]
20. Pan'ko, I.N.; Zazulyak, V.A. Four-point bending of a cylindrical specimen with an external circumferential notch. *Sov. Mater. Sci.* **1979**, *15*, 33–35. [CrossRef]
21. Baldo, N.; Rondinella, F.; Daneluz, F.; Vacková, P.; Valentin, J.; Gajewski, M.D.; Król, J.B. Stiffness Moduli Modelling and Prediction in Four-Point Bending of Asphalt Mixtures: A Machine Learning-Based Framework. *CivilEng* **2023**, *4*, 1083–1097. [CrossRef]
22. Denk, J.; Dallmeier, J.; Huber, O. Four-point-bending device for bending moment controlled cyclic reverse loading on plate materials and its application on AZ31B magnesium sheets. *Int. J. Fatigue* **2018**, *109*, 49–59. [CrossRef]
23. Sabuncuoglu, B.; Vanhee, F.; Willems, G.; Vandevelde, B.; Vandepitte, D. Evaluation of Fatigue Behavior of Lead-Free Solder Joints in Four-Point Bending Test by Finite-Element Modeling. *IEEE Trans. Compon. Packag. Manuf. Technol.* **2017**, *7*, 1957–1964. [CrossRef]
24. Wan, C.; Zhang, X.-N.; Yu, J.-M. Evaluation of fatigue performance of asphalt mixture using cooper NU-14 test system. *J. South China Univ. Technol.* **2009**, *37*, 52–56.
25. Yoshimoto, T.; Matsuo, T. Development of four-point bending fatigue test method using continuously hydrogen-charging pipe specimen. In Proceedings of the 12th International Conference on Damage Assessment of Structures (DAMAS), Kitakyushu, Japan, 10–12 July 2017.

26. Zhai, T.; Xu, Y.; Martin, J.; Wilkinson, A.; Briggs, G. A self-aligning four-point bend testing rig and sample geometry effect in four-point bend fatigue. *Int. J. Fatigue* **1999**, *21*, 889–894. [CrossRef]
27. Zou, X.; Ding, B.; Peng, Z.; Li, H. Damage analysis four-point bending fatigue tests on stone matrix asphalt using dissipated energy approaches. *Int. J. Fatigue* **2020**, *133*, 105453. [CrossRef]
28. Han, X.; Su, W.; Gong, J.; Xi, Z.; Zhang, J.; Cai, J.; Wang, Q.; Xie, H. Microstructure and dynamic mechanical properties epoxy/asphaltene composites. *J. Therm. Anal. Calorim.* **2022**, *147*, 2209–2219. [CrossRef]
29. Hao, Z.; Li, L.; Sheng, X.; Liu, P. Development and characterization of high temperature curing epoxy asphalt. *New Chem. Mater.* **2017**, *45*, 208–210+214.
30. Jing, F.; Wang, R.; Zhao, R.; Li, C.; Cai, J.; Ding, G.; Wang, Q.; Xie, H. Enhancement of Bonding and Mechanical Performance of Epoxy Asphalt Bond Coats with Graphene Nanoplatelets. *Polymers* **2023**, *15*, 412. [CrossRef]
31. Li, J.; Zhu, Y.; Yu, J. Study on Physical Properties, Rheological Properties, and Self-Healing Properties of Epoxy Resin Modified Asphalt. *Sustainability* **2023**, *15*, 689. [CrossRef]
32. Liu, H.; Li, X. High Temperature Rheological Properties of Waterborne Epoxy Emulsified Asphalt. *J. Chongqing Jiaotong Univ. Nat. Sci.* **2020**, *39*, 67–73.
33. Ishaq, M.A.; Giustozzi, F. Correlation between Rheological Fatigue Tests on Bitumen and Various Cracking Tests on As phalt Mixtures. *Materials* **2021**, *14*, 7839. [CrossRef] [PubMed]
34. Xue, L.-X. Viscoelastic characteristics of epoxy asphalt. *J. South China Univ. Technol.* **2008**, *36*, 71–75.
35. Xu, Z.; Liu, P.; Zeng, Q.; Xiong, Z.; Yang, X. Research on high temperature performance of epoxy asphalt based on rheology and molecular dynamics. *New Chem. Mater.* **2023**, *51*, 307–312+317.
36. Apostolidis, P.; Liu, X.; Erkens, S.; Scarpasa, T. Oxidative aging of epoxy asphalt. *Int. J. Pavement Eng.* **2022**, *23*, 1471–1481. [CrossRef]
37. Apostolidis, P.; Liu, X.; Marocho, P.; van de Ven, M.; Erkens, S.; Skarpas, T. Evaluation of epoxy modification in asphalt mastic. *Mater. Struct.* **2020**, *53*, 1–15. [CrossRef]
38. Guo, J.; Min, Z.-H.; Zhang, Z.-J.; Kong, D.-L. Dynamic mechanical property of epoxy resin asphalt in different crossing-density. *J. Change Univ.* **2011**, *31*, 26–30.
39. Zhang, X.; Ren, Y.; Chi, F. Study of viscous-elastic properties of epoxy asphalt mixtures by using dynamic frequency sweep. *J. Huazhong Univ. Sci. Technol.* **2009**, *37*, 102–105.
40. Luo, S.; Qian, Z.; Harvey, J. Research on Dynamic Modulus for Epoxy Asphalt Mixtures and Its Master Curve. *China J. Highw. Transp.* **2010**, *23*, 16–20.
41. Chen, L.; Qian, Z. Study on Dynamic Modulus of Epoxy Asphalt Mixture Based on Simple Performance Test. *J. Build. Mater.* **2013**, *16*, 341–344.
42. Chen, L.; Qian, Z.; Luo, S. Experimental study of dynamic modulus of thermosetting epoxy asphalt mixture for steel deck pavement. *J. Southeast Univ. (Engl. Ed.)* **2010**, *26*, 112–116.
43. Yao, B.; Cheng, G.; Wang, X.; Cheng, C. Characterization of the stiffness of asphalt surfacing materials on orthotropic steel bridge decks using dynamic modulus test and flexural beam test. *Constr. Build. Mater.* **2013**, *44*, 200–206. [CrossRef]
44. Yao, B.; Cheng, G.; Wang, X.; Cheng, C.; Liu, S. Linear viscoelastic behaviour of thermosetting epoxy asphalt concrete-Experiments and modeling. *Constr. Build. Mater.* **2013**, *48*, 540–547. [CrossRef]
45. *JTJ 052-2000*; Standard Test Methods of Bitumen and Bitu minous Mixtures for Highway Engineering. People's Communications Press: Beijing, China, 2000.
46. Buttlar, W.G.; Roque, R. Evaluation of empirical and theoretical models to determine asphalt mixture stiffnesses at low temperatures. In Proceedings of the 1996 Conference of the Association of Asphalt Paving Technologies, Baltimore, MD, USA, 18–20 March 1996; Asphalt Paving Technology: Cleveland, OH, USA, 1996.
47. Harran, G.; Shalaby, A. Dynamic modulus of asphalt concrete in uniaxial compression and indirect tension modes. In Proceedings of the Annual Conference of the Canadian Society for Civil Engineering 2007, Yellowknife, NT, USA, 6–9 June 2007.
48. He, Y.; Lv, S.; Xie, N.; Meng, H.; Lei, W.; Pu, C.; Ma, H.; Wang, Z.; Zheng, G.; Peng, X. Comparative Study on the Dynamic Response of Asphalt Pavement Structures: Analysis Using the Classic Kelvin, Maxwell, and Three-Parameter Solid Models. *Buildings* **2024**, *14*, 295. [CrossRef]
49. *Paving Technology 2003, 10–12 March 2003*; Association of Asphalt Paving Technologist: Lexington, KY, USA, 2003.
50. Li, X.; Liang, N.; Chen, L. Dynamic modulus and time-temperature equivalence equation of asphalt concrete. *J. Change Univ.* **2014**, *34*, 35–40.
51. Li, Z.L.; Qin, Y.; Sun, B.; Jia, C.L.; Zhang, W.J.; Yan, B.J.; Shi, Q.L. A fractional approach to the time-temperature dependence of dynamic viscoelastic behavior. *J. Mech. Sci. Technol.* **2019**, *33*, 139–147. [CrossRef]

buildings

MDPI

Article

Laboratory Performance and Micro-Characteristics of Asphalt Mastic Using Phosphorus Slag Powder as a Filler

Xiao Li [1,*], Xiaoge Tian [2] and Ronghua Ying [2]

1 Faculty of Civil Engineering, Changsha University of Science and Technology, Changsha 410114, China
2 School of Traffic and Transportation Engineering, Changsha University of Science and Technology, Changsha 410114, China; tianxiaoge@126.com (X.T.); yingronghua@vip.sina.com (R.Y.)
* Correspondence: lixiao@stu.csust.edu.cn

Abstract: To evaluate the possibility of using phosphorus slag powder instead of mineral powder as a filler in asphalt mastic, this study investigates the micro-characteristics of phosphorus slag powder and its viscoelastic mechanical properties in asphalt mastic. A systematic approach combining macro and micro test methods was used to analyze the physical and surface characteristics, void structure, and surface energy of phosphorus slag powder. The viscoelastic mechanical properties of phosphorus slag powder were evaluated using appropriate indexes. Meanwhile, the correlations between and limitations of various evaluation indexes and the high-temperature rheological properties were identified. The results demonstrate that phosphorus slag powder exhibits low density, small overall particle size, difficulty in forming agglomerates, developed pores, large specific surface area, and high surface energy, which is suitable for replacing mineral powder as a filler in asphalt mastic. The main factors affecting the viscoelastic properties of asphalt mastic are the particle size and dosage of phosphorus slag powder. Generally speaking, phosphorus slag powder asphalt mastic with particle sizes $\leq$ 18 μm exhibits the best performance. In practical engineering applications, the appropriate dosage (7%, 10%, 13%) can be selected based on different regions and specific design and construction requirements. Additionally, zero-shear viscosity (ZSV), non-recoverable creep compliance (J_{nr}), and creep recovery percentage (R) exhibit a strong correlation with the high-temperature rheological properties of asphalt mastic. At the same time, the rutting factor ($G^*/\sin\delta$) presents certain limitations.

Keywords: phosphorus slag powder; asphalt mastic; ZSV; micro-characteristics; rheological properties

check for updates

Citation: Li, X.; Tian, X.; Ying, R. Laboratory Performance and Micro-Characteristics of Asphalt Mastic Using Phosphorus Slag Powder as a Filler. *Buildings* **2024**, *14*, 2061. https://doi.org/10.3390/buildings14072061

Academic Editor: Huayang Yu

Received: 3 June 2024
Revised: 18 June 2024
Accepted: 25 June 2024
Published: 5 July 2024

1. Introduction

In recent years, nations worldwide have been dedicated to the pursuit of green, low-carbon, and sustainable development. The rational recycling and comprehensive utilization of industrial solid waste have emerged as practical and attractive trends. Phosphorus slag, a type of industrial waste from the phosphorus chemical industry, has become a significant environmental issue in China due to its large-scale stock and incremental stacking. Research data indicate that for every ton of yellow phosphorus produced, 8–12 tons of phosphorus slag are generated [1,2]. With an annual global demand of 1.5 million tons of yellow phosphorus, this results in 12–18 million tons of phosphorus slag, of which China alone produces up to 8 million tons annually, accounting for 44–67% of the global share, including a new stacking volume exceeding 1 million tons. However, the utilization rate of phosphorus slag remains below 50% [1,2]. Under the prolonged influence of natural environmental factors such as rain and strong winds, coupled with geological disasters like sudden rainstorms and earthquakes, pollutants within the phosphorus slag, including eutrophic elements like phosphorus and nitrogen, and heavy metals such as copper, lead, mercury, chromium, and arsenic, leach out [2]. Upon their infiltration into soil and water bodies, these elements exhibit resistance to degradation. Their accumulation adversely impacts the physical and chemical characteristics, biological attributes, and microbial community structure of the affected areas, ultimately leading to the disruption of ecological function and structure. Consequently, water, soil, and crops suffer serious pollution repercussions. Adjustments and rebalancing of the internal temperature field within phosphorus slag storage and incremental yards will further influence the viscosity and permeability of pore water. As a result, alterations in soil microstructure and composition will impact the diffusion and movement of contaminants. Subsequently, these substances will permeate the food chain through stepwise biological enrichment, impacting water quality, the ecological environment, and human health [1,2]. In accordance with 'The Targets of the 14th Five-year Plan for the Protection of Soil, Groundwater, and Rural Ecological Environment', the comprehensive and efficient utilization of phosphorus slag has emerged as an imperative issue necessitating attention. Consequently, numerous scholars have undertaken pertinent research in the realms of construction engineering, agricultural fertilizers, and environmental remediation materials.

In the realm of engineering applications for phosphorus slag, P.W Gao et al. have examined the microstructure and pore structure of cement concrete, where a portion of Portland cement is replaced with ultrafine phosphorus slag powder [3]. Y.Z. Peng et al. have explored the properties and microstructure of RPCs in reactive powder concrete containing high levels of phosphorus slag powder [4]. Jin Yang et al. have investigated the impacts of large-capacity ultrafine phosphorus slag powder (UPS) and superabsorbent polymers on the shrinkage behaviors, compressive strength, hydration products, and pore structure of cement mortars [5]. Ali Allahverdi has studied the influence of curing conditions on the physical and mechanical properties of chemically active phosphorus slag cement [6]. Existing research has demonstrated the viability and potential of utilizing phosphorus slag in the engineering sector, particularly in cement production. However, its feasibility as a filler substitute for mineral powder in modified asphalt preparation requires further investigation. Currently, modified asphalt represents a significant developmental direction in road engineering, with numerous scholars experimenting with materials such as solid waste and polymers to modify asphalt, yielding various findings. S.T. Lv has evaluated the properties of composite modified asphalt using bone glue and waste rubber powder [7]; X.H. Peng has modified rock asphalt and biological asphalt to assess the strength attributes of rock asphalt-modified biological asphalt mixtures [8]. Chao Zhang has analyzed the strength characteristics of rubberized asphalt mixtures under diverse stress loading paths [9]. J.N Liu et al. have produced an asphalt mixture by partially substituting limestone with iron oxide (F.O.) to study its self-healing capabilities under microwave heating [10]. G.Y. Tao et al. have discussed the potential of steel slag as a mineral filler in asphalt and mixtures [11]; Z.C. Li et al. have utilized fly ash and coal-liquefied slag instead of limestone ore powder to elucidate its effects on the rheological properties of asphalt mortar [12]. Fan Li has explored the physicochemical interactions between fly ash and limestone with asphalt mortar from a microscopic viewpoint [13]. Z.Y. Wei et al. have delved into the rheological behavior and mechanism of action of iron tailings asphalt mortar [14]. Y.J. Meng et al. have researched the rheological properties and microscopic mechanisms of asphalt mortar [15]. F. Russo et al. have assessed the rheological properties of asphalt mortar incorporating bottom ash as a filler in a waste-to-energy power plant [16]; Jiao Lin et al. have investigated the attributes of GFRP powder asphalt mortar derived from waste epoxy resin [17].

However, the current research on the modification of asphalt using phosphorus slag is limited, leaving a significant gap in its application to asphalt mastic. Studies by Y.P. Sheng et al. have explored the impact of phosphorus slag powder and polyester fibers on the performance characteristics of asphalt binders and mixtures [18]. G.P. Qian et al. have modified the surface of phosphorus slag powder and analyzed its effects and mechanisms on asphalt and asphalt mixture properties [19]. H.N. Yu evaluated how different contents and particle sizes of phosphorus slag affect asphalt performance [20]. S.C. Yi examined the properties of commonly used modifiers combined with phosphorus slag powder-modified asphalt after TM-P surface treatment [21]. Kun Wang conducted experiments on the surface modification of phosphorus slag powder and its influence on asphalt properties [22];

G.H. Nie performed performance tests on asphalt mixtures modified with phosphorus slag powder [23]. Given the existing research, it is evident that phosphorus slag can enhance asphalt performance to a certain degree. Its use in road engineering is not only feasible but also presents an excellent opportunity to reduce resource loss, address solid waste issues, and achieve both environmental and economic benefits. However, the current research on phosphorus slag predominantly centers on the modification of phosphorus slag asphalt binder and mixture using traditional asphalt and asphalt mixture forming methods. This involves modifying the phosphorus powder particles from a macroscopic perspective or incorporating additional materials such as common modifiers and polyester fibers, etc., to analyze their properties and modification effects. Nevertheless, there is currently no research on breaking the traditional asphalt binder molding method and using phosphorus slag powder as a filler instead of mineral powder for blending into matrix asphalt to create asphalt mastic. Additionally, there is no research focusing on the physical properties and mechanisms of phosphorous slag powder, as well as the performance and effects of modified asphalt mastic using phosphorous slag powder from a microscopic perspective by utilizing asphalt mastic theory.

In summary, researching phosphorus slag powder asphalt mastic is of practical significance for promoting engineering practice. To deeply explore the use of phosphorus slag powder in asphalt pavement, it is essential to investigate the physical properties and mechanisms of phosphorus slag powder, its influence on the properties of asphalt mastic, and the appropriate dosage. Based on these studies, Appropriate evaluation indexes should be selected to accurately characterize its high-temperature rheological properties, assess the rutting resistance of asphalt mixtures, and guide practical engineering applications. This study conducted both microscopic and macroscopic experiments on phosphorus slag powder from Yunnan Province. At the microscopic scale, the morphological characteristics and surface energy of phosphorus slag powder, including physical porosity, particle size, and surface energy, were analyzed using a laser particle size analyzer, nitrogen adsorption-specific surface area pore analyzer, scanning electron microscope (SEM), and contact angle analyzer. At the macro scale, the viscoelastic mechanical properties of phosphorus slag powder asphalt mastic were examined using conventional performance tests, temperature sweep tests, frequency sweep tests, multi-stress creep recovery (MSCR) tests, and bending beam rheometer (BBR) tests.

2. Materials and Methodologies

2.1. Raw Materials

In this experiment, the 70 grade matrix asphalt produced by Shell Xinyue Asphalt Co., Ltd. (Foshan, China) was selected, and its key technical indicators are listed in Table 1. The filler consists of ordinary limestone slag powder and phosphorus slag powder sourced from the Yunnan region. The phosphorus slag powder is divided into three distinct particle sizes (38–75 μm, 18–38 μm, ≤18 μm) using a cement fineness negative pressure sieve analyzer. The original photos of these three different particle sizes of phosphorus slag powder and mineral powder are displayed in Figure 1. In contrast, the technical indicators and particle size distribution of the four fillers are presented in Tables 2 and 3.

Table 1. Performance indexes of matrix asphalt.

Items	Unit	Test Results	Technical Requirements
Penetration (25 °C, 100 g, 5 s)	0.1 mm	69.8	60–80
Penetration index (PI)	--	−1.36	−1.5–+1.0
Ductility (5 cm/min, 15 °C)	cm	>150	≥100
Softening point	°C	47.2	≥46

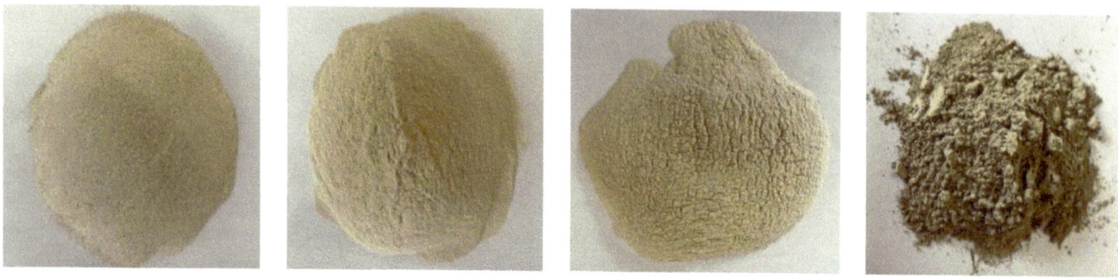

38-75 μm 18-38 μm ≤18 μm
(**a**) phosphorus slag powder (**b**) mineral powder

Figure 1. The original photos of phosphorus slag powder and mineral powder.

Table 2. Technical indicators of four types of fillers.

Items		**Unit**	**Phosphorus Slag Powder (μm)**			**Mineral Powder**
			38–75	**18–38**	**≤18**	
Apparent density		g/cm^3	2.69	2.65	2.63	2.66
Specific surface area	BET method	m^2/g	5.68	6.34	7.22	0.72
	Langmuir method		6.45	7.09	8.19	0.83
Hydrophilic coefficient		--		0.91		0.68

Table 3. Particle size distribution of the filler (unit: μm).

Type of Filler		χ_{10}	χ_{50}	χ_{90}	χ_{av}
Phosphorus slag powder	38–75	39.789	49.814	64.127	50.559
	18–38	20.168	26.047	31.127	26.032
	≤18	6.378	8.039	10.191	8.132
Mineral powder		0.421	0.789	1.478	0.890

Annotation: χ_{10}, χ_{50}, χ_{90} represent the accumulated analysis of the particles of 10%, 50%, and 90%. χ_{av} is the average particle size of the particle group.

2.2. Preparation of Asphalt Mastic

For this study, three types of phosphorus slag powder with varying dosages and particle sizes were selected to form the asphalt mastic of phosphorus slag powder. Following the "Test Regulations for Asphalt and Asphalt Mixtures in Highway Engineering" (JTG E20-2011) [24], the mineral powder asphalt mastic was prepared using conventional methods. Both the phosphorus slag powder and mineral powder were dried to constant weight and kept warm at a temperature of 140 °C to ensure that the filler remained dry. Subsequently, the 70 grade of matrix asphalt was heated to 135 °C and kept warm for two hours at a constant temperature to achieve an utterly fluid state.

In accordance with the test plan, the mass ratio of filler to asphalt was determined. Initially, 70 grades of matrix asphalt were added and stirred continuously with glass rods at high-speed shear. Subsequently, small amounts of phosphorus slag powder with different dosages and particle sizes were gradually mixed into the matrix asphalt multiple times, stirring evenly at different rates and durations until there were no bubbles on the surface. This process resulted in the preparation of the phosphorus slag powder asphalt mastic. During the mixing process, an electric furnace was placed directly beneath the vessel, maintaining the temperature between 145 °C and 165 °C to prevent uneven dispersion of the filler caused by the cooling of the asphalt.

2.3. Experimental Methods

The study's methodology encompasses five distinct phases, elaborated as follows:
Section A: Orthogonal Experiment Design

This phase employs an orthogonal array, denoted as $L_9(3^4)$, where a four-factor, three-level experimental schema was devised, leading to a total of nine experimental groups being executed.

Section B: Characterization of Micro-Void Structure and Surface Energy of Fillers

Comprehensive microstructural analyses, including specific surface area, porosity, micro-surface features, and surface energy assessments of the filling materials, were conducted using sophisticated techniques such as nitrogen adsorption-specific surface area analysis, scanning electron microscopy (SEM), and contact angle measurement.

Section C: Fundamental Properties of Asphalt Mastic

The fundamental indexes of asphalt mastic—penetration, softening point, and ductility—were examined.

Section D: High- and Low-Temperature Rheological Properties of Asphalt Mastic

Three distinct methodologies—frequency sweep, temperature sweep, and multi-stress creep recovery (MSCR) tests—were employed to assess the 60 °C zero-shear viscosity (ZSV), rutting parameter ($G^*/\sin\delta$), unrecoverable creep compliance (J_{nr}), and creep recovery percentage (R), thereby evaluating the high-temperature rheological behavior of phosphorus slag powder asphalt mastic. For low-temperature rheology, a bending beam rheometer (BBR) was utilized, with stiffness modulus (S) and creep rate (m) serving as evaluative criteria.

Section E: Fatigue Characteristics of Asphalt Mastic

Dynamic shear rheometer (DSR) was leveraged to investigate the fatigue performance of asphalt mastic, adopting fatigue parameters ($G^*\cdot\sin\delta$) for appraisal.

2.3.1. Orthogonal Experimental Design

This study employs a multifactorial, multi-level approach characterized by a large and complex test volume. To manage this complexity, the orthogonal design method of $L_9(3^4)$ was utilized for experimenting, and the standard deviation analysis method was employed for the analysis. The specific selection of factors and levels is detailed in Table 4, and the orthogonal design scheme can be seen in Table 5.

Table 4. Factor level.

Factors	Granularity A	Dosage B	Stirring Rate C	Stirring Time D
1	38–75 μm	7%	1000 r/min	20 min
2	18–38 μm	10%	2000 r/min	25 min
3	≤18 μm	13%	3000 r/min	30 min

Table 5. Orthogonal design scheme.

Factors		Granularity A	Dosage B	Stirring Rate C	Stirring Time D
Experiment number	1#	1	1	1	1
	2#	1	2	2	2
	3#	1	3	3	3
	4#	2	1	2	3
	5#	2	2	3	1
	6#	2	3	1	2
	7#	3	1	3	2
	8#	3	2	1	3
	9#	3	3	2	1

2.3.2. Microscopic Void Structure Properties and Surface Energy of the Filler

(1) Microscopic Surface Properties

Scanning electron microscopy (SEM) was employed to investigate the microscopic surface features of the filler. Specimens were prepared by applying a layer of conductive adhesive to a sample holder, upon which the filler powder was sprinkled. After the adhesive evaporated, the powder was secured, and any unadhered particles were removed using an ear suction ball. A conductive film was subsequently applied for testing.

(2) Microscopic Pore Structure

Nitrogen adsorption tests were conducted to assess the specific surface area of phosphorus slag powder with three distinct particle sizes. They were compared to mineral powder to elucidate the filler's pore structure characteristics. At the temperature of liquid nitrogen, the adsorption capacity of nitrogen on the solid surface is determined by the relative pressure (P/P_0). Within the range of 0.05 to 0.35, the adsorption conforms to the BET equation, serving as the basis for specific surface area determination. When P/P_0 exceeds 0.4, nitrogen begins to condense within micropores, leading to capillary condensation. Theoretical and experimental analyses were used to measure critical parameters such as pore volume and size distribution, and the pore types of phosphorus slag powder and mineral powder were identified based on the adsorption–desorption curves obtained from the tests.

(3) Surface Energy

The contact angle between the filler and a liquid with known surface free energy parameters was measured using the capillary rise method (wick method), recording the height and time of liquid ascent. The contact angle was determined by integrating the capillary rise theory with the Washburn equation, and the surface free energy of the filler was calculated using the Owens–Wendt method and the Young equation.

The contact angle measurement involved selecting a 10 cm section of a standard pipette (inner diameter of 3 mm) as the column tube. Approximately 1.8 g of powder sample was loaded into the tube, which was then placed at the bottom of a glass container with 2 mL of the test liquid. The impregnation time (s) and distance (cm) were recorded. The contact angle was calculated based on Equation (1). The test was conducted at 25 ± 1 °C, with each data point averaged over three replicates. In this study, a low surface energy liquid (n-hexane) was used to infiltrate the filler powder, and the wetting distance and time were documented. The contact angle between the solid and the liquid was measured using distilled water and ethylene glycol.

The expression of the Washburn equation is:

$$h^2/t = (\gamma_l R cos\theta)/2\eta \tag{1}$$

In the formula:

h—impregnation distance(cm);
t—impregnation time(s);
γ_l—Surface free energy of liquids;
R—The effective radius of the capillary;
θ—The angle of contact between the liquid and the solid;
η—The viscosity of the liquid.

The expression of the Young equation is as follows:

$$\gamma_{sl} = \gamma_s + \gamma_l cos\theta \tag{2}$$

During the ceremony:

γ_{sl}—Surface energy of solid-liquid interface;
γ_s—Solid surface free energy, $\gamma_s = \gamma_s^d + \gamma_s^p$;
γ_l—Liquid surface free energy, $\gamma_l = \gamma_l^d + \gamma_l^p$.

According to the Owens–Wendt method, the expression of free energy at the solid–liquid interface is:

$$\gamma_{sl} = \gamma_s + \gamma_l - 2\sqrt{\gamma_s{}^d \gamma_l{}^d} - 2\sqrt{\gamma_s{}^p \gamma_l{}^p} \tag{3}$$

In the formula:

γ_s^d—Solid dispersion component (London dispersion force);

$\gamma_l{}^d$—Liquid dispersion component (London dispersion Force);
$\gamma_s{}^p$—Solid polarity component γ^p (acid-base force);
$\gamma_l{}^p$—Liquid polarity component γ^p (acid-base force). The rest is the same as above.

Combined with Young's equation, the expression of the free energy at the solid–liquid interface is as follows:

$$\gamma_l(1 + cos\theta) = 2\sqrt{\gamma_s{}^d\gamma_l{}^d} + 2\sqrt{\gamma_s{}^p\gamma_l{}^p} \tag{4}$$

Transform Equation (4) into $y = mx + b$ the form:

$$\frac{(1 + cos\theta)}{2} \times \frac{\gamma_l}{\sqrt{\gamma_l{}^d}} = \sqrt{\gamma_s{}^d} + \sqrt{\gamma_s{}^p} \times \sqrt{\frac{\gamma_l{}^p}{\gamma_l{}^d}} \tag{5}$$

Based on the surface free energy of the known liquid and its components, as well as the contact angle between the known liquid and the filler, the surface free energy of the filler can be calculated by Equation (5). According to the characteristics of the formula, y or $\frac{(1+cos\theta)}{2} \times \frac{\gamma_l}{\sqrt{\gamma_l{}^d}}$ is plotted against x or $\sqrt{\frac{\gamma_l{}^p}{\gamma_l{}^d}}$ to fit the function, the slope m^2 represents the polarity, and b^2 represents the dispersive component.

2.3.3. Basic Properties of Asphalt Mastic

According to the "Test Regulations for Highway Engineering Asphalt and Asphalt Mixture" (JTG E20-2011) [24], the penetration degree, softening point, and 10 °C ductility of phosphorus slag powder asphalt mastic were experimentally studied.

2.3.4. High- and Low-Temperature Rheological Properties of Asphalt Mastic

(1) High-Temperature Rheological Test

To evaluate the high-temperature rheological characteristics of phosphorus slag powder asphalt mastic, temperature sweep, frequency sweep, and multiple stress creep recovery (MSCR) tests were conducted using a dynamic shear rheometer (DSR). The evaluation indexes used were the rutting factor (G*/sin δ), zero shear viscosity (ZSV), non-recoverable creep compliance (J_{nr}), and creep recovery percentage (R).

Temperature sweep test: Conducted within the range of 40 °C to 82 °C, with increments of 6 °C and a loading frequency of 10 rad/s.

Frequency sweep test: Conducted at 60 °C, covering a frequency range of 0.1 to 100 rad/s.

MSCR test: Performed at 60 °C, with stress levels of 0.1 kPa and 3.2 kPa, a creep time of 1 s, and a recovery strain time of 9 s.

(2) Low-Temperature Rheological Test

A bending beam rheometer (BBR) was used to evaluate the low-temperature rheological properties of the asphalt mastic. The evaluation indexes were creep stiffness (S) and creep rate (m), with test temperatures set at −6 °C, −12 °C, −18 °C, and −24 °C.

2.3.5. Fatigue Properties of Asphalt Mastic

The fatigue factor (G*·sin δ) was obtained using a dynamic shear rheometer (DSR) within the temperature range of 15 °C to 40 °C to evaluate the fatigue characteristics of the asphalt mastic.

3. Results and Discussion

3.1. Microscopic Void Structure Characteristics and Surface Energy Analysis of Phosphorus Slag Powder

The scanning electron microscope (SEM) image, displayed in Figure 2, reveals the microstructural details of phosphorus slag powder. Under SEM, phosphorus slag powder particles showcase an irregular shape and a rough surface. The particles are relatively uniform in size, with prominent mosaic and self-locking abilities. Noteworthy are the visible pores and micropores, as well as the transparent gap pores between particles, which indicate a well-developed pore structure. When the particle size of phosphorus slag powder decreases, the particles become more petite and more uniform. At the same time, the gap pores between them increase. This is quite different from mineral powder particles, which are smaller and have more pores of varying sizes but are less microporous and not as tightly connected.

(I) Magnify 200× (II) Magnify 10,000×
(a)
(I) Magnify 200× (II) Magnify 10,000×
(b)
(I) Magnify 200× (II) Magnify 10,000×
(c)
(I) Magnify 200× (II) Magnify 10,000×
(d)

Figure 2. Scanning electron microscope (SEM) of the filler; (**a**) 38–75 μm phosphorus slag powder; (**b**) 18–38 μm phosphorus slag powder; (**c**) ≤18 μm phosphorus slag powder; (**d**) mineral powder.

Figure 3 and Table 6, respectively, depict the adsorption–desorption curves and pore volume of phosphorus slag powder. The adsorption and desorption curve of phosphorus slag powder fits the Type IV(a) adsorption isotherm, characteristic of a mesoporous structure (2–50 nm). The trend in the middle- and low-pressure areas is relatively gentle, escalating sharply in the high-pressure area. This sharp increase results in a hysteresis loop caused by capillary condensation in the medium- and high-pressure areas, suggesting that phosphorus slag powder contains slit pores with active adsorption capabilities. In stark contrast, the mineral powder displays no significant hysteresis and virtually lacks an internal pore structure.

Notably, the nitrogen adsorption capacity of 18–38 μm and ≤18 μm phosphorus slag powder is significantly higher than that of 38–75 μm phosphorus slag powder and mineral powder. The pore volume is larger, and the pores are more developed, which is more conducive to the adsorption of asphalt on the filler.

The contact angle between the filler and each liquid, as well as the surface free energy of the filler and its components, are presented in Tables 7 and 8. Generally, the contact angle of phosphorus slag powder is smaller than that of mineral powder, which indicates that phosphorus slag powder has more excellent wettability. Additionally, smaller particle sizes result in a more substantial wetting capacity of phosphorus slag powder, suggesting improved adsorption between phosphorus slag powder and asphalt.

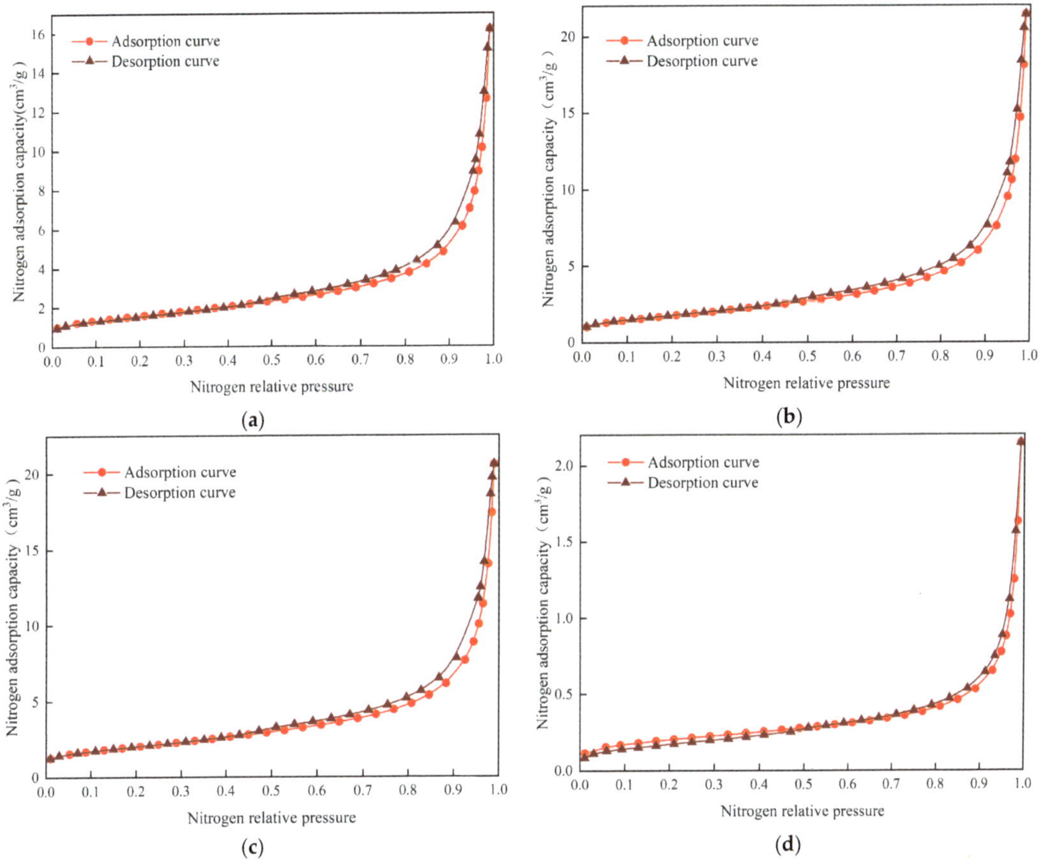

Figure 3. Nitrogen adsorption–desorption curves of phosphorus slag powder and mineral powder; (**a**) 38–75 μm phosphorus slag powder adsorption–desorption curve; (**b**) 18–38 μm phosphorus slag powder adsorption–desorption curve; (**c**) ≤18 μm phosphorus slag powder adsorption–desorption curve; (**d**) mineral powder adsorption–desorption curve.

Table 6. Packing adsorption–desorption pore volume (unit: cm^3/g).

Type of Filler	Phosphorus Slag Powder (μm)			Mineral Powder
	38–75	18–38	≤18	
BJH adsorption	0.025	0.033	0.031	0.003
BJH desorption	0.025	0.033	0.032	0.003

Table 7. Contact angle between the packing material and each liquid (unit: °).

Type of Filler		Average Value	
		Steaming Water	Ethylene Glycol
Phosphorus slag powder	38–75 μm	25.9	0
	18–38 μm	29.6	0
	≤18 μm	33.3	0
Mineral powder		71.7	48.1

Table 8. Surface free energy and its components of packing material (unit: m J/m^2).

Type of Filler		Surface Free Energy		
		γ	γ^d	γ^p
Phosphorus slag powder	38–75 μm	71.8	5.66	66.14
	18–38 μm	67.94	6.94	61
	≤18 μm	63.96	8.59	55.37
Mineral powder		33.68	19.11	14.57

Based on the findings, the surface energy of phosphorus slag powder is higher than that of mineral powder, with the dispersion component constituting the majority of phosphorus slag powder regardless of particle size. However, for mineral powder, the polar and dispersion components are nearly equal. This indicates that phosphorus slag powder is a non-polar material with intermolecular solid forces, while the intermolecular forces of mineral powder are equivalent to its surface dispersion force.

3.2. Analysis of Basic Characteristics of Asphalt Mastic

The test results for the conventional performance indexes of phosphorus slag powder asphalt mastic are presented in Figure 4. The test values are found to be within the specified requirements. Notably, the penetration degree of the mastic decreases as the particle size diminishes or the content of phosphorus slag powder increases, suggesting an enhanced resistance to shear deformation. The ductility value rises with a decrease in particle size. It falls with an increase in content, indicating superior plastic deformation capacity when the particle size is small, or the content is low. The softening point exhibits an upward trend with reduced particle size or increased content of phosphorus slag powder, implying that smaller particles or higher dosages facilitate the adsorption and bonding between asphalt and phosphorus slag powder, thereby benefiting the high-temperature performance of the mastic.

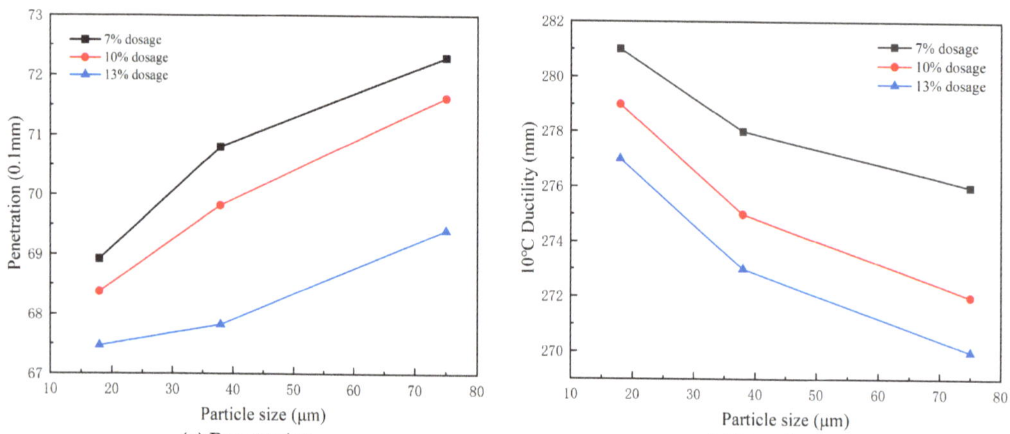

(**a**) Penetration (**b**) Ductility

Figure 4. *Cont.*

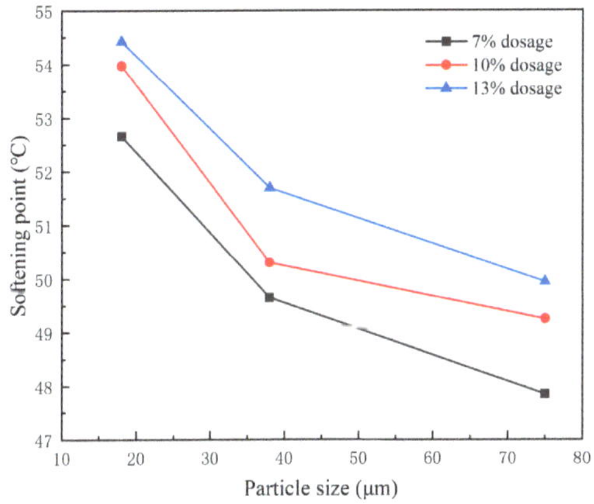

(**c**) Softening point

Figure 4. Three indexes of phosphorus rock powder asphalt mastic.

3.3. Analysis of Rheological Properties of Asphalt Mastic

3.3.1. Rheological Properties at High Temperatures

(1) Analysis of Temperature Scanning Test Results

As illustrated in Figure 5, the particle size and dosage of phosphorus slag powder are key factors influencing the rutting factor (G*/sin δ) of asphalt mastic, with particle size having the most significant impact on high-temperature performance. Theoretical analysis suggests that when the phosphorus slag powder particle size is ≤18 μm, the content is 10%, the stirring rate is 3000 r/min, and the stirring time is 30 min, the asphalt mastic with phosphorus slag powder can exhibit excellent high-temperature rheological properties.

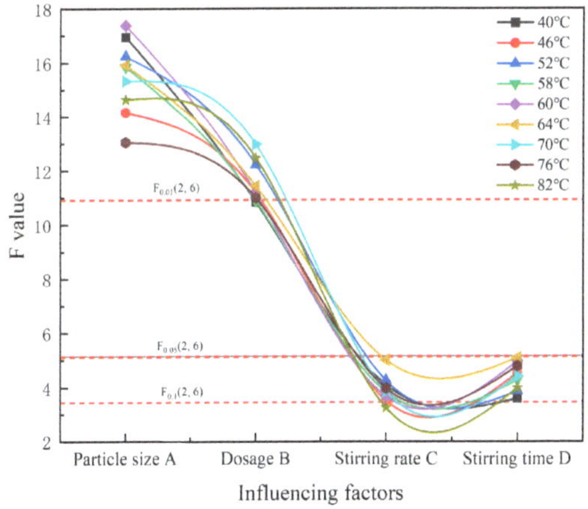

Figure 5. G*/sin δ standard deviation analysis diagram.

Similarly, Figure 6 demonstrates that temperature significantly affects the high-temperature performance of the mastic, with the rutting factor (G*/sin δ) decreasing in a curvilinear fashion as temperature increases, levelling off as it approaches 60 °C. At a constant temperature and dosage, smaller phosphorus slag powder particle sizes correlate with better high-temperature performance and a reduced likelihood of rutting. Under identical temperature and particle size conditions, an optimal phosphorus slag powder content exists, with a 10% content yielding the best results in this experiment, aligning with the theoretical analysis.

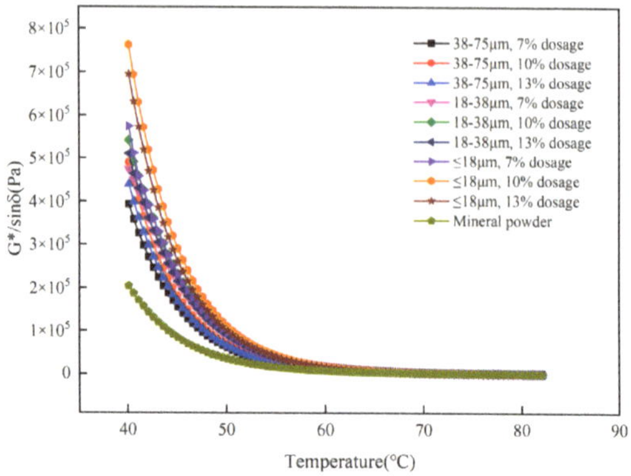

Figure 6. Diagram of G*/sin δ as a function of temperature.

(2) Analysis of Frequency Sweep Test Results

As indicated in Table 9, the zero-shear viscosity (ZSV) was fitted and derived computationally from the Carreau rheological model. Generally, the ZSV of phosphorus slag powder asphalt mastic obtained through the primary curve fitting is lower than that derived from the formula. However, the results and the trend of change remain broadly consistent. The primary factors influencing the ZSV of asphalt mastic are particle size and dosage.

Table 9. Zero-shear viscosity (ZSV) of asphalt mastic (unit: Pa·s).

Trial Number	Formula Derivation			Carreau Rheological Model Fitting Curves	
	K	*m*	ZSV	ZSV	R^2
1#	71.26	0.17	13,058	9908	0.99071
2#	32.4	0.02	14,577	14,461	0.97284
3#	74.96	0.003	13,854	13,529	0.99569
4#	49.53	0.02	14,787	14,727	0.98679
5#	48.9	0.014	18,901	18,775	0.9823
6#	50.04	0.007	17,636	17,670	0.97922
7#	48.18	0.007	19,842	19,760	0.99244
8#	36.4	0.04	29,761	25,980	0.97734
9#	37.7	0.008	21,508	21,097	0.98369
10#	44.07	0.003	7795	7806	0.99612

Notably, the ZSV of phosphorus slag powder asphalt mastic with particle sizes of 18–38 μm and ≤18 μm is predominantly higher than the average value and significantly exceeds the ZSV of mineral powder asphalt mastic. For a given particle size, increasing the dosage has a significant impact on the ZSV of asphalt mastic, achieving a peak in high-temperature performance at a 10% dosage. Furthermore, when using the exact dosage, smaller particle sizes lead to improved high-temperature performance of phosphorus slag powder asphalt mastic.

(3) Analysis of the Results of the Multi-Stress Creep Recovery Test (MSCR)

As shown in Table 10 and Figures 7 and 8, the J_{nr} value of phosphorus slag powder asphalt mastic at 60 °C is relatively low, not exceeding 6.5 kPa^{-1}. This indicates that the high-temperature performance of phosphorus slag powder asphalt mastic is satisfactory. Notably, the J_{nr} value is the lowest at a 10% dosage under the same particle size, suggesting that the high-temperature performance is optimal at this dosage. Additionally, for a given dosage, a smaller particle size of phosphorus slag powder results in a lower J_{nr} value of asphalt mastic, indicating better high-temperature performance.

Table 10. MSCR index test results.

Trial Number	Unit	1#	2#	3#	4#	5#	6#	7#	8#	9#	10#
$J_{nr0.1}$	kPa^{-1}	4.94	4.33	4.35	3.54	3.01	3.20	3.14	2.93	2.99	3.24
$J_{nr3.2}$	kPa^{-1}	5.53	4.66	6.09	4.01	3.25	4.03	3.38	3.55	3..90	3.97
$J_{nr\text{-}diff}$	--	12.04	7.75	39.94	13.35	8.05	25.93	7.81	20.99	30.43	22.50
$R_{0.1}$	%	49.36	57.81	51.3	60.25	69.58	66.26	69.12	75.89	71.07	62.01
$R_{3.2}$	%	37.53	49.65	40.42	51.04	58.99	54.75	60.05	68.91	65.43	47.55

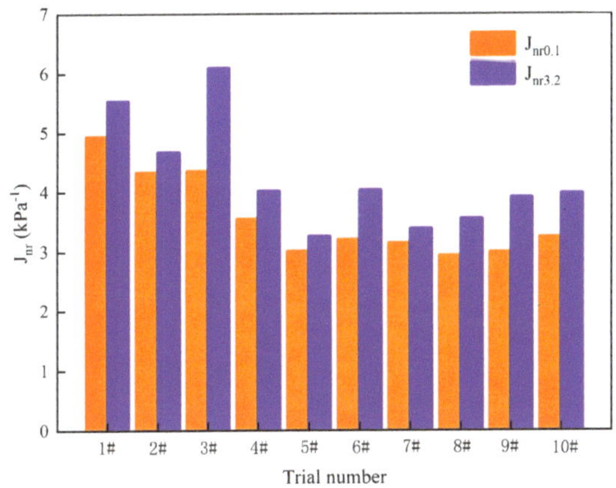

Figure 7. J_{nr} comparison diagram.

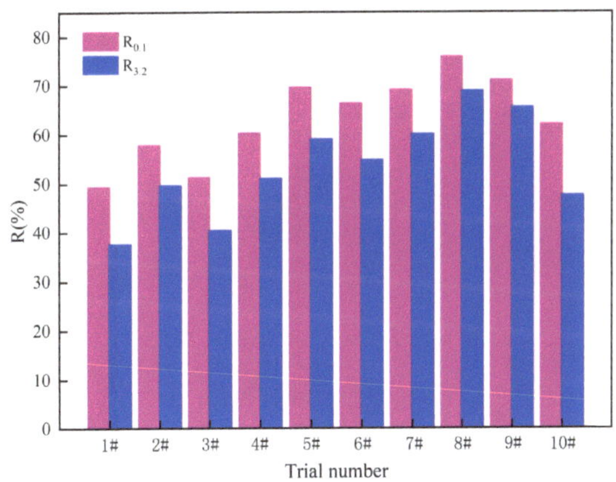

Figure 8. R comparison diagram.

The $J_{nr\text{-}diff}$ does not exhibit any significant change characteristics, making it difficult to determine the variation in stress sensitivity of the asphalt mastic. The creep deformation recovery rate (R) increases as the particle size of phosphorus slag powder decreases, indicating better delayed elastic recovery performance. Compared to fillers with larger particle sizes, this demonstrates higher elastic composition and resistance to viscosity and flow deformation at high temperatures.

(4) Analysis and Discussion of Different High Temperature Evaluation Indexes

By comparing the high-temperature rheological properties of asphalt mastic obtained from three different tests, it is evident that all three indexes lead to similar conclusions. These indexes can be used to assess the high-temperature rheological properties of asphalt mastic to some extent. Among them, zero-shear viscosity (ZSV) is an inherent property of the material that remains unaffected by external factors. The intermittent loading mode of the multiple stress creep recovery (MSCR) test effectively simulates actual pavement conditions and better reflects the natural stress conditions of asphalt materials in use. These two evaluation indicators are highly reliable and have a reasonable correlation with the high-temperature performance of asphalt mastic, accurately reflecting their high-temperature rheological characteristics.

However, the rutting factor ($G^*/\sin\delta$), obtained under continuous loading, fails to simulate the actual stress conditions of asphalt pavements accurately. It also does not account for delayed elasticity and recovery ability, separating the elastic and viscous components of asphalt materials, thus failing to describe their viscoelastic rheological behavior accurately. Therefore, it is recommended to use ZSV or MSCR test indexes as the evaluation indicators for the high-temperature rheological properties of phosphorus slag powder asphalt mastic.

3.3.2. Analysis of Rheological Properties at Low Temperatures

As shown in Figures 9 and 10, the dosage and particle size of phosphorus slag powder are the main factors influencing the properties of asphalt mastic, while the preparation process (stirring rate and stirring time) has a less significant impact. Theoretical analysis indicates that with a phosphorus slag powder particle size of ≤18 μm, a dosage of 13%, a stirring rate of 2500 r/min, and a stirring time of 25 min, asphalt mastic can maintain better low-temperature crack resistance.

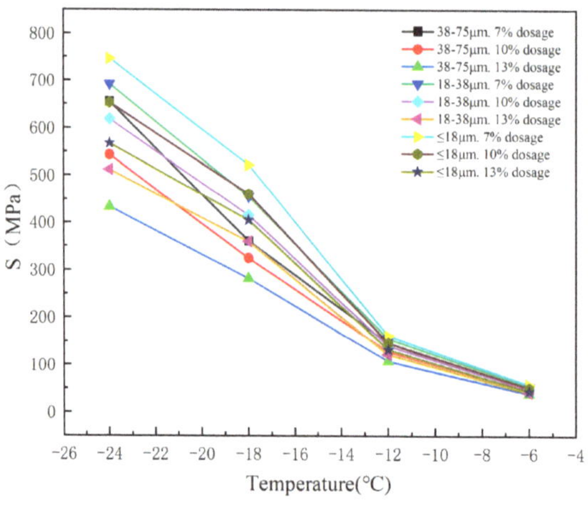

Figure 9. S with temperature change relationship diagram.

Based on the test results, it was observed that under consistent temperature and particle size conditions, an increase in phosphorus slag powder dosage results in a decrease in the stiffness modulus (S value) of asphalt mastic, as well as an increase in the creep rate (m value). This suggests that a higher deformation capacity of asphalt mastic contributes to improved crack resistance at low temperatures, although it has a relatively minor impact on stress relaxation capacity.

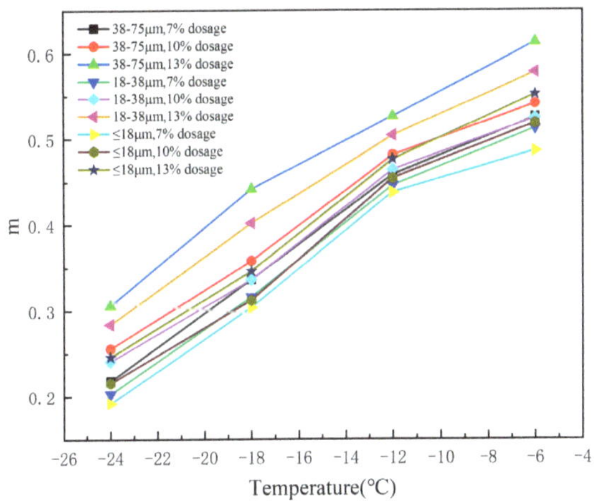

Figure 10. m with temperature change relationship diagram.

3.4. *Analysis of Fatigue Characteristics*

As depicted in Figure 11, regardless of the filler type, the fatigue factor (G* sin δ) of asphalt mastic consistently declines with increasing temperature. Notably, asphalt mastic containing mineral powder exhibits a higher fatigue factor than those with phosphorus slag powder across all temperatures, reaching up to 7000 kPa. This indicates superior fatigue resistance in asphalt mastic with phosphorus slag powder compared to those with mineral powder.

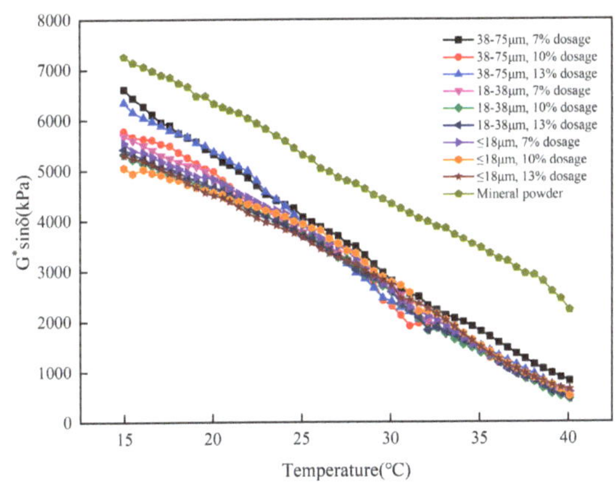

Figure 11. G*·sin δ with temperature change diagram.

Furthermore, the fatigue resistance of asphalt mastic is significantly influenced by the particle size and dosage of phosphorus slag powder. A smaller particle size enhances fatigue resistance under equivalent dosage conditions, and asphalt mastic with ≤18 μm phosphorus slag powder demonstrates heightened temperature sensitivity, which is beneficial for fatigue resistance. Under constant particle size, the fatigue limit temperature exhibits a 'V-shaped trend, with the lowest fatigue limit temperature observed in asphalt mastic containing 10% phosphorus slag powder.

4. Conclusions

In this paper, the feasibility of using phosphorus slag powder as a filler in asphalt mastic, replacing mineral powder, was studied from a microscopic perspective. The high- and low-temperature rheological properties, as well as fatigue characteristics of phosphorus slag powder asphalt mastic, were evaluated. Furthermore, different evaluation indexes for the high-temperature rheological properties of asphalt mastic were compared and analyzed. Based on the study results, the following conclusions were drawn:

(1) Phosphorus slag powder exhibits low density, small overall particle size, difficulty in forming agglomerates, developed pores, large specific surface area, and high surface energy. These properties facilitate bonding between phosphorus slag powder and asphalt, making it suitable as a filler in asphalt mastic and mixtures. However, an appropriate dosage and small particle size of phosphorus slag powder should be selected.

(2) The main factors affecting the viscoelastic properties of asphalt mastic are the particle size and dosage of phosphorus slag powder. The stirring rate and stirring time have little influence on its properties. Among these factors, particle size has the most significant effect on high-temperature performance and fatigue resistance, while dosage has the most apparent effect on low-temperature performance.

(3) Generally speaking, phosphorus slag powder asphalt mastic with particle sizes $\leq$18 μm exhibits the best performance. Phosphorus slag powder asphalt mastic shows good fatigue resistance at a dosage of 7%, good high-temperature rheological properties at a dosage of 10%, and good low-temperature rheological properties at a dosage of 13%. In practical engineering applications, the appropriate dosage can be selected based on different regions and specific design and construction requirements.

(4) High-temperature evaluation indexes: Comparing and analyzing different high-temperature evaluation indexes for asphalt mastic, the test indexes obtained by ZSV and MSCR show a good correlation with the high-temperature performance of asphalt mastic, whereas $G^*/\sin\delta$ has certain limitations and is not able to accurately reflect the high-temperature performance of asphalt mastic. In this paper, it is recommended that ZSV or J_{nr} and R be used to represent the high-temperature rheological properties of asphalt mastic.

(5) This study has important theoretical significance and research value, as it reveals the characteristic change mechanisms of phosphorus slag powder asphalt mastic and optimizes evaluation indexes. However, this paper focuses only on the study of different high-temperature evaluation indexes of asphalt mastic and the physical surface properties of the filler. The chemical properties of the filler and the evaluation indexes for low-temperature and fatigue resistance characteristics of asphalt mastic merit further exploration.

Author Contributions: Conceptualization, R.Y.; methodology, X.T.; formal analysis, X.L.; investigation, X.L.; data curation, X.L.; writing—original draft preparation, X.L.; writing—review and editing, X.L. All authors have read and agreed to the published version of the manuscript.

Funding: This research received no external funding.

Data Availability Statement: The data analyzed in this study are subject to the following licenses/restric the restrictions can be received from the first author. Requests to access these datasets should be directed to lixiao@stu.csust.edu.cn.

Conflicts of Interest: The authors declare no conflict of interest.

References

1. Liu, X.; Liu, X.; Zhang, Z. Recycling and comprehensive utilization of yellow phosphorus slag in building materials: A review. *Constr. Build. Mater.* **2023**, *396*, 132384. [CrossRef]
2. Chen, X.; Zhou, X.; Deng, S.S.; Peng, Z.L.; Fan, Z.Y.; Gao, Z.F.; Lu, L. Review and prospect of research on the release and diffusion of solid waste pollutants from phosphate rock. *J. Yangtze River Res. Inst.* **2024**, 1–9.
3. Gao, P.; Lu, X.; Yang, C.; Li, X.; Shi, N.; Jin, S. Microstructure and pore structure of concrete mixed with superfine phosphorous slag and superplasticizer. *Constr. Build. Mater.* **2008**, *22*, 837–840. [CrossRef]
4. Peng, Y.; Zhang, J.; Liu, J.; Ke, J.; Wang, F. Properties and microstructure of reactive powder concrete having a high content of phosphorous slag powder and silica fume. *Constr. Build. Mater.* **2015**, *101*, 482–487. [CrossRef]
5. Yang, J.; Huang, J.; He, X.; Su, Y.; Oh, S.K. Shrinkage properties and microstructure of high volume ultrafine phosphorous slag blended cement mortars with superabsorbent polymer. *J. Build. Eng.* **2020**, *29*, 101121. [CrossRef]
6. Allahverdi, A.; Pilehvar, S.; Mahinroosta, M. Influence of curing conditions on the mechanical and physical properties of chemically-activated phosphorous slag cement. *Powder Technol.* **2016**, *288*, 132–139. [CrossRef]
7. Lv, S.; Tan, L.; Peng, X.; Hu, L.; Cabrera, M.B. Experimental investigation on the performance of bone glue and crumb rubber compound modified asphalt. *Constr. Build. Mater.* **2021**, *305*, 124734. [CrossRef]
8. Peng, X.; Yuan, J.; Wu, Z.; Lv, S.; Zhu, X.; Liu, J. Investigation on strength characteristics of bio-asphalt mixtures based on the time-temperature equivalence principle. *Constr. Build. Mater.* **2021**, *309*, 125132. [CrossRef]
9. Zhang, C.; Yu, H.; Zhu, X.; Yao, D.; Peng, X.; Fan, X. Unified Characterization of Rubber Asphalt Mixture Strength under Different Stress Loading Paths. *J. Mater. Civ. Eng.* **2024**, *36*, 04023498. [CrossRef]
10. Liu, J.; Wang, Z.; Jia, H.; Jing, H.; Chen, H.; Zhou, L.; Yuan, L.; Hoff, I. Characteristics and properties of asphalt mortar containing FO filler. *Constr. Build. Mater.* **2023**, *392*, 132039. [CrossRef]
11. Tao, G.; Xiao, Y.; Yang, L.; Cui, P.; Kong, D.; Xue, Y. Characteristics of steel slag filler and its influence on rheological properties of asphalt mortar. *Constr. Build. Mater.* **2019**, *201*, 439–446. [CrossRef]
12. Li, Z.; Zhang, X.; Liu, F.; Su, J.; Zhao, Q.; Liu, L. Effect of replacement of limestone mineral powder with fly ash and direct coal liquefaction residue on the rheological properties of asphalt mastic. *Constr. Build. Mater.* **2024**, *412*, 134803. [CrossRef]
13. Li, F.; Yang, Y. Experimental investigation on the influence of interfacial effects of limestone and fly ash filler particles in asphalt binder on mastic aging behaviors. *Constr. Build. Mater.* **2021**, *290*, 123184. [CrossRef]
14. Wei, Z.; Jia, Y.; Wang, S.; Zhou, Z.; Zhang, Z.; Wang, X.; Huang, X.; Gao, Y. Influence of iron tailing filler on rheological behavior of asphalt mastic. *Constr. Build. Mater.* **2022**, *352*, 129047. [CrossRef]
15. Meng, Y.; Lai, J.; Ling, L.; Zhang, C.; Chen, J.; Zhu, J. Preparation of an eco-friendly de-icing filler and its effects on the performance of different asphalt mastic. *Constr. Build. Mater.* **2023**, *364*, 129967. [CrossRef]
16. Russo, F.; Veropalumbo, R.; Oreto, C.; Cassese, D.; Papa, B.; Malvezzi, S. Reusing bottom ash as a filler from a Waste-to-Energy plant for making asphalt mastic. *Case Stud. Constr. Mater.* **2022**, *17*, e01406. [CrossRef]
17. Lin, J.; Guo, Z.; Hong, B.; Xu, J.; Fan, Z.; Lu, G.; Wang, D.; Oeser, M. Using recycled waste glass fiber reinforced polymer (GFRP) as filler to improve the performance of asphalt mastic. *J. Clean. Prod.* **2022**, *336*, 130357. [CrossRef]
18. Sheng, Y.; Zhang, B.; Yan, Y.; Chen, H.; Xiong, R.; Geng, J. Effects of phosphorus slag powder and polyester fiber on performance characteristics of asphalt binders and resultant mixtures. *Constr. Build. Mater.* **2017**, *141*, 289–295. [CrossRef]
19. Qian, G.; Wang, K.; Bai, X.; Xiao, T.; Jin, D.; Huang, Q. Effects of surface modified Phosphorus slag powder on performance of asphalt and asphalt mixture. *Constr. Build. Mater.* **2018**, *158*, 1081–1089. [CrossRef]
20. Yu, H.; Zhu, X.; Qian, G.; Gong, X.; Nie, X. Evaluation of phosphorus slag (PS) content and particle size on the performance modification effect of asphalt. *Constr. Build. Mater.* **2020**, *256*, 119334. [CrossRef]
21. Yi, S.C. Experimental Study on the Performance of Commonly Used Modifier/Phosphorus Slag Powder Composite Modified Asphalt. Master's Thesis, Changsha University of Science and Technology, Changsha, China, 2018.
22. Wang, K. Experimental Study on Surface Modification of Phosphorus Slag Powder and Its Modified Asphalt Properties. Master's Thesis, Changsha University of Science and Technology, Changsha, China, 2016.
23. Nie, X.H. Experimental Study on the Properties of Phosphorus Slag Powder Modified Asphalt and Mixture. Master's Thesis, Changsha University of Science and Technology, Changsha, China, 2014.
24. *JTG E20-2011*; Test Regulations for Asphalt and Asphalt Mixtures for Highway Engineering. Research Institute of Highway, Ministry of Transport: Beijing, China, 2011; 373P, A374.

buildings

Article

Comparative Study on the Dynamic Response of Asphalt Pavement Structures: Analysis Using the Classic Kelvin, Maxwell, and Three-Parameter Solid Models

Yonghai He [1], Songtao Lv [2,3,4], Nasi Xie [2,3,4,*], Huilin Meng [1], Wei Lei [1], Changyu Pu [1], Huabao Ma [1], Ziyang Wang [2,3,4,5], Guozhi Zheng [2,3,4] and Xinghai Peng [2,3,4]

1 Hebei Provincial Communications Planning, Design and Research Institute Co., Ltd., Shijiazhuang 050090, China; yonghaihe@126.com (Y.H.); smenghl@163.com (H.M.); hbleiwei@163.com (W.L.); puchangyu_1980@163.com (C.P.); pushumhb@163.com (H.M.)
2 School of Traffic and Transportation Engineering, Changsha University of Science & Technology, Changsha 410114, China; lst@csust.edu.cn (S.L.); wzyang@seu.edu.cn (Z.W.); zhengguozhi@stu.csust.edu.cn (G.Z.); pengxinghsi@stu.csust.edu.cn (X.P.)
3 National Engineering Research Center of Highway Maintenance Technology, Changsha 410114, China
4 National Key Laboratory of Green and Long-Life Road Engineering in Extreme Environment, Changsha 410114, China
5 School of Transportation, Southeast University, Nanjing 211189, China
* Correspondence: xns@stu.csust.edu.cn

Abstract: This study addressed the complex problems of selecting a constitutive model to objectively characterize asphalt mixtures and accurately determine their viscoelastic properties, which are influenced by numerous variables. Inaccuracies in model or parameter determination can result in significant discrepancies between the calculated and measured results of the pavement's structural dynamic response. To address this, the research utilized the physical engineering principles of asphalt pavement structure to perform dynamic modulus tests on three types of high-content rubberized asphalt mixtures (HCRAM) within the surface layer. The research aimed to investigate the influencing factors of the dynamic modulus and establish a comprehensive master curve. This study also critically evaluated the capabilities of three viscoelastic models—the three-parameter solid model, the classical Maxwell model, and the classical Kelvin model—in depicting the dynamic modulus of HCRAM. The findings indicated a negative correlation between the dynamic modulus of the asphalt mixture and temperature, while a positive association exists between the loading frequency and temperature, with the impact of the loading frequency diminishing as the temperature increases. Notably, the three-parameter solid model was identified as the most accurate in describing the viscoelastic properties of the HCRAM. Furthermore, the dynamic response calculations revealed that most indexes in the surface layer's dynamic response are highest when evaluated using the three-parameter viscoelastic model, underscoring its potential to enhance the pavement performance's predictive accuracy. This research provides valuable insights into optimizing the material performance and guiding the pavement design and maintenance strategies.

Keywords: dynamic modulus; viscoelastic model; moving load; dynamic response; numerical simulation

Citation: He, Y.; Lv, S.; Xie, N.; Meng, H.; Lei, W.; Pu, C.; Ma, H.; Wang, Z.; Zheng, G.; Peng, X. Comparative Study on the Dynamic Response of Asphalt Pavement Structures: Analysis Using the Classic Kelvin, Maxwell, and Three-Parameter Solid Models. *Buildings* **2024**, *14*, 295. https://doi.org/10.3390/buildings14010295

Academic Editor: Pengfei Liu

Received: 28 December 2023
Revised: 19 January 2024
Accepted: 20 January 2024
Published: 22 January 2024

1. Introduction

One of the most often utilized road materials worldwide is asphalt pavement, with higher durability, comfort, and other advantages [1]. With the acceleration of urbanization and transportation demand growth, asphalt pavement application is still promising. In the future, with the advancement of technology and the application of new materials, many researchers will devote themselves to the study of asphalt pavements in terms of materials, design, construction, and maintenance to meet the ever-increasing transportation demand [2–5].

In terms of materials, in order to enhance the pavement performance and make the materials more ecologically and energy-efficient, and ultimately to achieve the goal of a "carbon neutral" vision [6], the application of rubberized asphalt (RA) and its mixtures have been a prominent topic of study, with numerous researchers conducting in-depth investigations on it [7–9]. Rubberized asphalt is widely used in stress-absorbing layers and rutting-resistant sections of asphalt pavements due to its exemplary environmental performance, rutting resistance, and durability, thus prolonging the pavement's service life [10]. For rubberized asphalt, a rubber powder content of over 25% can become a high-content rubberized asphalt, and a large amount of research has proven the feasibility of high-content rubberized asphalt [11,12]. In recent years, it has been found that high-content rubberized asphalt (HCRA) has better performance in terms of high-temperature rutting resistance and fracture resistance compared to ordinary RA [13,14]. Also, its asphalt mixtures have good rutting and fatigue resistance [15–18]. However, few domestic and international studies have been conducted on the viscoelastic characteristics of high-content rubberized asphalt mixtures (HCRAM). Therefore, this study selected HCRAM as the research object, and their viscoelastic characteristics were examined using the dynamic modulus (DM) test.

Since asphalt mixture is a common viscoelastic material, it will display various mechanical properties at different temperatures and frequencies [19,20]. In order to investigate the viscoelastic properties, many scholars have carried out DM studies on asphalt mixtures [21,22]. Currently, the research on the DM of asphalt mixtures in various countries mainly includes test methods, influencing factors, and prediction of the DM. The DM of asphalt mixtures can be measured utilizing a variety of domestic and international test techniques [23,24]. The commonly used methods in China are the DM test method in "Standard Test Methods of Bitumen and Bituminous Mixtures in Highway Engineering" (JTG E20-2011) and the Superpave Simple Performance Tester test method (SPT) [25,26]. In addition to indoor testing, on-site estimation of the dynamic modulus is also crucial. Andreas Loizos et al. [27] conducted on-site tests on the experimental road section and non-destructive testing using a drop hammer deflectometer and ground penetrating radar to study the viscoelastic response of the road surface. In addition, regarding the investigation of factors affecting the DM, Wang, H et al. [28] investigated the impacts of migration, temperature, and loading frequency on the DM and phase angle of asphalt mixtures. Zhou et al. [29] found that the void ratio has the most significant combined effect on the DM and phase angle.

Regarding the analysis of DM forecasting, Witzack [30] developed a DM prediction model for densely mixed hot mix asphalt mixtures. Using the Witzack model, Wei et al. [31] fitted the Sigmoidal model to the DM test findings. They found that the Sigmoidal model could better respond to the viscoelastic mechanical characteristics of asphalt mixtures. Dana Daneshvar and Ali Behnood [32] created a DM prediction model utilizing the random forest algorithm and confirmed its superiority. In summary, to study the viscoelastic characteristics of HCRAM, this work measured the DM of asphalt mixtures by indoor DM tests and plotted the master curve utilizing the Sigmoidal to accurately predict the DM.

Stress state analysis and structural dynamic response calculation in asphalt pavement are essential for the design, construction, and maintenance [33]. Choosing a constitutive model that can adequately characterize the material and precisely estimate its characteristics is essential to obtaining a real and accurate stress state and structural dynamic response [34]. Researchers have proposed a number of viscoelastic constitutive models to explain the viscoelastic properties of asphalt mixes [19], such as the classical Maxwell and its generalized model, the classical Kelvin and its generalized model [35], and the three-parameter solid model [36]. However, the majority of nations still design asphalt pavements using the layered elastic system mechanics theory under vertical stresses due to computational simplicity [37]. This method needs to consider the influence of horizontal forces generated by driving loads on the pavement and accurately describes the viscoelastic behavior of road materials under repeated loads, which can easily lead to consistency between the calculated and measured results of the dynamic response of the road structure. Consequently, in this study, the viscoelastic constitutive model is employed to determine the internal mechanical response of the pavement structure to objectively describe the structural and mechanical response features of pavement.

Exploring the response of pavement structures is extremely necessary for pavement design or performance evaluation. However, this study cannot solely focus on pavement structure or material characteristics. Instead, the two should be combined to study the performance of asphalt pavement under their combined action. A helpful technique for bridging the gap between experimental and pavement research is numerical simulation [38,39]. With the application of numerical simulation methods, more and more scholars are using finite element software to study mechanical responses. Ji et al. [40] used ANSYS to simulate the pavement structure under different driving loads. They studied the connection between the tensile stress, shear stress, and pavement's asphalt layer's thickness. Z Dong et al. [41] established a large axle load model using Abaqus-6.14 software. They proposed that for semi-rigid mobile heavy-duty asphalt pavements, the vertical strain at the top of the roadbed and the vertical stress of the pavement must be considered in the design. In addition, the viscoelastic constitutive model of asphalt mixtures can be implemented into Abaqus-6.14 software through user-defined subroutines (UMAT) to evaluate the response of asphalt structures [42]. This study chooses to use UMAT to compute the constitutive model of HCRAM. It conducts dynamic response analysis of pavement under viscoelastic behavior.

2. Study Objectives and Scope

Based on the above background, this study aimed to characterize the dynamic response of HCRA pavement structures realistically and objectively. This study relied on the physical engineering of the pavement structure of the Jingde Expressway, considering its surface materials' viscoelastic properties. Dynamic modulus tests and analysis of factors affecting the DM were conducted on three types of HCRAM, and a dynamic modulus master curve (DMMC) was established utilizing the Sigmoidal and studied the viscoelastic properties of high content (30%, 40%, and 50%) rubber asphalt mixtures. Further, the variability in the description of the DM of the HCRAM by the classical Maxwell model, the classical Kelvin model, and the three-parameter solid model was evaluated according to the theory of viscoelasticity of the HCRAM concerning the established master curves. Finally, the UMAT subroutine was invoked to implement the viscoelastic intrinsic model numerically in Abaqus. At the same time, the pavement structure's dynamic response was calculated, and the dynamic response and differences of pavement structures based on three viscoelastic models were studied, objectively describing the structural and mechanical response properties of the pavement and providing the accurate basis for pavement structure design.

3. Materials and Methods

3.1. Raw Material and Mix Design

3.1.1. Asphalt

The HCRA employed in this study had 30%, 40%, and 50% rubber doping. The asphalt was provided by Hebei Provincial Communications Planning, Design and Research Institute Co., Ltd. in Shijiazhuang, China, which is consistent with the asphalt materials used in China's Jingde Highway's upper, middle, and lower layers, respectively. The PG grading performance is presented in Table 1, and the related indexes are displayed in Table 2 [43].

Table 1. RA PG grading properties.

Test Items	30% RA	40% RA	50% RA
PG classification	88-34	88-34	82-28
Rutting factor (88 °C)	1683.8	1421.6	1638
Rutting factor after aging (88 °C)	3429.5	2767.2	1672
Fatigue limit temperature (°C)	13	10	16
Creep strength (−24 °C)	201.9205	166.9801	343
Creep rate	0.3054	0.3147	0.335

Table 2. Performance test results of RA.

Test Items		30% RA	40% RA	50% RA
Penetration 25 °C, 100 g, 5 s (0.1 mm)		62	66	58
Ductility 5 cm/min, 5 °C (cm)		16	18	12.8
Softening point (°C)		72.5	72	78.7
Dynamic viscosity 180 °C (Pa·s)		2.8	3.5	2.96
Elastic recovery 25 °C (%)		84	81	71
48 h softening point difference (°C)		2.0	0.5	6.4
0.1 Kpa recovery rate (%)		98.67	99.43	92.08
3.2 Kpa recovery rate (%)		90.79	92.55	86.45
Residue after	−0.29	−0.3	0.3	T 0610-2011
TFOT (163 °C,	77	76	86.2	T 0604-2011
85 min)	11	12	10.4	T 0605-2011

3.1.2. Aggregates

The aggregate used in this study is consistent with the research of Wang et al. [44], which is limestone, and the filler is limestone mineral powder. Various characteristics indicators of aggregates were examined, and aggregates' physical and mechanical performance indicators were finally obtained. Relevant literature can be found in [44].

3.1.3. Mix Design

The HCRAM gradation design method used in this paper mainly refers to the "Technical Specifications for Construction of Highway Asphalt Pavements" (JTG F40-2004), the ARHM-13 (The maximum nominal particle size of the asphalt rubber hot mixture is 13), ARHM-20, and ARHM-25 gradation curve presented in Figure 1. The asphalt aggregate ratio of the ARHM-13, ARHM-20, and ARHM-25 corresponding to the finalized 30%, 40%, and 50% blends of RA is 5.22%, 4.79%, and 4.36%, respectively.

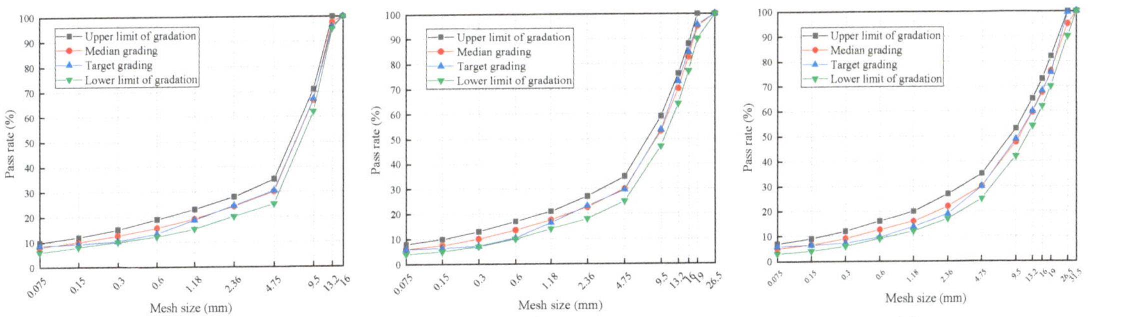

Figure 1. Mineral grading curve of HCRAM. (**a**) ARHM-13, (**b**) ARHM-20, (**c**) ARHM-25.

3.2. *Experiment Approach*

3.2.1. Preparation of Test Specimens

After being mixed according to the desired gradation, the limestone aggregate was dried in an oven maintained at 110 °C until it attained a constant weight. For more than two hours, the asphalt and aggregate were placed in an oven set at 180 °C, and an indoor mixer was used to mix the HCRAM under the optimal oil-stone ratio conditions. Following demolding, a cylindrical sample measuring 150 mm ± 2 mm in height and 100 mm ± 2 mm in diameter was obtained. The samples were made using the rotary compaction method. The resulting HCRAM specimen is displayed in Figure 2.

Figure 2. HCRAM specimens.

3.2.2. Dynamic Modulus Test Methods

This study used a multifunctional material testing machine, MTS Landmark 370.25, for uniaxial compression the DM testing. The schematic diagram of the testing instrument and the uniaxial compression DM test are shown in Figures 3 and 4, respectively. The selected temperatures were 15 °C, 30 °C, 45 °C, and 60 °C. Table 3 shows the number of repeated loads and frequencies. The test was conducted according to the uniaxial compression DM test method in the "Standard Test Methods of Bitumen and Bituminous Mixtures in Highway Engineering" (JTG E20-2011).

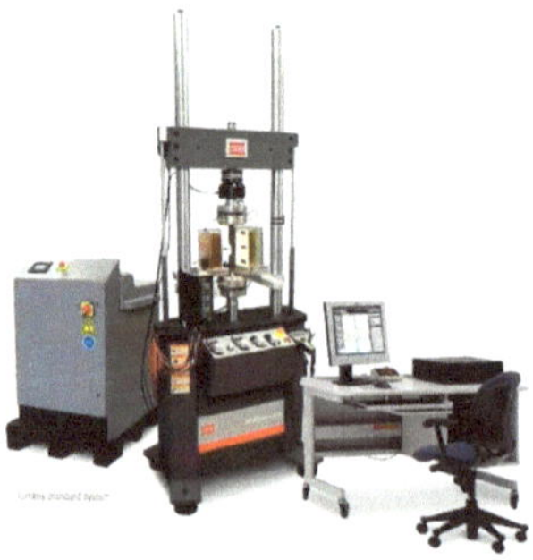

Figure 3. MTS machine.

Figure 4. Uniaxial compression DM test.

Table 3. RA PG grading properties.

Frequency	Repetitions	Frequency	Repetitions
25	200	1	20
10	200	0.5	15
5	100	0.1	15

3.3. Numerical Simulation Methods

3.3.1. Establishment of Finite Element Model

The Jingde Expressway test road in China's pavement structure was replicated in the pavement structure model. The UMAT subroutine was used to set the parameters for the viscoelastic surface layer material. Elastic materials included the base course and soil foundation, and the material characteristics were directly determined by the internal software program. The pavement materials and structural design parameters are displayed in Table 4. The model dimensions were 9 m in length, 2.75 m in width, and 5 m in depth. The construction basis of the model was as follows: due to the symmetrical pavement structure, considering computer performance and computational efficiency, only half of the pavement model was established in ABAQUS-6.14 software. A load band was drawn in advance to affect the exact condition of the pavement, with a length of 6 m and a width of 0.23 m. The load band is presented in Figure 5.

Table 4. Pavement structure combination and material parameters.

Structure Layer No	Horizon	Material Type	Model	Density (kg/m^3)	Compressive Modulus (MPa)	Tensile Modulus (MPa)	Pressure Poisson's Ratio
1	Upper layer	ARHM-13	Viscoelastic model	0.04	2400	-	0.25
2	Middle layer	ARHM-20	Viscoelastic model	0.08	2400	-	0.25
3	Lower layer	ARHM-25	Viscoelastic model	0.1	2400	-	0.25
4	Upper base course	Inorganic binding material stabilizer	Linear elastic model	0.18	2300	10,500	0.25
5	Medium base course	Inorganic binding material stabilizer	Linear elastic model	0.18	2300	10,500	0.25
6	Lower base	Inorganic binding material stabilizer	Linear elastic model	0.18	2300	8500	0.25
7		Soil foundation	Linear elastic model	-	1800	280	0.40

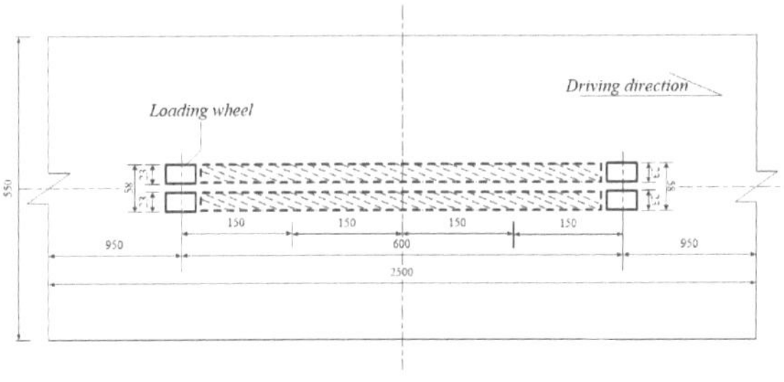

Figure 5. Pavement load band under moving load.

Considering computer performance, the grid division method in the x-direction was as follows: the grid width of the moving load band was set to 2.3 cm. The load was carried to the pavement's center, and the grid width was set to 0.2 cm. The remaining grids were divided according to the principle of uniform distribution. The grid division method in the y-direction ensured that the grid shape approximates a square with a fixed grid length of 2.3 cm. The grid division method in the z-direction was as follows: near the upper part of the model, for example by the asphalt surface layer, the grid was encrypted, and the grid height was set to 1 cm. The height of the grid at the grassroots level was set to 3 cm. The grid of the soil base layer gradually became sparse according to the principle of equal distribution.

The boundary conditions of the finite element model directly affected the convergence of the model. The boundary conditions of the finite element model in this study were as follows: on the X = 0 m and X = 6 m planes, the displacement in the x-direction was limited, that is, $U_1 = 0$ on this vertical plane; on the Y = 0 m plane, the displacement in the y-direction was limited, i.e., $U_2 = 0$ on this vertical plane; on a plane with Y = 2.75 m, the displacement was set as a symmetrical boundary; at the bottom of the model, on a plane of Z = 5 m, the displacement was set as an infinite mesh.

Due to the tire width of the loading device being 0.23 m, the moving load was set to a rectangle with a length of 0.23 m and a width of 0.23 m. The method of applying moving loads was referred to the literature [45].

3.3.2. Numerical Implementation of Viscoelastic Constitutive Model

ABAQUS provides standard finite element analysis programs and has good openness, which can generate non-standard analysis programs by providing user subroutine interfaces to meet user needs. Through the user material subroutine, users can define any supplementary material model. Not only can any material constants be read as data, but ABAQUS also provides storage functionality for any state variables related to the solution at each material count point for application in these subroutines.

This work used UMAT to numerically apply the viscoelastic constitutive model of HCRAM, defining the asphalt pavement surface material as a viscoelastic material. The UMAT subroutine can be found in Appendix A (Table A1).

3.3.3. Validation Method for Numerical Simulation Effectiveness

The methods for verifying the effectiveness of numerical simulations can be divided into two types: direct verification and indirect verification. Direct verification refers to comparing the measured data from laboratories or actual engineering with the model calculated by numerical simulations. Indirect verification involves comparing the results of theoretical calculations or analytical solutions with numerical simulation results. This study adopts the second method, which uses the calculation results of Bisar3 software to verify the effectiveness of the numerical simulation method.

Because Bisar3 software is based on an elastic layered system calculation method, it is necessary to degrade the constitutive model in Abaqus-6.14 software from a viscoelastic model to an elastic model and then calculate the same road surface model to compare the calculation results of the two.

4. Results and Discussion

4.1. Analysis of DM Test Results for Asphalt Mixtures

4.1.1. Analysis of Factors Affecting the DM

(1) The impact of loading frequency

Figure 6 displays the trend of the DM of the HCRAM with frequency. From Figure 6, the DM of the HCRAM was observed to increase with frequency and to have a tendency of fast growth in the early stage and slow growth in the later stage. Especially when the loading rate increased from 0.1 Hz to 1 Hz, the modulus of the HCRAM increased roughly linearly. However, the growth rate was prolonged when the frequency increased from 10 Hz to 25 Hz. It can be inferred that the HCRAM's DM eventually reached a certain limit value and fails to grow indefinitely with an increase in loading rate.

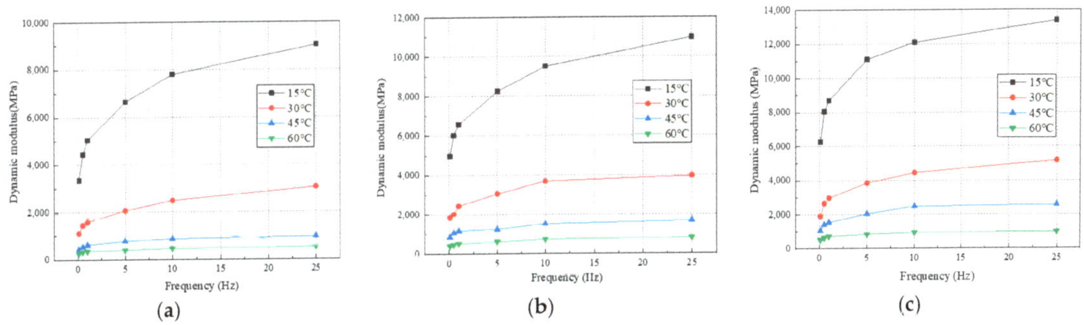

Figure 6. Trend of DM of HCRAM with frequency. (**a**) ARHM-13, (**b**) ARHM-20, (**c**) ARHM-25.

The aforementioned phenomenon can be elucidated from the standpoint of the viscoelastic properties of asphalt mixtures [46], mainly because when the loading frequency is low, the HCRAM mainly exhibit viscous characteristics. When the frequency is high, the asphalt mixture mainly exhibits elastic properties, which means that when subjected to high-frequency loads, the asphalt mixture is no longer an apparent viscoelastic material. At this point, the impact of the loading rate on the DM is no longer as significant as that at low frequencies. The explanation for the hysteresis of asphalt mixture deformation is that when the asphalt mixture is subjected to external loads [47], the stress–strain changes generated inside the specimen will exhibit a certain degree of hysteresis. If the frequency is reduced, the viscosity characteristics of the asphalt mixture are apparent, and the subsequent phenomenon will be more pronounced; that is, the deformation is more minor, and the DM is larger. Conversely, if the loading frequency is reduced, the deformation is more significant, and the DM is smaller. To some extent, this explains that asphalt is more prone to deformation and rutting under low-speed traffic loads, and the pavement is more prone to cracking under high-speed traffic loads [48].

Figure 7 shows the variation trend of the phase angle of each asphalt mixture as loading frequency. This indicates that the phase angle variation trend with frequency differs with temperature. As the loading frequency increases at test temperatures of 15 °C and 30 °C, the phase angle progressively decreases. This trend can indicate that when the temperature is low, the viscous properties of the asphalt mixture decrease, and the elastic properties increase with the rise of loading frequency. At both 45 °C and 60 °C test temperatures, the phase angle of the asphalt mixture rises proportionally with the increase in frequency. This indicates that at elevated temperatures, the viscosity of the asphalt mixture intensifies while its elasticity diminishes with the rise in loading frequency.

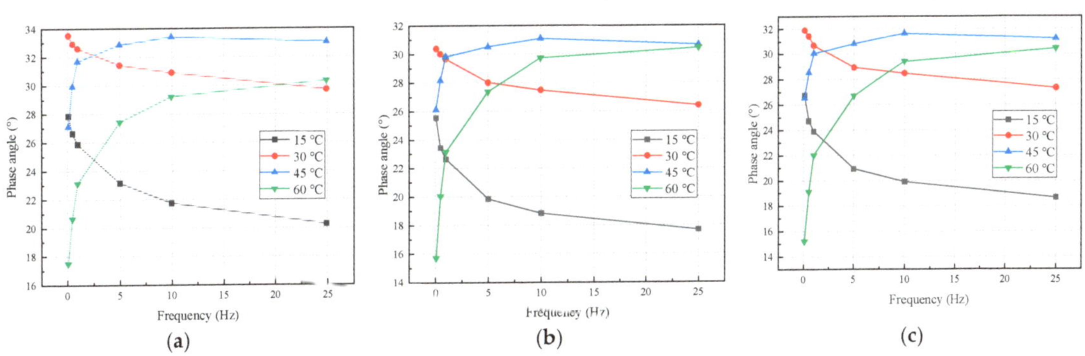

Figure 7. The trend of phase angle variation of HCRAM with loading frequency. (**a**) ARHM-13, (**b**) ARHM-20, (**c**) ARHM-25.

Figure 8 shows the variation trend of the storage modulus of the HCRAM as frequency. According to Figure 8a, it is evident that the trend of the storage modulus changing with the loading frequency is similar to that of the DM, with a faster rate of change in the early stage and a slower trend in the later stage. The storage modulus characterizes the elasticity of the HCRAM, which can store and release energy when subjected to dynamic loads. At 25 Hz, the maximum DM is measured. This implies that the energy contained in the asphalt mixture is unable to be completely dissipated when subjected to high frequencies and low temperatures. The elastic properties exhibited by the asphalt mixture are apparent, allowing the stored modulus to reach its maximum value under these conditions. On the contrary, under high temperature and low-frequency conditions, the elastic performance of the material is weak, resulting in a decrease in the DM.

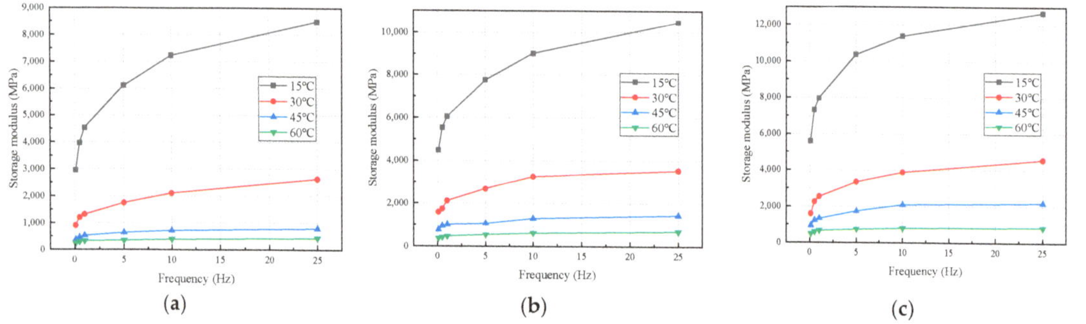

Figure 8. Trend of storage modulus of HCRAM with loading frequency. (**a**) ARHM-13, (**b**) ARHM-20, (**c**) ARHM-25.

Figure 9 shows the variation trend of the loss modulus as frequency. This demonstrates that the trends of the DM, storage modulus, and loss modulus with frequency are all the same. The loss modulus characterizes the HCRAM's viscous properties. Under certain temperature conditions and high loading frequencies, the viscosity and elasticity of asphalt mixtures increase, increasing the energy loss. This is manifested as a rise in the loss modulus of asphalt mixtures under dynamic loads.

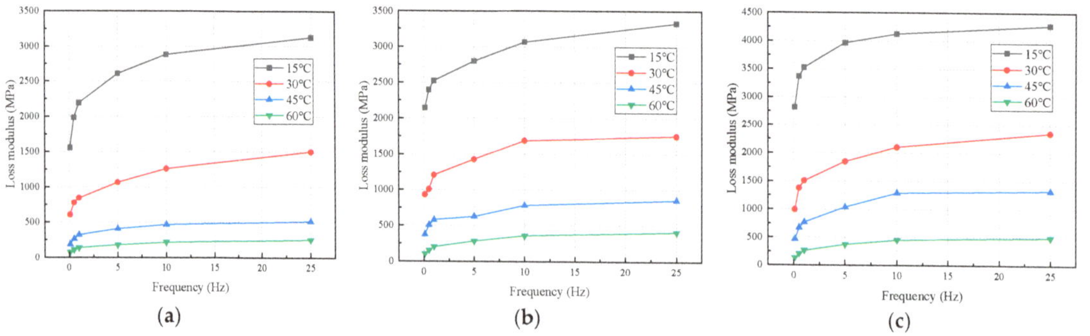

Figure 9. Trend of loss modulus of HCRAM with loading frequency. (**a**) ARHM-13, (**b**) ARHM-20, (**c**) ARHM-25.

(2) The influence of temperature

Figure 10 shows the variation trend of the DM and other experimental data of the ARHM-13 with the temperature. The DM trend of ARHM-13 with a temperature fluctuation is presented in Figure 10a. In Figure 10a, it can be observed that, under various frequencies, as the temperature rises, the HCRAM's DM drastically drops. For instance, the DM of the

ARHM-13 is 66%, 89%, and 94% lower at 30 °C, 45 °C, and 60 °C when the frequency is 25 Hz compared to when it is at 15 °C. As the temperature rises, the DM of the asphalt mixture decreases, and its rutting resistance decreases. This phenomenon mostly occurs because, at high temperatures, the asphalt in the asphalt mixture mainly exhibits viscoelastic properties. As the temperature rises, the viscosity of the asphalt reduces, and its rutting resistance performance decreases. The pavement is prone to permanent shear deformation and rutting when subjected to external forces.

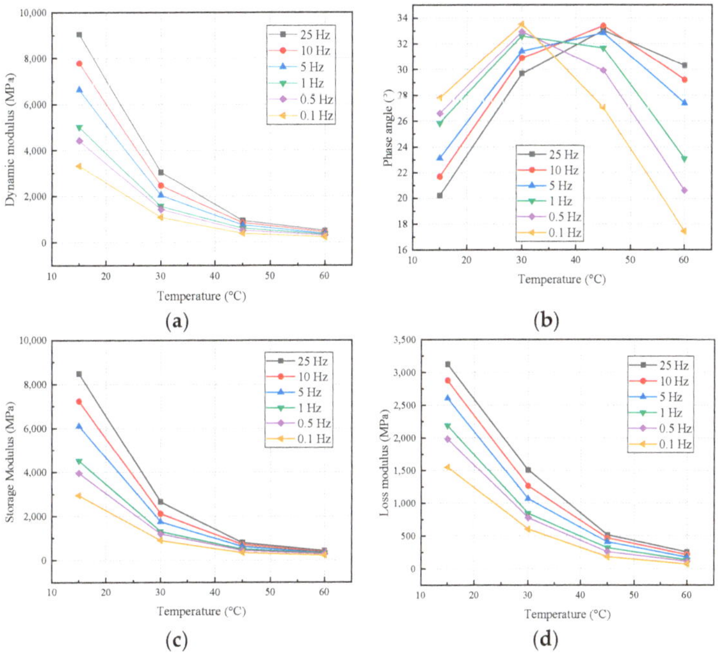

Figure 10. The trend of DM, etc., of ARHM-13 with temperature (15 °C). (**a**) dynamic modulus, (**b**) phase angle, (**c**) storage modulus, (**d**) loss modulus.

Figure 10b shows the trend of the phase angle with the temperature. As the figure illustrates, the variation of the phase angle with the loading frequency is more complicated. The phase angle change mainly reflects the viscoelasticity of the polymer. At low temperatures, by increasing the frequency, the phase angle change first rises and then falls, while at high temperatures, it shows an upward trend when the frequency is increased. When the loading frequency is 0.1 Hz, 0.5 Hz, and 1 Hz, the phase angle peaks at 30 °C; when the loading frequency is 5 Hz, 10 Hz, and 25 Hz, the phase angle reaches their peak at 45 °C. The primary cause for this changing pattern is that, under low temperatures and high-frequency loads, the asphalt binder exhibits significant elastic properties, with solid performance and increased viscosity with rising temperatures. Since the characteristics of the asphalt binder have a greater influence on the asphalt mixture's performance, the phase angle rises as the temperature rises. In theory, as a composite material, the phase angle of the asphalt mixture is equal to the phase angle of the matrix material. The asphalt mixture's phase angle is unaffected by aggregates. However, under high temperature and low-frequency action, the asphalt binder becomes soft, and the embedding force between the mineral skeleton increases, which exceeds the influence of the asphalt bonding force. As minerals are elastic materials, their phase angle is zero, so the phase angle will decrease and further stiffen [49].

Figure 10c,d demonstrate the temperature-dependent trends of the storage and loss modulus of ARHM-13. It is not difficult to find from the figures that their trends are similar to those of the DM, decreasing with increasing temperatures. This phenomenon is because the storage and loss modulus are related to the molecular chain movement of asphalt. The frictional resistance between the asphalt molecules is greatly affected by temperature, and temperature changes cause changes in asphalt performance.

4.1.2. Establishment of the Dynamic Modulus Master Curve

The DM was fitted and analyzed in this work using a Sigmoidal model [50], whose formula is as follows:

$$\lg(|E^*|) = \delta + \frac{\alpha}{1 + e^{\beta + \gamma \cdot \lg(\omega_r)}} \tag{1}$$

where $|E^*|$ is the DM; δ is the logarithm of the minimum dynamic modulus; β, γ are the parameter that describes the shape of the sigmoidal function; $\delta + \alpha$ is the logarithm of the maximum dynamic modulus; ω_r is frequency reduction.

The DM obtained at different temperatures can be converted from the loading frequency ω, which is the logarithm of the minimum dynamic modulus temperatures to the corresponding reduced frequency ω_r at the reference temperature through a shift factor $\lg\alpha_T$. As a result, the DM value at the converted frequency can finally be obtained. The relationship between $\lg\alpha_T$, ω, and ω_r is shown as follows.

$$\omega_r = \alpha_T \cdot \omega \tag{2}$$

$$\lg\omega_r = \lg\alpha_T + \lg\omega \tag{3}$$

The DMMC is established using a sigmoidal model based on the DM test results at different temperatures and frequencies, with $|E^*|$ as the vertical axis and ω_r as the horizontal axis. For ease of comparison, 30 °C was selected as the reference temperature. The DM values at the other three temperatures were translated using Origin-2018 software to obtain the shift factor, and then the fitted sigmoidal function was used to obtain the DMMC. Figure 11 presented the fitting results of the DMMC for each HCRAM, as presented in Table 5. The R^2 values are all higher than 0.99, indicating a very good fitting effect. That is, the model is highly consistent with the experimental results.

As demonstrated by Figure 11, the DMMC slope is comparatively small at both lower and higher frequencies., indicating that the DM tends to stabilize at this time. The change trend is not apparent and will tend to have a limited value. The shift factor is a function of temperature, and its amplitude can indicate the degree of temperature dependence of asphalt mixtures. Table 5 demonstrates that, starting from the gradation of the mixture and the rubber content, in terms of temperature dependence, ARHM-13 has 30% rubber content, <ARHM-20 has 40% rubber content, and <ARHM-25 has 50% rubber content. This result indicates that, as the rubber content and gradation increase, the temperature dependence of asphalt mixtures also increases.

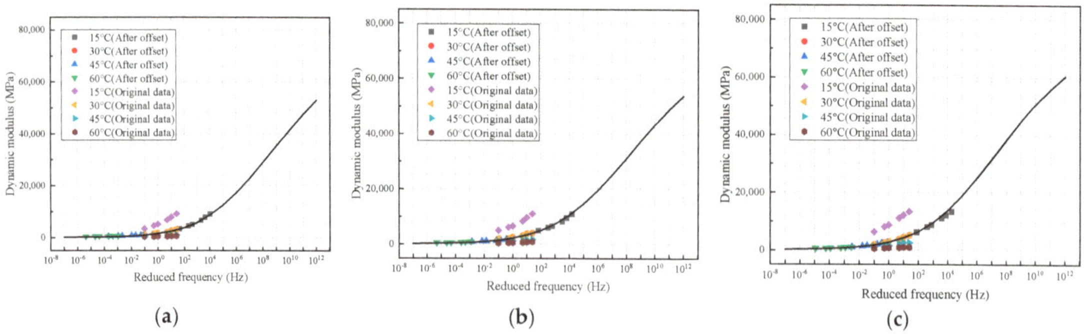

Figure 11. DMMC of asphalt mixture. (**a**) ARHM-13, (**b**) ARHM-20, (**c**) ARHM-25.

Table 5. DMMC of various HCRAM sigmoidal model.

Types	Temperature (°C)	Shift Factor	Dynamic Modulus Main Curve Model	R^2
ARHM-13	15 30 45 60	2.62 0 −2.67 −4.37	$\lg(\lvert E^*\rvert) = 1.9298 + \frac{2.9669}{1+e^{0.2725-0.2551\lg(\omega_r)}}$	0.9990
ARHM-20	15 30 45 60	2.81 0 −1.85 −4.31	$\lg(\lvert E^*\rvert) = 2.0630 + \frac{2.8337}{1+e^{0.1589-0.2449\lg(\omega_r)}}$	0.9943
ARHM-25	15 30 45 60	2.88 0 −1.48 −3.93	$\lg(\lvert E^*\rvert) = 2.2018 + \frac{2.6949}{1+e^{0.1321-0.2701\lg(\omega_r)}}$	0.9971

Figure 12 shows the DMMC of three types of HCRAM at different temperatures. It depicted that there is no substantial disparity in the DMMC of the three types of HCRAM, especially in the high-frequency and low-frequency regions, where the three master curves tend to overlap. The rate at which the DM changes from low frequency to high frequency steadily slows down as the temperature rises, suggesting that the asphalt mixture is less impacted by frequency as the temperature rises. At constant temperature, the master curve's position rises with increasing asphalt mixture gradation, suggesting that the ARHM-25 asphalt mixture has a higher DM.

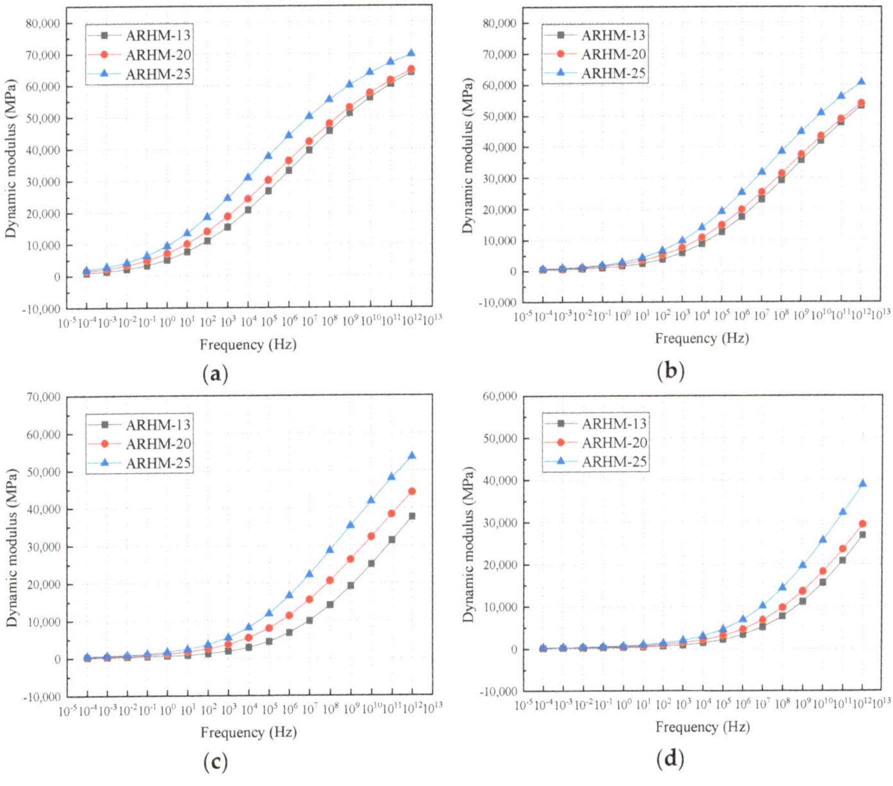

Figure 12. Comparison of DMMC for several HCRAM. (**a**) 15 °C, (**b**) 30 °C, (**c**) 45 °C, (**d**) 60 °C.

4.2. *Establishment of Intrinsic Relationships and Parameter Fitting for Viscoelastic Models*

4.2.1. Establishment of Constitutive Relations for Viscoelastic Models

(1) Classic Kelvin model

The Kelvin model comprises a spring and a sticky pot in parallel. The model structure diagram is displayed in Figure 13. In this connection method, the strain of the two components is equal, and the total stress is equal to the sum of the stresses of the spring and the adhesive pot. The constitutive equation is represented by Equation (4).

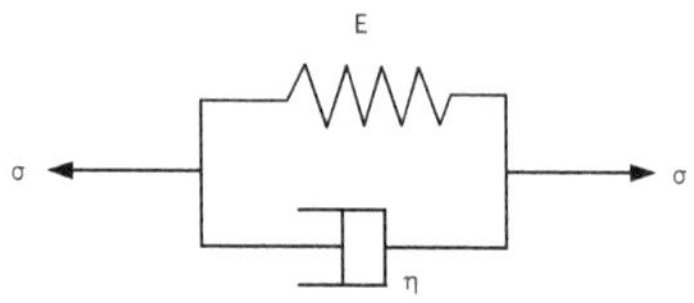

Figure 13. Classic Kelvin model.

$$\sigma = \sigma_1 + \sigma_2 = E\varepsilon + \eta\varepsilon' \tag{4}$$

By applying Fourier transform to the Kelvin constitutive model, Equation (5) can be obtained:

$$\sigma(\omega) = E\varepsilon(\omega) + \eta(i\omega)\varepsilon(\omega) \tag{5}$$

where: i is an imaginary singular number; ω is the frequency.

The complex modulus of the Kelvin model can be obtained by performing a Fourier transform on Equation (5) and then using the definition of complex modulus, as shown in Equation (6):

$$Y^*(i\omega) = \frac{\sigma(\omega)}{\varepsilon(\omega)} = E + \eta(i\omega) = Y_1(\omega) + iY_2(\omega) \tag{6}$$

From this, the expression for the dynamic modulus can be obtained as shown in Equation (7):

$$|Y^*(i\omega)| = \sqrt{E^2 + \eta^2\omega^2} \tag{7}$$

(2) Classic Maxwell model

The Maxwell model comprises a spring and a sticky pot connected in series. The model structure is shown in Figure 14. Assuming that under stress, the strains generated by the spring and the adhesive pot are ε_1 and ε_2, respectively, the total deformation of the component is the sum of their deformations. The constitutive equation is represented by Equation (8).

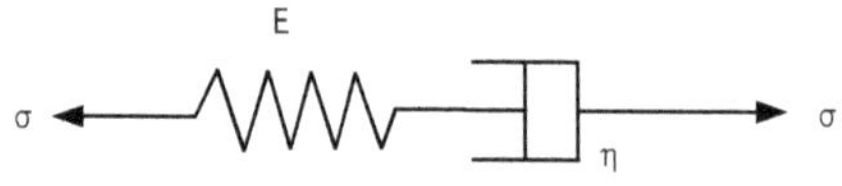

Figure 14. Classic Maxwell model.

$$\varepsilon = \varepsilon'_1 + \varepsilon'_2 = \frac{\dot{\sigma}}{E} + \frac{\sigma}{\eta} \tag{8}$$

Using the same method to process the Classic Maxwell model yields the expression for dynamic modules as shown in Equation (9):

$$|Y^*(i\omega)| = \sqrt{\frac{(E\eta^2\omega^2)^2 + (E^2\eta\omega)^2}{(E^2 + \eta^2\omega^2)^2}} \tag{9}$$

(3) Three-parameter solid model

A Kelvin model coupled in series with a spring element makes up the three-parameter solid model. The model structure is presented in Figure 15. The constitutive equation is represented by Equation (10).

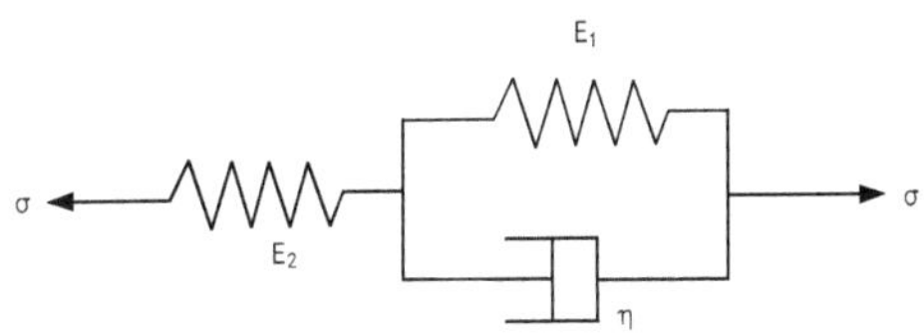

Figure 15. Three-parameter solid model.

$$\varepsilon = \varepsilon_1 + \varepsilon_2 \tag{10}$$

For the Kelvin model:

$$\frac{\sigma}{E_1} = \varepsilon_1 + \frac{\eta_1}{E_1}\dot{\varepsilon}_1 \tag{11}$$

For elastic components:

$$\frac{\sigma}{E_2} = \varepsilon_2 \tag{12}$$

After sorting out the above equations and taking their derivatives, Equation (13) can be obtained:

$$\sigma + \frac{\eta_1}{E_1+E_2}\dot{\sigma} = \frac{E_1E_2}{E_1+E_2}\varepsilon + \frac{\eta_1 E_2}{E_1+E_2}\dot{\varepsilon} \tag{13}$$

Using the same method to process the three-parameter solid model yields the expression for dynamic modules as shown in Equation (14):

$$|Y^*(i\omega)| = \sqrt{\frac{\left(\frac{E_1E_2}{E_1+E_2} + E_2\left(\frac{\eta_1}{E_1+E_2}\right)^2\omega^2\right)^2 + \left(\left(E_2 - \frac{E_1E_2}{E_1+E_2}\right)\cdot\frac{\eta_1\omega}{E_1+E_2}\right)^2}{\left(1+\left(\frac{\eta_1}{E_1+E_2}\right)^2\omega^2\right)^2}} \tag{14}$$

(4) Jacobian Matrix and UMAT

For linear elastic models, there is a relationship between the elements in the stiffness matrix and the elastic modulus as follows:

$$\begin{cases} c_{11} = c_{22} = c_{33} = \frac{1-\mu}{(1-2\mu)(1+\mu)}E \\ c_{12} = c_{13} = c_{23} = \frac{\mu}{(1-2\mu)(1+\mu)}E \\ c_{44} = c_{55} = c_{66} = \frac{1}{2(1+\mu)}E \end{cases} \tag{15}$$

Assuming that the Poisson's ratio of the classical viscoelastic model is a constant value, the relationship between the stiffness matrix after the Laplace transform and the corresponding values of the elastic modulus is similar to that of the linear elastic model, with:

$$\begin{cases} \overline{c}_{11}(s) = \overline{c}_{22}(s) = \overline{c}_{33}(s) = \frac{1-\mu}{(1-2\mu)(1+\mu)}\overline{E}(s) \\ \overline{c}_{12}(s) = \overline{c}_{13}(s) = \overline{c}_{23}(s) = \frac{\mu}{(1-2\mu)(1+\mu)}\overline{E}(s) \\ \overline{c}_{44}(s) = \overline{c}_{55}(s) = \overline{c}_{66}(s) = \frac{1}{2(1+\mu)}\overline{E}(s) \end{cases} \tag{16}$$

For common viscoelastic models, the corresponding values of their elastic modulus are presented in Table 6.

Table 6. Corresponding values of elastic modulus of classical viscoelastic models.

	Classic Kelvin Model	Classic Maxwell Model	Three Parameter Solid Model
$\overline{E}(s)$	$E+\eta s$	$\frac{E\eta s}{E+\eta s}$	$\frac{E_2(E_1+\eta s)}{E_1+E_2+\eta s}$

Based on the above assumptions, the constitutive relationship of the three-parameter solid model can be inferred as follows:

$$\begin{cases} \left(1+\frac{\eta}{E_1+E_2}d\right)\sigma_{ii} = \frac{E_2(E_1+\eta d)}{(1+\mu)(E_1+E_2)}\varepsilon_{ii} + \frac{3\mu E_2(E_1+\eta d)}{(1-2\mu)(1+\mu)(E_1+E_2)}\varepsilon_h \\ \left(1+\frac{\eta}{E_1+E_2}d\right)\sigma_{ij} = \frac{E_2(E_1+\eta d)}{2(1+\mu)(E_1+E_2)}\gamma_{ij} \end{cases} \tag{17}$$

$$\begin{cases} \Delta\sigma_{ii} = A\Delta\varepsilon_{ii} + B\Delta\varepsilon_h + C\varepsilon_{ii} + D\varepsilon_h - E\sigma_{ii} \\ \Delta\sigma_{ij} = \frac{1}{2}A\Delta\gamma_{ij} + \frac{1}{2}C\gamma_{ij} - E\sigma_{ij} \end{cases} \tag{18}$$

where $\begin{cases} A = \frac{E_2(E_1\Delta t+2\eta)}{(1+\mu)[(E_1+E_2)\Delta t+2\eta]}, B = \frac{3\mu E_2(E_1\Delta t+2\eta)}{(1+\mu)(1-2\mu)[(E_1+E_2)\Delta t+2\eta]} \\ C = \frac{2E_1E_2\Delta t}{(1+\mu)[(E_1+E_2)\Delta t+2\eta]}, D = \frac{6\Delta t\mu E_1E_2}{(1+\mu)(1-2\mu)[(E_1+E_2)\Delta t+2\eta]} \\ E = \frac{2(E_1+E_2)\Delta t}{(E_1+E_2)\Delta t+2\eta} \end{cases}$.

Its Jacobian matrix is as follows:

$$\begin{cases} c_{11} = \frac{(1-\mu)E_2(E_1\Delta t+2\eta)}{(1+\mu)(1-2\mu)[(E_1+E_2)\Delta t+2\eta]} \\ c_{12} = \frac{E_2(E_1\Delta t+2\eta)}{(1+\mu)(1-2\mu)[(E_1+E_2)\Delta t+2\eta]} \\ c_{44} = \frac{E_2(E_1\Delta t+2\eta)}{2(1+\mu)[(E_1+E_2)\Delta t+2\eta]} \end{cases} \tag{19}$$

When $E_2 \to \infty$, the three-parameter solid model degenerates to the Kelvin model, and the relationship between the stress increment and the strain increment and the Jacobian matrix are as follows:

$$\begin{cases} \Delta\sigma_{ii} = A\Delta\varepsilon_{ii} + B\Delta\varepsilon_h + C\varepsilon_{ii} + D\varepsilon_h - E\sigma_{ii} \\ \Delta\sigma_{ij} = \frac{1}{2}A\Delta\gamma_{ij} + \frac{1}{2}C\gamma_{ij} - E\sigma_{ij} \end{cases} \tag{20}$$

where $\begin{cases} A = \frac{E_1\Delta t+2\eta}{(1+\mu)\Delta t}, B = \frac{3\mu(E_1\Delta t+2\eta)}{(1+\mu)(1-2\mu)\Delta t} \\ C = \frac{2E_1}{1+\mu}, D = \frac{6\mu E_1}{(1+\mu)(1-2\mu)} \\ E = 2 \end{cases}$.

Its Jacobian matrix is:

$$\begin{cases} c_{11} = \frac{(1-\mu)(E_1\Delta t+2\eta)}{(1+\mu)(1-2\mu)\Delta t} \\ c_{12} = \frac{\mu(E_1\Delta t+2\eta)}{(1+\mu)(1-2\mu)\Delta t} \\ c_{44} = \frac{E_1\Delta t+2\eta}{2(1+\mu)\Delta t} \end{cases} \tag{21}$$

When $E_1 = 0$, the three-parameter solid model degenerates to the Maxwell model, and the relationship between the stress increment and the strain increment and the Jacobian matrix are as follows:

$$\begin{cases} \Delta\sigma_{ii} = A\Delta\varepsilon_{ii} + B\Delta\varepsilon_h - C\sigma_{ii} \\ \Delta\sigma_{ij} = \frac{A}{2}\Delta\gamma_{ij} - C\sigma_{ij} \end{cases} \tag{22}$$

where $\begin{cases} A = \frac{2E_2\eta}{(1+\mu)(E_2\Delta t+2\eta)}, B = \frac{6\mu E_2\eta}{(1+\mu)(1-2\mu)(E_2\Delta t+2\eta)} \\ C = \frac{2E_2\Delta t}{E_2\Delta t+2\eta} \end{cases}$.

Its Jacobian matrix is:

$$\begin{cases} c_{11} = \frac{2(1-\mu)E_2\eta}{(1+\mu)(1-2\mu)(E_2\Delta t+2\eta)} \\ c_{12} = \frac{2\mu E_2\eta}{(1+\mu)(1-2\mu)(E_2\Delta t+2\eta)} \\ c_{44} = \frac{E_2\eta}{(1+\mu)(E_2\Delta t+2\eta)} \end{cases} \quad (23)$$

According to the above content, the code of the UMAT subroutine is shown in the Appendix A (Table A1).

4.2.2. Evaluation of Several Classical Viscoelastic Models for the DM Description of Asphalt Mixtures

This work used the DM calculation formula in Section 4.2.1 to fit the DM calculated from the DMMC of the three HCRAM established in Section 4.1.2. Figure 16 shows the fitting curve of the DM of ARHM-13 at diverse temperatures, and the viscoelastic parameters obtained by fitting each asphalt mixture are listed in Table 7.

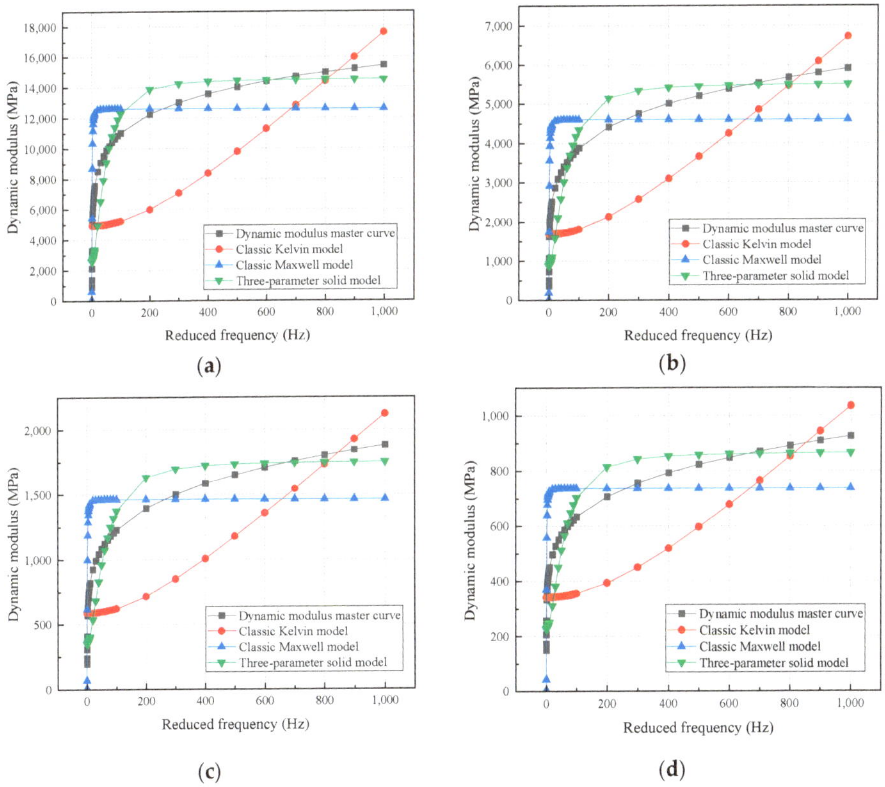

Figure 16. Fitting curve between ARHM-13 DMMC and viscoelastic model. (**a**) 15 °C, (**b**) 30 °C, (**c**) 45 °C, (**d**) 60 °C.

Figure 16 illustrates how the DM determined by the three-parameter solid model is the most similar to the test data of the three models, followed by Maxwell model, while the Kelvin model has the most significant difference. This result is also verified in Table 7. It demonstrates that the viscoelastic parameters and correlation coefficients were obtained by fitting the three models in different HCRAM. In contrast, the three-parameter solid model has the largest correlation coefficient, so the fitting effect of the model is better, which aligns most closely with the outcomes of the DM test. The constitutive equation of the three-parameter solid model can comprehensively describe the instantaneous elasticity, creep, stress relaxation, and other behaviors of viscoelastic materials, showing the characteristics of solids. Therefore, using the three-parameter solid model to describe the viscoelastic mechanical property of the HCRAM is most appropriate.

Table 7. Viscoelastic model parameters of various HCRAM.

Mixture	Temperature (°C)	Fitting Parameters of Viscoelastic Models									
		Classic Kelvin Model			Classic Maxwell Model			Three-Parameter Solid Model			
		E	η	R^2	E	η	R^2	E_1	E_2	η	R^2
ARHM-13	15	4929.89	2.69	0.46	12,602.79	957.69	0.85	3115.43	14,552.65	43.08	0.93
	30	1683.62	1.04	0.54	4604.64	299.60	0.82	1031.70	5526.33	13.03	0.94
	45	589.37	0.32	0.55	1461.25	108.33	0.75	433.68	1755.65	4.24	0.94
	60	340.99	0.16	0.53	736.44	67.92	0.63	305.63	867.06	2.45	0.93
ARHM-20	15	6659.82	3.21	0.43	15,638.63	1395.49	0.82	4740.86	17,698.89	61.16	0.91
	30	2345.27	1.31	0.51	5999.82	440.16	0.81	1553.55	7061.87	18.81	0.93
	45	1169.86	0.64	0.53	2906.24	218.61	0.78	835.69	3451.10	8.86	0.94
	60	522.61	0.24	0.52	1123.76	105.48	0.64	466.49	1314.37	3.86	0.93
ARHM-25	15	8771.20	4.18	0.41	20,531.81	1861.68	0.83	6191.55	23,092.45	83.16	0.91
	30	2933.94	1.70	0.51	7734.54	537.87	0.82	1856.92	9148.62	23.51	0.94
	45	1634.33	0.95	0.54	4249.58	295.72	0.79	1095.27	5088.36	12.33	0.94
	60	693.00	0.33	0.54	1557.14	134.24	0.68	586.32	1843.44	4.97	0.94

4.3. Numerical Simulation Validation and Result Analysis

Using Bisar3 software and Abaqus-6.14 software to calculate the mechanical response of the pavement structure in Table 4, we determined that the load area is the unit load, and the load center, half of the load circle radius, and the load circle circumference are taken as the mechanical response analysis points, referred to as points 1, 2, and 3, respectively. Figure 17 shows a comparison of calculated mechanical response values. From Figure 17, we see that the calculation results of Abaqus are smaller than those of Bisar. However, the differences in longitudinal stress calculated by the two software are minimal, indicating that the difference between the analytical and numerical solutions is insignificant. Therefore, the numerical simulation method used in this article is practical and applicable.

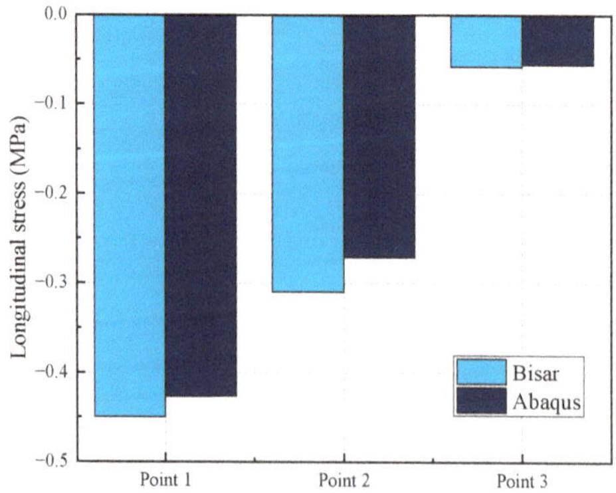

Figure 17. Comparison of longitudinal stress.

4.4. Dynamic Response Analysis of Pavement Structures and Its Differential Analysis

This study mainly analyzed the structural dynamic mechanical response of the pavement of Jingde Expressway under the conditions of driving speed of 10.8 km/h, vertical load of 0.7 MPa, horizontal load of 0.3 MPa, and temperature of 15 °C. The mechanical response indexes include equivalent stress, longitudinal stress, deflection, and longitudinal strain, and the variation laws along the depth direction, longitudinal along the surface, and transverse to the pavement were studied. As illustrated in Figure 18, several distinctive aspects of mechanical response analysis were identified here to aid in the study. Based on the mechanical response findings determined by the three-parameter solid model, this chapter first examined the dynamic mechanical response of the Jingde Expressway pavement structure and then analyzed the differences of the results calculated by different viscoelastic models. To facilitate the difference analysis, only feature point A was selected for comparative analysis.

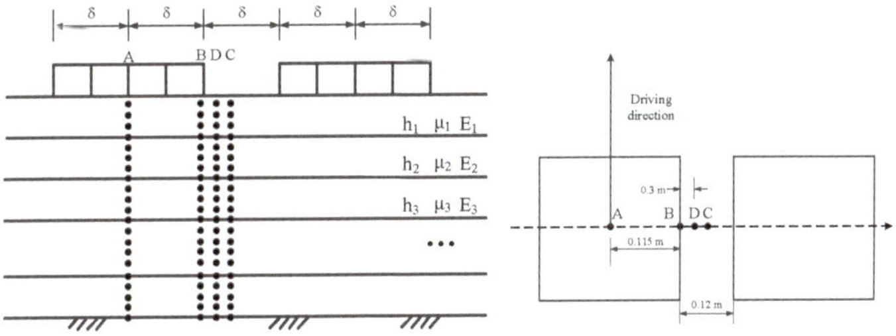

Figure 18. Schematic diagram of mechanical response calculation points.

Figure 19 shows the cloud map results of three viscoelastic models. Figure 19 shows that the mechanical indicators of the pavement structure calculated based on the classical Kelvin model, classical Maxwell model, and three-parameter solid model have roughly similar changes in different directions, but there are still slight differences. In addition, the differences between the calculation results are also relatively significant. Therefore, in the following section, the differences between the sizes of the mechanical indicators will be emphasized.

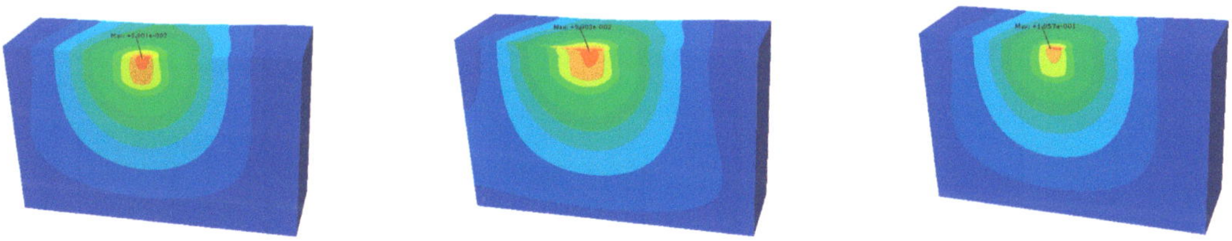

(a) (b) (c)

Figure 19. Calculation cloud map results of three viscoelastic models. (**a**) Classic Kelvin model, (**b**) Classic Maxwell model, (**c**) Three parameter solid model.

4.4.1. Equivalent Stress Dynamic Response

The pavement equivalent stress is an essential parameter in pavement structure design and evaluation. Accurately calculating the pavement equivalent stress is essential to ensure the safety and durability of pavement structure [51]. Therefore, the equivalent stress dynamic response was selected for analysis, and its calculation method is as follows [52]:

$$\sigma_e = \sqrt{\left[(\sigma_1 - \sigma_2)^2 + (\sigma_2 - \sigma_3)^2 + (\sigma_3 - \sigma_1)^2\right]/2} \tag{24}$$

where σ_1, σ_2, σ_3 is the first, second, and third principal stresses.

Figure 20 shows the equivalent stress curve based on the three-parameter solid model. Figure 20a shows the variation curve of equivalent stress along the depth direction. It manifests that the base layer differs somewhat, whereas the four characteristic points on the surface layer differ significantly. The maximum value of equivalent stress at each characteristic point is in the order of A > B > D > C from large to small, and the calculated values are 0.59 MPa, 0.52 MPa, 0.17 MPa, and 0.31 MPa, respectively. Therefore, for the pavement structure, the equivalent stress at the wheel load center is highest, and the equivalent stress at the wheel gap center is the lowest, meaning that damage is more likely to happen at the wheel load center of the pavement. Figures 20b and 20c, respectively, show the variation curves of equivalent stress along the longitudinal and transverse directions of the pavement. It is evident that each layer's maximum equivalent stress ratio is arranged from large to tiny as follows: base layer bottom, surface layer bottom, and surface layer. This phenomenon occurs because of the asphalt surface layer being in direct contact with the traffic load and being subject to the impact, vertical, and horizontal forces of that load. Furthermore, the base layer is composed of a stiff material, while the surface layer is composed a viscoelastic material. In contrast, the overall stiffness of the surface layer is more minor, so the equivalent stress of the surface layer is the largest.

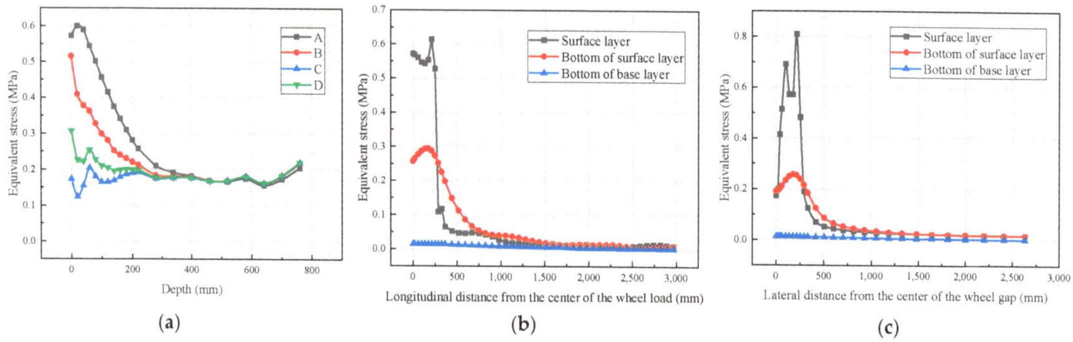

Figure 20. Equivalent stress variation curve based on a three-parameter solid model. (**a**) Along depth direction, (**b**) Longitudinal along the pavement, (**c**) Horizontally along the pavement.

Figure 21 compares the equivalent stress time history curves based on different models. In Figure 21, the abscissa is the time axis, and 2 s is when the load completes a movement on the belt. Under the action of moving load, the equivalent stress of different viscoelastic models is the same with time, and the maximum value appears when t = 1 s. Based on Figure 22, it can be observed that the equivalent stress calculated by the three viscoelastic models at point A of the surface layer from large to small are the three-parameter solid model, classical Kelvin model, classical Maxwell model, and the equivalent stress response values are 0.57 MPa, 0.56 MPa, and 0.55 MPa, respectively. The reason for this phenomenon is that, among the viscoelastic model fitting parameters calculated by using the DMMC, the modulus of the classical Maxwell model is the largest, the stronger with regard to the stiffness, the stronger with regard to the resistance to the load, and the smaller with regard to the equivalent stress.

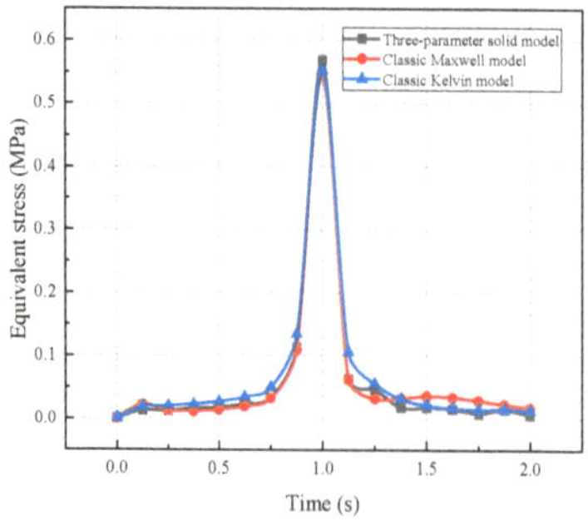

Figure 21. Comparison diagram of equivalent stress time history curves based on different models (point A of surface layer).

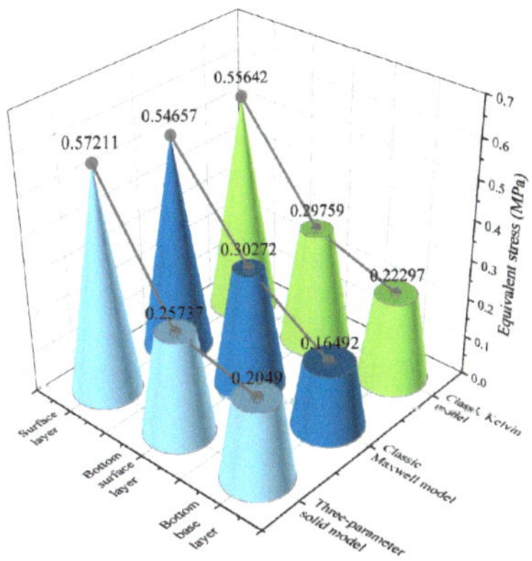

Figure 22. Comparison diagram of equivalent stress based on different models (characteristic point A of different horizons).

4.4.2. Longitudinal Stress Dynamic Response

Figure 23 shows the longitudinal stress curve based on the three-parameter solid model. According to Figure 23a, the equivalent stress is the same as that the maximum longitudinal stress of each characteristic point in the pavement structure is it arranged from large to small as A > B > D > C, and the maximum calculated longitudinal stress of point A is −0.39 MPa, which is 1.63 times, 2.2 times, and 2.0 times of points B, C, and D, respectively. Further, it is discovered that the pavement is in a state of compression, and the longitudinal stress at points A and B generally decreases along the pavement depth. In contrast, the longitudinal stress at points C and D first increases and then decreases. However, most of the base is in tension, and the longitudinal stress increases along the depth of the pavement. Figures 23b and 23c show the variation curves of longitudinal stress along the pavement's longitudinal and transverse directions, respectively. The figure illustrates that the equivalent stress is the same; the highest longitudinal stress appears in the surface layer and is located at the longitudinal distance of 173 mm from the center of the wheel load, and the response value is −0.65 MPa. It indicates that the most unfavorable point of the pavement is not necessarily located at the characteristic point specified in the "Specifications for Design of Highway Asphalt Pavement". Simultaneously, it also proves that it is more reasonable to calculate the most unfavorable point using the global search method in China's "Design Standard of Durable Asphalt Pavement for Highway with Structural Life Increment".

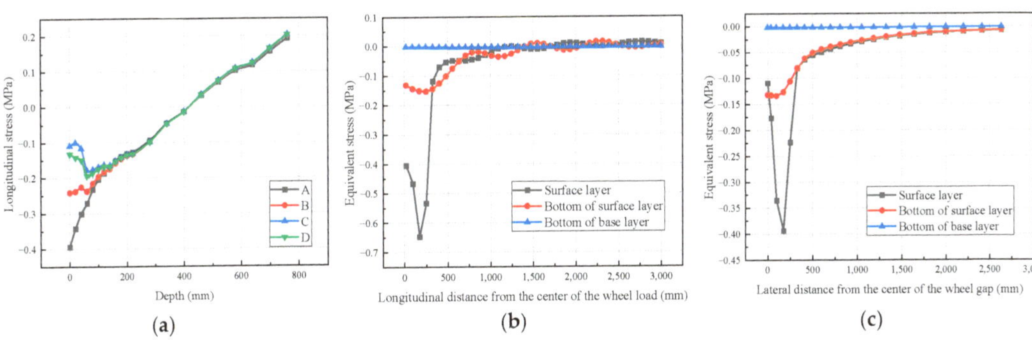

Figure 23. Longitudinal stress variation curve based on a three-parameter solid model. (**a**) Along depth direction, (**b**) Longitudinal along the pavement, (**c**) Horizontally along the pavement.

Figure 24 is a comparison diagram of the longitudinal stress time history curves based on the different models. From an analysis of Figure 24, it is apparent that the time history curves of longitudinal stress calculated by the three models are the same. In the process of the load movement, the longitudinal stresses calculated by the three models show positive and negative alternating changes, that is, tension-compression alternating phenomenon. This phenomenon is because, within 0 s–0.875 s, most of the surface layer is in a tension state and has also changed to a compression state. It is preliminarily considered that the load has just begun to move and has not yet stabilized, so it fluctuates wildly. Within 0.875 s–1 s, that is, before the arrival of the wheel, the asphalt concrete is under pressure. As the tire approaches, the longitudinal compressive stress gradually increases and reaches its maximum when the tire arrives. As the tire moves away from the tire, the longitudinal compressive stress gradually decreases and then transits to the tensile stress.

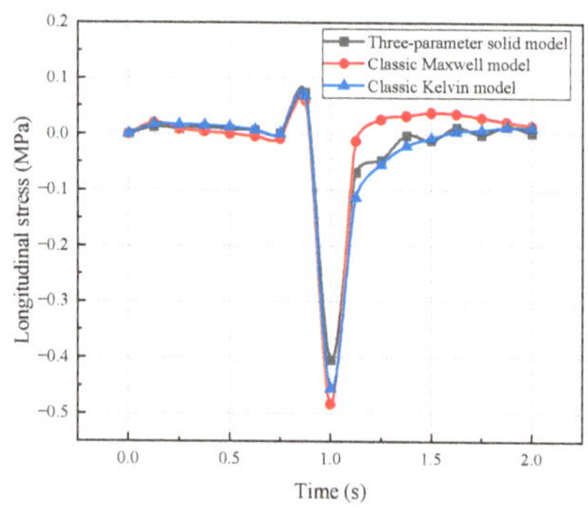

Figure 24. Comparison diagram of longitudinal stress time history curves based on different models (point A of surface layer).

Figure 25 depicts the comparison of longitudinal stresses of the different viscoelastic models. The figure clearly illustrates that the maximum longitudinal stress at point A of the surface layer calculated by the three viscoelastic models is ranked from large to small as follows: Maxwell, Kelvin, and three-parameter solid model, and the response sizes are −0.47 MPa, −0.44 MPa, and −0.39 MPa, respectively. Combined with Figure 23a, the longitudinal stress based on the classical Maxwell model is the fastest with the depth change. This is because this model can better reflect the instantaneous elastic deformation and relaxation phenomenon than other models, and its properties are closer to liquid.

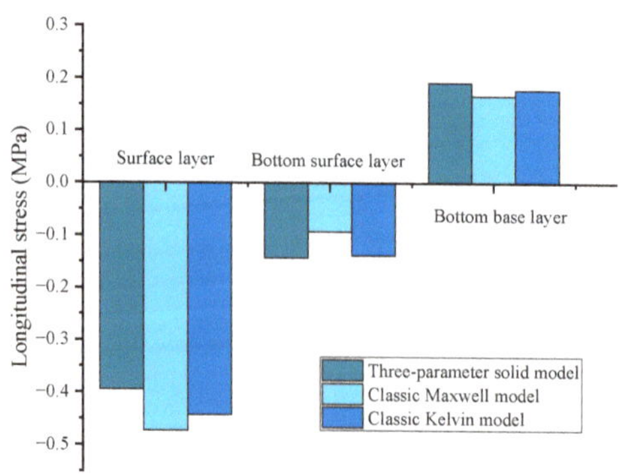

Figure 25. Comparison diagram of longitudinal stress based on different models (characteristic point A of different horizons).

4.4.3. Dynamic Response of Deflection

Deflection is an essential index for the pavement design and construction. This paper studied the dynamic response of the deflection (vertical displacement). Figure 26 demonstrates the deflection curve based on the three-parameter solid model. As shown from Figure 26, the wheel load center of the surface layer is the location of the maximum deflection value, which is 0.1 mm., indicating that this position is more prone to pavement settlement or deformation than the other three characteristic points.

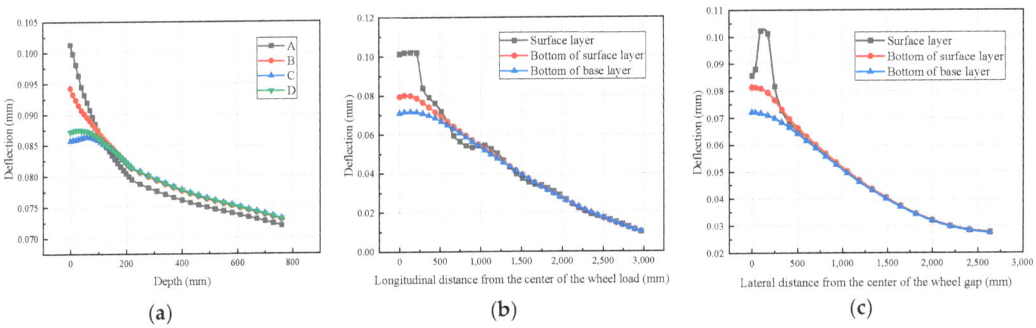

Figure 26. Deflection curve based on three parameter solid model. (**a**) Along depth direction, (**b**) Longitudinal along the pavement, (**c**) Horizontally along the pavement.

Figure 27 is a comparison chart of the bending time history curves based on the different models. In Figure 26, the peak value of the curve appears at t = 1 s; that is, the load is just at point A. In addition, we can see that in 1 s–2 s, the load progressively shifts away from point A, the slowest rate of change is found in the deflection response based on the Maxwell model, and the three-parameter solid model has the fastest rate of change. In other words, the residual value of the deflection response calculated by the classical Maxwell model is the largest after the movement of a load. Figure 28 is a comparison diagram of the deflection based on the different models. It is evident from the diagram that the different viscoelastic models obtain the maximum deflection response of the different layers, and the three-parameter solid model calculates the maximum deflection of the surface layer. The base course bottom and surface layer bottom maximum values are computed by the classical Maxwell model. The surface layer, surface layer bottom, and base layer bottom have maximum deflection values of 0.1 mm, 0.0799 mm, and 0.072 mm, respectively.

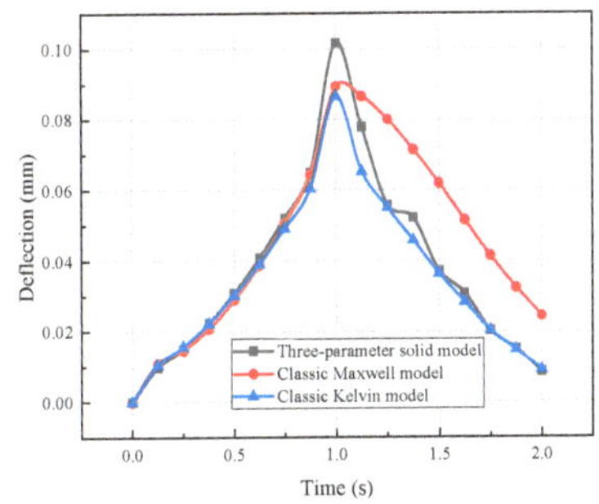

Figure 27. Comparison diagram of deflection time history curves based on different models (point A of surface layer).

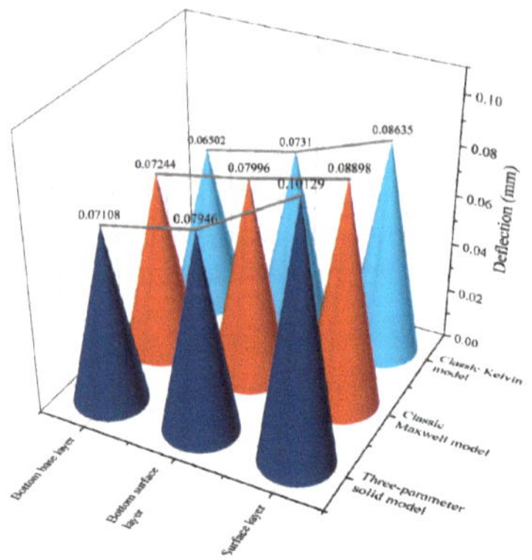

Figure 28. Comparison diagram of deflection based on different models (characteristic point A of different horizons).

4.4.4. Longitudinal Strain Dynamic Response

Strain refers to the degree of deformation of pavement under external load. The longitudinal strain of asphalt pavement is a crucial design index. Figure 29 shows the longitudinal strain curve based on the three-parameter solid model. From Figure 29a, it is demonstrated that the greatest longitudinal strain at the four characteristic points is arranged from large to small as C > D > A > B, and the response values are −41.89 με, −40.43 με, −23.49 με, −28.14 με. This phenomenon is because the characteristic point C is the wheel gap center of the road, which is squeezed by the wheel, resulting in the superposition effect of two-wheel actions [45], so the longitudinal strain of point C is considerable. Furthermore, it is apparent that the asphalt pavement is subject to the longitudinal compressive strain, with a progressive reduction in the longitudinal strain as depth increases. The longitudinal strain progressively grows as the depth deepens and reaches its greatest value at the base's bottom. Figure 29b,c demonstrate that along the pavement's longitudinal direction, the surface layer's longitudinal strain has a peak value of 134.6 με at 210 mm from the wheel load center. Along the longitudinal and transverse directions of the pavement, the variation trend of longitudinal strain is the same. With the direction away from the load, the response value gradually tends to zero, and more speed is placed on the longitudinal strain attenuation rate of each structural layer compared to on the transverse strain attenuation rate.

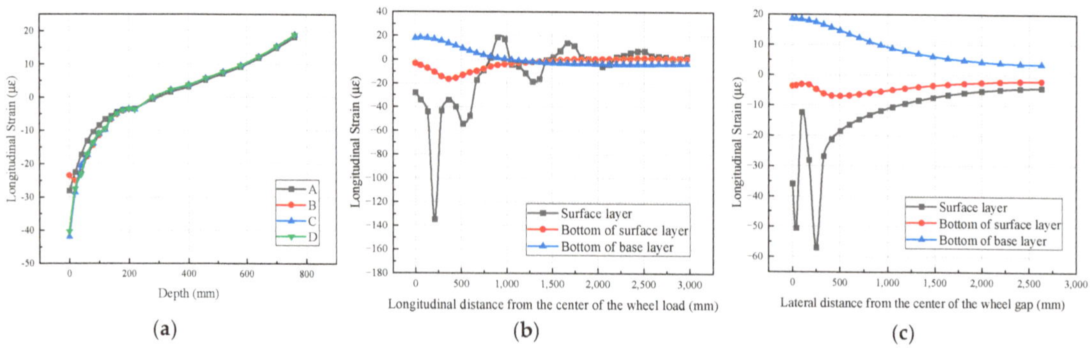

Figure 29. Longitudinal strain curve based on three-parameter solid model. (**a**) Along depth direction, (**b**) Longitudinal along the pavement, (**c**) Horizontally along the pavement.

Figure 30 is a comparison diagram of the longitudinal strain time history curves based on different models. Combined with Figure 24, the longitudinal strain exhibits a similar temporal change pattern as the longitudinal stress, and there will be tension compression alternation during the load movement. The difference is that the peak value of the curve does not appear at t = 1 s. This is because the damping effect is considered in the analysis of moving load, and the existence of damping affects the dynamic time history response, resulting in the hysteresis of the strain response. A comparison diagram of the longitudinal strain based on the several models is shown in Figure 31. From Figures 30 and 31, it is evident that the longitudinal strain at point A of the surface layer based on the three-parameter solid model is the largest, which is 1.007 and 2.223 times that of the Maxwell and the Kelvin model, respectively. Moreover, its attenuation rate in the asphalt pavement is the slowest. The reason for this is that the asphalt pavement surface is defined as viscoelastic material in the finite element calculation, and the properties of the three-parameter solid model are closer to solids. Compared with other models, the strain change rate is slower when the depth is increased.

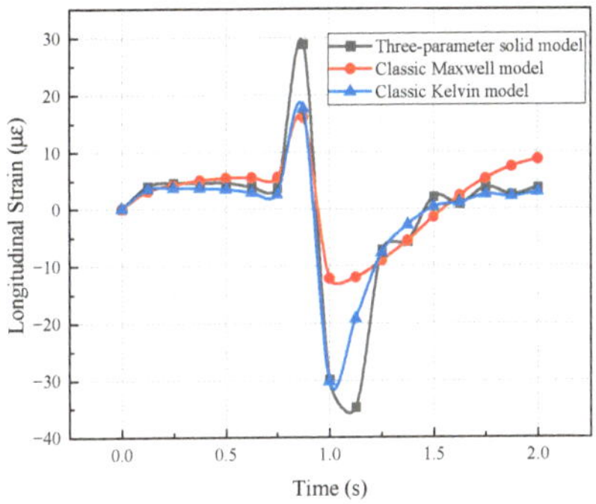

Figure 30. Comparison diagram of longitudinal strain time history curves based on different models (point A of surface layer).

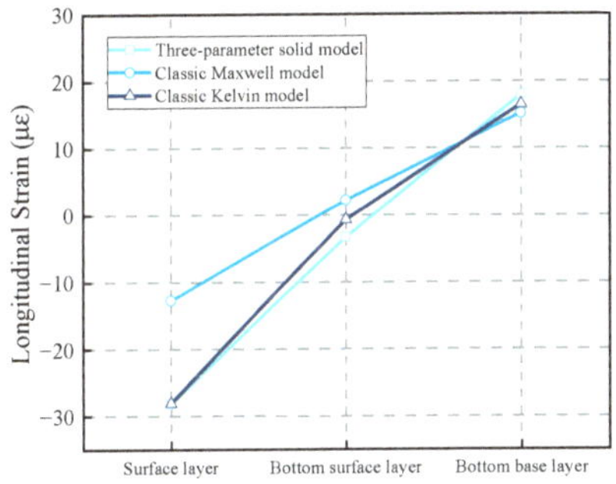

Figure 31. Comparison diagram of longitudinal strain based on different models (characteristic point A of different horizons).

5. Conclusions

This study selected the pavement structure and materials of the Jingde Expressway in China as the research object, explored the viscoelastic characteristics of the HCRAM, and studied the mechanical response and differences of the asphalt pavement structures utilizing three viscoelastic models of the HCRAM. The following are the primary conclusions:

(1) The relationship between temperature, frequency, and the DM was obtained through DM testing of the HCRAM. Specifically, the DM had a negative correlation with temperature and a positive correlation with loading frequency. The relationship between the storage and loss modulus and temperature and frequency had a similar pattern to that of the DM. The phase angle exhibited complicated variations with the frequency and temperature.

At low temperatures, the phase angle decreased as the loading frequency increased. At high temperatures, the phase angle of the HCRAM increased as the loading frequency increased. This reveals that under the action of high-frequency and low-temperature loads, the asphalt mixtures exhibited significant elastic characteristics, and their viscosity rose proportionally with the rise in temperature.

(2) Utilizing the DM test results, a DMMC was established based on the Sigmoidal model. The master curve demonstrated that the asphalt mixture's reliance on temperature intensified as the rubber fraction and gradation increased. The rate at which the DM changed from low frequency to high frequency steadily slowed down as temperature rose, indicating that the impact of the loading frequency on the asphalt mixture decreased as the temperature increased.

(3) This study found that the three-parameter solid model most accurately described the viscoelastic dynamic mechanical behavior of HCRAM. From the results of mechanical response calculations, we found that the equivalent stress, longitudinal stress, and bending response of characteristic point A were all the largest, with the response values of 0.59 MPa, −0.39 MPa, and 0.1 mm, respectively. Under the comparison of three viscoelastic models, most of the response indexes in the surface layer calculations were all the largest in the three-parameter viscoelastic model.

(4) This study systematically investigated the dynamic mechanical response of road surfaces under different viscoelastic models. However, there has yet to be research on the fatigue and high and low-temperature performance of high-content rubber asphalt mixtures, nor has the influence of temperature, load, and speed on the dynamic mechanical response of pavement structures been considered. In addition, it is necessary to compare and analyze the numerical simulation results with the accelerated loading test data. This will be carried out in subsequent work, and further consideration of the above research content is needed.

Author Contributions: Conceptualization, Y.H., S.L. and N.X.; methodology, S.L. and N.X.; Software, Y.H., S.L., N.X. and H.M. (Huabao Ma); formal analysis, W.L., C.P. and H.M. (Huilin Meng); investigation, H.M. (Huilin Meng) and X.P.; resources, H.M. (Huabao Ma) and G.Z.; data curation, Y.H.; writing—original draft, Y.H., S.L., X.P., H.M. (Huabao Ma), W.L., C.P., H.M. (Huilin Meng) and G.Z.; writing—review and editing, Y.H., S.L., Z.W. and X.P.; visualization, W.L. and C.P.; supervision, X.P.; project administration, Y.H.; Funding acquisition, Y.H. and S.L. All authors have read and agreed to the published version of the manuscript.

Funding: The research was funded by the National Science Fund for Distinguished Young Scholars (No. 52225806), the National Natural Science Foundation of China (No. 52078063), the Science and Technology Project of Hebei Provincial Department of Transportation (No. TH-201918), and the Postgraduate Scientific Research Innovation Project of Changsha University of Science and Technology (No. CXCLY2022012 and No. CLSJCX22016).

Data Availability Statement: Some or all data, models, or code generated or used during the study are available from the corresponding author by request.

Conflicts of Interest: My contributions to this work and manuscript were made independently without any requirement, guidance or input by my employer. I received no financial compensation from any source for the contributions I made to this scientific work and manuscript.

Appendix A Appendix A

The UMAT subroutine is shown below, and the input parameters are shown in Table A1.

Table A1. The UMAT subroutine and the input parameters (the "*" symbol is equivalent to "×").

```
SUBROUTINE UMAT(STRESS, STATEV, DDSDDE, SSE, SPD, SCD,
    1 RPL, DDSDDT, DRPLDE, DRPLDT,
  2 STRAN, DSTRAN, TIME, DTIME, TEMP, DTEMP, PREDEF, DPRED, CMNAME,
    3 NDI, NSHR, NTENS, NSTATV, PROPS, NPROPS, COORDS, DROT, PNEWDT,
    4 CELENT, DFGRD0, DFGRD1, NOEL, NPT, LAYER, KSPT, KSTEP, KINC)
```

Table A1. *Cont.*

```
C
      INCLUDE 'ABA_PARAM.INC'
C
      CHARACTER*80 CMNAME
      DIMENSION STRESS(NTENS), STATEV(NSTATV),
     1 DDSDDE(NTENS, NTENS),
     2 DDSDDT(NTENS), DRPLDE(NTENS),
     3 STRAN(NTENS), DSTRAN(NTENS), TIME(2), PREDEF(1), DPRED(1),
     4 PROPS(NPROPS), COORDS(3), DROT(3,3), DFGRD0(3,3), DFGRD1(3,3)
      DIMENSION DSTRES(6), D(3,3)
C
C   EVALUATE NEW STRESS TENSOR
C
      EV = 0
      DEV = 0
      DO K1 = 1, NDI
            EV = EV + STRAN(K1)
            DEV = DEV + DSTRAN(K1)
      END DO
C
      TERM1 = PROPS(2)*(0.5*DTIME*PROPS(1) + PROPS(3))/(1 + PROPS(4))/
     1 (0.5*DTIME*(PROPS(1) + PROPS(2)) + PROPS(3))
      TERM2 = 3*PROPS(4)*PROPS(2)*(0.5*DTIME*PROPS(1) + PROPS(3))/
     1 (1 + PROPS(4))/(1−2*PROPS(4))/(0.5*DTIME*(PROPS(1) + PROPS(2))
     2 + PROPS(3))
        TERM3 = DTIME*PROPS(1)*PROPS(2)/(1 + PROPS(4))/(0.5*DTIME*
     1 (PROPS(1) + PROPS(2)) + PROPS(3))
      TERM4 = 3*DTIME*PROPS(4)*PROPS(1)*PROPS(2)/(1 + PROPS(4))/
     1 (1−2*PROPS(4))/(0.5*DTIME*(PROPS(1) + PROPS(2)) + PROPS(3))
      TERM5 = DTIME*(PROPS(1) + PROPS(2))/(0.5*DTIME*(PROPS(1)+
     1 PROPS(2)) + PROPS(3))
C
C
C
      DO K1 = 1, NDI
            DSTRES(K1) = TERM1*DSTRAN(K1) + TERM2*DEV/3 + TERM3*
     1      STRAN(K1) + TERM4*EV/3−TERM5*STRESS(K1)
            STRESS(K1) = STRESS(K1) + DSTRES(K1)
      END DO
C
C
C
      I1 = NDI
      DO K1 = 1, NSHR
            I1 = I1 + 1
            DSTRES(I1) = TERM1/2*DSTRAN(I1) + TERM3/2*STRAN(I1)−
     1      TERM5*STRESS(I1)
            STRESS(I1) = STRESS(I1) + DSTRES(I1)
      END DO
C
C
C
      DO K1 = 1, NTENS
            DO K2 = 1, NTENS
                  DDSDDE(K2,K1) = 0
            END DO
      END DO
```

Table A1. *Cont.*

```
C
        DO K1 = 1, NDI
                DDSDDE(K1,K1) = TERM1 + TERM2/3
        END DO
C
        DO K1 = 2, NDI
                N2 = K1−1
                DO K2 = 1, N2
                        DDSDDE(K2,K1) = TERM2/3
                        DDSDDE(K1,K2) = TERM2/3
                END DO
        END DO
C
        I1 = NDI
        DO K1 = 1, NSHR
                I1 = I1 + 1
                DDSDDE(I1,I1) = TERM1/2
        END DO
        RETURN
        END
```

References

1. Lv, S.; Hu, L.; Xia, C.; Wang, X.; Borges Cabrera, M.; Guo, S.; Chen, J. Development of Fatigue Damage Model of Asphalt Mixtures Based on Small-Scale Accelerated Pavement Test. *Constr. Build. Mater.* **2020**, *260*, 119930. [CrossRef]
2. Xie, N.; Peng, X.; He, Y.; Lei, W.; Pu, C.; Meng, H.; Ma, H.; Tan, L.; Zhao, P. Investigation on Performance and Durability of Bone Glue and Crumb Rubber Compound Modified Asphalt and Its Mixture. *Case Stud. Constr. Mater.* **2023**, *19*, e02437. [CrossRef]
3. Wang, J.; Lv, S.; Liu, J.; Peng, X.; Lu, W.; Wang, Z.; Xie, N. Performance Evaluation of Aged Asphalt Rejuvenated with Various Bio-Oils Based on Rheological Property Index. *J. Clean. Prod.* **2023**, *385*, 135593. [CrossRef]
4. Peng, X.; Lv, S.; Huang, T.; Zheng, C.; Jiang, M.; Yuan, J.; Hu, L. Performance Check of Asphalt Mixture Layer Based on Yield Criterion and Normalized Fatigue Equation. *J. Mater. Civ. Eng.* **2022**, *34*, 04022073. [CrossRef]
5. Yuan, J.; Lv, S.; Peng, X.; Liu, K. Comprehensive Properties Evaluation of Modified Bio-Asphalt Mixture Based on Comparison Matrix. *J. Mater. Civ. Eng.* **2024**, *36*, 04023616. [CrossRef]
6. Peng, X.; Xie, N.; Xia, C.; Zhou, X.; Zhao, P.; Ma, S.; Zhang, C.; Lv, S. Laboratory Evaluation of Different Bio-Oil Recycled Aged Asphalts: Conventional Performances and Microscopic Characteristics. *J. Clean. Prod.* **2023**, *428*, 139442. [CrossRef]
7. Lv, S.; Tan, L.; Peng, X.; Hu, L.; Liu, H.; Yuan, J.; Wang, S. Fatigue Resistance Design of Rubberized Asphalt Mixture Pavement under Three-Dimensional Stress State. *Constr. Build. Mater.* **2021**, *307*, 125138. [CrossRef]
8. Wang, Q.-Z.; Wang, N.-N.; Tseng, M.-L.; Huang, Y.-M.; Li, N.-L. Waste Tire Recycling Assessment: Road Application Potential and Carbon Emissions Reduction Analysis of Crumb Rubber Modified Asphalt in China. *J. Clean. Prod.* **2020**, *249*, 119411. [CrossRef]
9. Chen, S.; Cao, J.; Zheng, C.F.; Guo, X.D. Rheological Characteristics of Tire Rubber-Modified Asphalt Following Thermal Variation. *Geofluids* **2021**, *2021*, 9928562. [CrossRef]
10. Wang, S.; Zhao, X.; Wang, Q. Rheological and Structural Evolution of Rubberized Asphalts under Weathering. *J. Mater. Civ. Eng.* **2017**, *29*, 04017180. [CrossRef]
11. Zhang, Z.; Yang, W.; Ma, Y.; Zhang, C.; Wang, S. Effect of Dynamic Vulcanization on the Performance of High Content Rubber Modified Asphalt. *J. Appl. Polym. Sci.* **2023**, *140*, e53889. [CrossRef]
12. Xu, G.; Yao, Y.; Ma, T.; Hao, S.; Ni, B. Experimental Study and Molecular Simulation on Regeneration Feasibility of High-Content Waste Tire Crumb Rubber Modified Asphalt. *Constr. Build. Mater.* **2023**, *369*, 130570. [CrossRef]
13. Zhu, Y.; Xu, G.; Ma, T.; Fan, J.; Li, S. Performances of Rubber Asphalt with Middle/High Content of Waste Tire Crumb Rubber. *Constr. Build. Mater.* **2022**, *335*, 127488. [CrossRef]
14. Wang, G.; Wang, X.; Lv, S.; Qin, L.; Peng, X. Laboratory Investigation of Rubberized Asphalt Using High-Content Rubber Powder. *Materials* **2020**, *13*, 4437. [CrossRef]
15. Sun, Y. Performance Evaluation and Mixture Design Technology of High Content Rubber Modified Asphalt. Master's Thesis, Hebei University of Technology, Tianjin, China, 2019.
16. Zhao, Z.; Xiao, F.; Toraldo, E.; Crispino, M.; Ketabdari, M. Effect of Crumb Rubber and Reclaimed Asphalt Pavement on Viscoelastic Property of Asphalt Mixture. *J. Clean. Prod.* **2023**, *428*, 139422. [CrossRef]
17. Huang, J.; Sun, Y. Viscoelastic Analysis of the Damping Asphalt Mixtures (DAMs) Made with a High Content of Asphalt Rubber (AR). *Adv. Civ. Eng.* **2020**, *2020*, 8826926. [CrossRef]
18. Zhang, C.; Yu, H.; Zhu, X.; Yao, D.; Peng, X.; Fan, X. Unified Characterization of Rubber Asphalt Mixture Strength under Different Stress Loading Paths. *J. Mater. Civ. Eng.* **2024**, *36*, 04023498. [CrossRef]

19. Chen, S.; Xu, L.; Jia, S.; Wang, J. Characterization of the Nonlinear Viscoelastic Constitutive Model of Asphalt Mixture. *Case Stud. Constr. Mater.* **2023**, *18*, e01902. [CrossRef]
20. Han, D.; Xia, X.; Zhao, Y.; Xie, Y.; Zhou, Z. Investigation of the Time-Domain Linear Viscoelastic Response of Warm Mix Asphalt Mixture via Discrete and Continuous Spectrum. *Case Stud. Constr. Mater.* **2023**, *19*, e02290. [CrossRef]
21. Chen, H.; Barbieri, D.M.; Zhang, X.; Hoff, I. Reliability of Calculation of Dynamic Modulus for Asphalt Mixtures Using Different Master Curve Models and Shift Factor Equations. *Materials* **2022**, *15*, 4325. [CrossRef]
22. Kuchiishi, A.K.; Castorena, C.; Kim, Y.R.; Underwood, B.S. Ruggedness Evaluation of Dynamic Modulus Test Using Small-Scale Test Specimens for Asphalt Mixture Performance Tester. *Transp. Res. Rec. J. Transp. Res. Board* **2023**, *2677*, 387–400. [CrossRef]
23. Witzcak, M.W. *Simple Performance Test for Superpave Mix Design*; Transportation Research Board: Washington, DC, USA, 2002.
24. Qin, X.; Ma, L.; Wang, H. Comparison Analysis of Dynamic Modulus of Asphalt Mixture: Indirect Tension and Uniaxial Compression Test. *Transp. A Transp. Sci.* **2019**, *15*, 165–178. [CrossRef]
25. Dong, Y.; Tan, Y., Yang, L. Evaluation of Performance on Crumb-Rubber-Modified Asphalt Mixture. *J. Test. Eval.* **2012**, *40*, 20120186. [CrossRef]
26. Li, S.; Fan, M.; Xu, L.; Tian, W.; Yu, H.; Xu, K. Rutting Performance of Semi-Rigid Base Pavement in RIOHTrack and Laboratory Evaluation. *Front. Mater.* **2021**, *7*, 590604. [CrossRef]
27. Loizos, A.; Gkyrtis, K.; Plati, C. Modelling Asphalt Pavement Responses Based on Field and Laboratory Data. In *Accelerated Pavement Testing to Transport Infrastructure Innovation*; Chabot, A., Hornych, P., Harvey, J., Loria-Salazar, L.G., Eds.; Springer International Publishing: Cham, Switzerland, 2020; Volume 96, pp. 438–447.
28. Wang, H.; Zhan, S.; Liu, G. The Effects of Asphalt Migration on the Dynamic Modulus of Asphalt Mixture. *Appl. Sci.* **2019**, *9*, 2747. [CrossRef]
29. Zhou, Z.; Zhou, Y.; Liu, Z. Analysis of influencing factors on dynamic modulus of permeable asphalt mixture. *Mater. Guide* **2022**, *36*, 121–127.
30. Fonseca, O.S.; Witczak, M.W. A Prediction Methodology for the Dynamic Modulus Of In-Place Aged Asphalt Mixtures (with Discussion). *J. Assoc. Asph. Paving Technol.* **1996**, *65*, 532–572.
31. Wei, J.; Cui, S.; Hu, J. Experimental study on dynamic modulus of asphalt mixture. *J. Build. Mater.* **2008**, *11*, 657–661.
32. Ali, Y.; Irfan, M.; Ahmed, S.; Khanzada, S.; Mahmood, T. Investigation of Factors Affecting Dynamic Modulus and Phase Angle of Various Asphalt Concrete Mixtures. *Mater. Struct.* **2016**, *49*, 857–868. [CrossRef]
33. Liu, P.; Xing, Q.; Wang, D.; Oeser, M. Application of Linear Viscoelastic Properties in Semianalytical Finite Element Method with Recursive Time Integration to Analyze Asphalt Pavement Structure. *Adv. Civ. Eng.* **2018**, *2018*, 9045820. [CrossRef]
34. Wang, T.; Dong, Z.; Xu, K.; Ullah, S.; Wang, D.; Li, Y. Numerical Simulation of Mechanical Response Analysis of Asphalt Pavement under Dynamic Loads with Non-Uniform Tire-Pavement Contact Stresses. *Constr. Build. Mater.* **2022**, *361*, 129711. [CrossRef]
35. Han, D.; Xia, X.; Liu, G.; Xi, Y.; Zhao, Y. Improved Parameter Acquisition of Generalized Maxwell Model in Asphalt Mixtures by Continuous Relaxation Spectrum. *Constr. Build. Mater.* **2023**, *408*, 133775. [CrossRef]
36. Lagos-Varas, M.; Movilla-Quesada, D.; Arenas, J.P.; Raposeiras, A.C.; Castro-Fresno, D.; Calzada-Pérez, M.A.; Vega-Zamanillo, A.; Maturana, J. Study of the Mechanical Behavior of Asphalt Mixtures Using Fractional Rheology to Model Their Viscoelasticity. *Constr. Build. Mater.* **2019**, *200*, 124–134. [CrossRef]
37. Jiang, D.; Wang, D.; Zhang, R.; Feng, D.; Yi, J. Pavement Mechanics Calculation Based on the Analytical Solution of Layered Viscoelastic System under Horizontal and Vertical Loads. *J. Eng. Mech.* **2022**, *148*, 04022059. [CrossRef]
38. Wang, W.; Yan, G.; Zhao, K.; Wang, L. Numerical Simulation and Experimental Measurements of Dynamic Responses of Asphalt Pavement in Dry and Saturated Conditions under Full-Scale Accelerated Loading. *Appl. Sci.* **2022**, *12*, 12291. [CrossRef]
39. Deng, Y.; Shi, X.; Zhang, Y.; Chen, J. Numerical Modelling of Rutting Performance of Asphalt Concrete Pavement Containing Phase Change Material. *Eng. Comput.* **2023**, *39*, 1167–1182. [CrossRef]
40. Ji, H.; Xie, H.; Liu, Z. Mechanical response analysis of semi-rigid base asphalt pavement based on finite element ANSYS. *Sci. Technol. Eng.* **2021**, *21*, 5092–5097.
41. Dong, Z.; Cao, L.; Tan, Y.; Wu, S. Simulation analysis of three-dimensional strain dynamic response of asphalt pavement under moving loads. *China Civ. Eng. J.* **2009**, *42*, 133–139.
42. Ban, H.; Im, S.; Kim, Y.-R. Nonlinear Viscoelastic Approach to Model Damage-Associated Performance Behavior of Asphaltic Mixture and Pavement Structure. *Can. J. Civ. Eng.* **2013**, *40*, 313–323. [CrossRef]
43. Yuan, S. Research on Fatigue Characteristics of High Content Rubber Asphalt and Mixtures Based on Permanent Pavement. Master's Thesis, Chongqing Jiaotong University, Chongqing, China, 2023.
44. He, Y.; Lv, S.; Wang, Z.; Ma, H.; Lei, W.; Pu, C.; Meng, H.; Xie, N.; Peng, X. Research on the Modulus Decay Model under a Three-Dimensional Stress State of Asphalt Mixture during Fatigue Damage. *Buildings* **2023**, *13*, 2570. [CrossRef]
45. Xie, N.; Lv, S.; He, Y.; Lei, W.; Pu, C.; Meng, H.; Ma, H.; Peng, X. Load Response and Fatigue Life of Cement-Stabilized Macadam Base Structure Considering Dynamic and Static Load Differences and Tension-Compression Modulus Differences. *Constr. Build. Mater.* **2023**, *394*, 132060. [CrossRef]
46. Wang, G.; Wang, X.; Yan, Z.; Qin, L.; Gao, Z. Analysis of the Influence of Temperature Field on the Dynamic Modulus of Rubber Asphalt Pavement. *Front. Mater.* **2020**, *7*, 586457. [CrossRef]
47. Wang, P. Research on the Characterization Method of Dynamic Modulus of Asphalt Mixtures under Different Stress States and Temperature Frequency Conditions. Master's Thesis, Changsha University of Science and Technology, Changsha, China, 2023.

48. Zhao, H.; Gao, W.; Cui, S.; Zhang, W.; Ma, S.; Miao, B.; Su, C.; Zhang, P.; Wang, X.; Wei, J. Study on the Dynamic Performance of PU and SBS-Modified Asphalt Mixtures with Dense Gradation. *Coatings* **2023**, *13*, 1143. [CrossRef]
49. Hua, Y. Research on Vehicle Road Interaction Based on Fractional Derivative Viscoelasticity Theory. Doctor's Thesis, Nanjing Forestry University, Nanjing, Chian, 2011.
50. Chen, Y.; Wang, Y.; Guo, S.; Zhao, J.; Feng, D.; Yi, J. Research on Aging Performance of Hot Mix Fully Recycled Epoxy Asphalt Mixture with 100% RAP Utilization-Take the Surface Course as the Application Target. *Constr. Build. Mater.* **2023**, *392*, 131430. [CrossRef]
51. Yuan, J.; Lv, S.; Cheng, H.; Peng, X. Application of Miller Yield Criterion to Develop a Unified Fatigue Law of Asphalt Mixture for Different Loading Modes. *Fatigue Fract. Eng. M* **2023**, *46*, 2639–2650. [CrossRef]
52. Fan, X.; Liu, C.; Lv, S.; Ge, D.; Liu, J.; Pan, Q. Unified Fatigue Characterization of Asphalt Mixture under Multi-Field Coupling Condition: Stress State, Frequency, and Temperature. *Constr. Build. Mater.* **2022**, *353*, 129027. [CrossRef]

buildings

MDPI

Article

Assessment of Intelligent Unmanned Maintenance Construction for Asphalt Pavement Based on Fuzzy Comprehensive Evaluation and Analytical Hierarchy Process

Gensheng Hu [1], Gongzuo Shi [1], Runhua Zhang [2,3], Jianfeng Chen [1], Haichang Wang [2,3] and Junzhe Wang [2,3,*]

1 Zhejiang Communication Investment Expressway Operation Management Co., Ltd., Hangzhou 310020, China; 13757144113@163.com (G.H.); shigongzuo@163.com (G.S.); 15180182073@163.com (J.C.)
2 China Academy of Transportation Sciences, Beijing 100029, China; 17801128125@163.com (R.Z.); 15901377101@163.com (H.W.)
3 Testing Technology (Beijing) Co., Ltd. of China Academy of Transportation Sciences, Beijing 100029, China
* Correspondence: wangjunzhe@motcats.ac.cn

Abstract: Conventional human-involved maintenance methods for asphalt pavement pose significant challenges when applied to high-traffic road sections, often leading to congestion and safety risks, as well as reduced maintenance efficiency. In recent years, explorations into unmanned construction technology for newly constructed expressways have yielded beneficial and encouraging results. However, its application in road maintenance in more complex environments still needs to be expanded. In this study, an intelligent unmanned maintenance technology for asphalt pavement was applied to the Lilong Highway in Zhejiang Province, China, and the compactability, thickness, surface smoothness, permeability coefficient, and constructure depth of maintenance road sections were measured. Then, based on fuzzy comprehensive mathematics and the analytic hierarchy process, a comprehensive evaluation was performed on the intelligent unmanned maintenance technology, considering the aspects of road quality, safety, application, and socio-economic benefits. The results show that the road quality of intelligent unmanned maintenance technology can meet the road specification requirements. In addition, the membership degree of unmanned maintenance technology in the excellent grade is the highest, reaching 0.805, and the quantified value for the overall evaluation of the application effectiveness of unmanned maintenance technology is 92.10. This means that the final comprehensive evaluation result of unmanned maintenance technology is rated as excellent. The research findings provide decision-makers with valuable insights into the unmanned automation maintenance challenges faced by asphalt pavement, enabling them to implement appropriate measures to elevate the maintenance standards of road transportation.

Keywords: road engineering; asphalt pavement; unmanned maintenance; fuzzy comprehensive evaluation; analytical hierarchy process

Citation: Hu, G.; Shi, G.; Zhang, R.; Chen, J.; Wang, H.; Wang, J. Assessment of Intelligent Unmanned Maintenance Construction for Asphalt Pavement Based on Fuzzy Comprehensive Evaluation and Analytical Hierarchy Process. *Buildings* **2024**, *14*, 1112. https://doi.org/10.3390/buildings14041112

Academic Editor: Pengfei Liu

Received: 25 March 2024
Revised: 10 April 2024
Accepted: 14 April 2024
Published: 16 April 2024

1. Introduction

With the rapid growth of highway mileage and the increase in road service life, highways have entered a period where construction and maintenance are equally important [1–4]. Common drawbacks associated with conventional engineering construction and maintenance methods include low productivity and high costs, largely attributed to heavy reliance on manual labor. With the rapid advancement of digitalization, automation, artificial intelligence, big data, and other technologies, the landscape of engineering construction and maintenance is gradually shifting towards intelligent systems [5–7]. Consequently, as a vital component of transportation infrastructure, pavement construction and maintenance urgently require the adoption of new technologies, techniques, and solutions to alleviate the workload of workers and enhance construction efficiency.

In road construction, effective paving and compaction techniques will ensure road quality and traffic safety. In the development of intelligent pavement paving and compaction technology, scholars have conducted extensive research. For pavers, Liang and Guan [8] devised an intelligent automatic leveling control system for asphalt pavers, leveraging ultrasonic distance-measuring technology. This innovative system automatically adjusts the speed of the hydraulic cylinder in response to road elevation deviations and autonomously verifies the accuracy of the measured data. Consequently, it mitigates human effects on achieving a smooth road surface during construction. Qi et al. [9] developed a specialized paving control device and control method for the construction of curved sections with super-large lateral inclinations. They programmed a reference control line into the paver's operational control device and adjusted the machine's lateral displacement deviation to align its data with the travel route, thus enabling precise control over the forward trajectory of the paving machine. For rollers, Chen et al. [10] proposed an intelligent control system for asphalt pavement compaction quality. Through on-site measurements and analysis, they investigated and verified the correlation and applicability of two proposed indicators for controlling compaction quality with compactness and smoothness. Lu et al. [11] proposed an intelligent technology for roller path tracking and mapping in pavement compaction. Based on laboratory and field tests, the proposed technology can achieve comparable accuracy to GPS for rollers' lateral position estimation and a satisfactory accuracy for longitudinal position estimation. Meanwhile, in addition to intelligently optimizing equipment parameters, the devices themselves have been endowed with more intelligent functionalities. For instance, 3D intelligent paving technology involves the use of non-contact electronic guidance systems to guide pavers in laying surfaces without the need for strings. In terms of controlling the direction and elevation of the paver, global positioning system (GPS) and laser radar technology have replaced strings, increasing accuracy to the millimeter level [12,13]. Researchers in France and other countries have delved into "Computer-Integrated Road Construction Projects" supported by relevant research initiatives in the European Union. By amalgamating high-precision positioning technology and laser guidance technology, they achieved precise positioning guidance for road rollers and pavers [14]. Ye et al. [15] conducted comparative tests between the 3D paving control system and the conventional paving control system using specific engineering examples. They found that the use of 3D intelligent paving technology reduced the measurement workload by approximately 40%, increased paving accuracy by around 30%, and improved work efficiency by approximately 50%.

Furthermore, current research and development efforts for paving and compaction equipment are moving towards unmanned vehicle fleets. This technology for unmanned construction involves construction machinery operated from a distance, along with supporting equipment for remotely controlling the operation of this machinery. Bian et al. [16] proposed a path-tracking control method based on the fuzzy algorithm to solve the path-tracking error of unmanned vibration rollers during automatic rolling operations and confirmed the effectiveness and superiority of this method. Xie R. [17] provided a detailed introduction to the unmanned paving and compaction intelligent construction system. This system utilizes the BeiDou Navigation Satellite System and, with the assistance of a 3D laser leveling system, automatically plans the travel paths of pavers and compactors for newly constructed asphalt pavement. Wang et al. [18] analyzed the statistical results of road condition indicators for unmanned construction sections and indicated that unmanned construction machinery based on road trajectory data can ensure construction accuracy within 2 cm and achieve a pass rate of over 90%. Compared to conventional manual construction methods, they found a 50% reduction in the need for auxiliary personnel, resulting in improved construction quality and efficiency.

In summary, unmanned intelligent road construction technology has promising prospects. However, the technology is currently in the research stage and is only used in new road engineering constructions on a small scale, and its application in highway maintenance projects still needs to be assessed. Compared with a newly constructed highway project, a maintenance project in operating expressway experiences more complex situations and the quality requirements of pavement surface conditions are much stricter. Specifically, the current maintenance of highways faces two prominent issues: (1) Maintenance operations on high-traffic sections are difficult to carry out, leading to congestion and lower construction efficiency. Organizing traffic during maintenance operations is complex, and there are significant safety risks for workers involved in these activities. (2) The level of intelligence in maintenance equipment is not high. Some maintenance machinery and equipment have low levels of automation, resulting in many maintenance operations still being labor-intensive. The degree of reliance on information technology for quality management is not high enough to meet current maintenance needs. In this study, unmanned intelligent road construction technology is used in the maintenance construction of a practically engineered structure. The engineering technical indicators of asphalt pavements after intelligent unmanned maintenance and conventional manual maintenance are investigated and compared. Finally, based on fuzzy comprehensive mathematical methods and the analytic hierarchy process, an evaluation index system for the effectiveness of unmanned maintenance technology is established, considering road quality, safety, technology application, and socio-economic benefits.

2. Project Overview

2.1. Study Area

In this study, to comprehensively evaluate the application effectiveness of intelligent unmanned road construction technology in maintaining road sections, a local road to be repaired in Jinyun County, Zhejiang Province, was selected as the study area to complete the field tests. The road is situated in the Lishui section of the Longli Expressway. The Longli Expressway spans a total length of 85,056 km and was constructed in accordance with bidirectional four-lane expressway standards. The roadbed maintains a standard width of 24.5 m, and the mainline road surface, during construction, comprises a typical structure with three asphalt concrete layers (4 cm top layer + 6 cm middle layer + 8 cm bottom layer), a water-stable crushed stone base layer, and a sub-base layer. The expressway was completed and opened to traffic in December 2006.

Table 1 outlines the specific details of the road sections maintained using intelligent unmanned maintenance technology in this study. Sections 7 and 8 are specifically designated for completing unmanned maintenance work at night. In addition, to further assess and validate the maintenance quality of sections under unmanned maintenance technology, the road sections utilizing conventional manual maintenance methods are selected, as shown in Table 2. The expressway's "up" and "down" directions are determined based on the expressway milepost, with "up" denoting the direction of increasing expressway mileposts and "down" indicating the direction of decreasing highway mileposts.

Table 1. Description of unmanned maintenance sections.

Road Section Number	Direction	Starting Milepost	Ending Milepost	Length/km
W1	Up	K445 + 050	K445 + 444	0.394
W2	Down	K461 + 306	K460 + 606	0.7
W3	Down	K460 + 460	K459 + 760	0.7
W4	Down	K466 + 308	K465 + 991	0.317
W5	Up	K431 + 803	K432 + 503	0.7
W6	Up	K432 + 494	K433 + 250	0.756
W7	Up	K475 + 244	K475 + 977	0.733
W8	Down	K94 + 422	K93 + 928	0.494

Table 2. Description of conventional maintenance sections.

Road Section Number	Direction	Starting Milepost	Ending Milepost	Length/km
C1	Up	K415 + 282	K415 + 975	0.693
C2	Up	K437 + 296	K438 + 171	0.875
C3	Up	K443 + 622	K445 + 031	1.409
C4	Up	K445 + 527	K446 + 179	0.652
C5	Up	K474 + 619	K475 + 241	0.622
C6	Down	K467 + 230	K466 + 308	0.922
C7	Down	K463 + 136	K462 + 046	1.09
C8	Down	K438 + 605	K437 + 960	0.645

2.2. *Intelligent Unmanned Maintenance Technology*

In these maintenance activities, the asphalt overlayers are used for strengthening the existing pavement structure, correcting surface defects, and improving pavement serviceability. The unmanned maintenance process for the asphalt overlayers primarily involves the paving and compaction of the asphalt mixture. The test road in this study is equipped with two unmanned asphalt pavers and five unmanned rollers. Additionally, two asphalt pavers and five double steel-wheel rollers were mobilized as emergency equipment to ensure the smooth progress of maintenance on the test section.

During the paving process, two unmanned automatic telescopic pavers operate in tandem, with adjacent lanes overlapping by a width of 0.1–0.2 m. The pavers are positioned and pre-adjusted to the desired curvature 30–60 min before paving. The smoothing plates were preheated to 120 °C prior to the paving process. According to the Technical Specification for Construction of Highway Asphalt Pavements (JTG F40-2004) [19], both pavers maintain consistent compaction hammer vibration frequencies, and the paving temperature of the asphalt mixture must not be kept lower than 160 °C. In the initial 10–20 m from the starting point of the pavement section, manual operation mode is employed for paving, and manual leveling is conducted for joint areas. Once the paving machine stabilizes, it transitions into unmanned paving mode. The loose thickness is determined based on the measured loose coefficient. The paving machine uniformly, slowly, and continuously paves the material. The paving speed is controlled within the range of 2–4 m/min, avoiding arbitrary changes in speed or mid-process stops to enhance smoothness and reduce material segregation. At 10–20 m from the end point of the paving section, the system switches to manual control mode and remains under manual control until it passes the joint.

Simultaneously, the compaction of the asphalt mixture is undertaken. For the initial 10–20 m of the joint starting section, the compaction of the asphalt pavement is manually controlled. After the initial compaction is completed, it transitions to unmanned compaction mode to start the unmanned compaction operation. During the initial compaction, the roller closely follows the paver, applying static pressure while moving forward and vibratory pressure while moving backward. A relatively short initial compaction zone is maintained to quickly complete the compaction, minimizing heat loss. Each paver is followed by two rollers. Following the completion of paving by the first paver, an approximately 0.5 m joint is left without compaction temporarily. This serves as a reference surface for the movement of the floating screeds of the second paver. Once the second paver completes its paving, compaction is then applied to the joint. The roller's secondary compaction is carried out immediately after the initial compaction, without arbitrary pauses. The recommended length for the roller's compaction is 50–80 m. After the secondary compaction, on-site technicians use a three-meter straightedge to inspect the smoothness. If any suboptimal points are identified, the roller is directed for spot vibratory compaction by a manually operated small roller. The final compaction is conducted immediately after the secondary compaction and should be carried out in segmented rolling, until all roller marks are eliminated. According to the JTG F40-2004, the travel speeds for initial compaction, secondary compaction, and final compaction are 2–3 km/h, 3–5 km/h, and 3–6 km/h, respectively.

2.3. *Performance Tests*

Following the completion of pavement construction on the test section, the assessments of road quality were conducted to compare the maintenance quality of different maintenance technologies. Firstly, based on the Field Test Methods of Highway Subgrade and Pavement (JTG 3450-2019) [20], core samples were collected on-site from the unmanned maintenance section and the conventional maintenance section. Then, the performances of the test road, including compactability, thickness, smoothness, permeability coefficient, and construction depth, were measured.

2.3.1. Compactability

To obtain the compaction degree of the asphalt pavement, the real-time density ρ_s of the core samples is obtained using the test method in the Standard Test Methods of Bitumen and Bituminous Mixtures for Highway Engineering (JTG E20-2011) [21]. Simultaneously, the standard density ρ_0 is determined according to the JTG F40-2004. Finally, compactability K is calculated by Equation (1) according to the T 0924-2019 in the JTG 3450-2019.

$$K = \frac{\rho_s}{\rho_0} \tag{1}$$

The compactability of two maintenance sections is shown in Figure 1. It can be observed that, upon completion of paving and compaction on the test section, the compactability for each section under conventional maintenance mode can meet the specification requirements. However, the unmanned maintenance sections achieve a compactability meeting the specified standards at a rate of 75%, suggesting room for improvement in the qualification rate. Interestingly, there is notable variability in data within the starting and ending points of unmanned maintenance Sections 3 and 6. This is attributed to the transition between manual and unmanned maintenance modes in these regions, introducing a certain level of impact on data stability. After excluding data with significant variability within the starting and ending regions, the compactability across various unmanned maintenance sections meets the specified standards at a rate of 88%. In addition, the compactability for unmanned maintenance sections after two months greatly exceeds the design values, meeting the specification requirements, which indicates the feasibility of the unmanned maintenance technology.

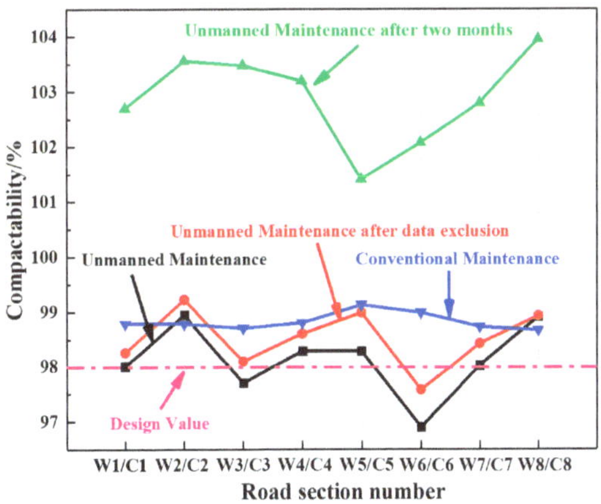

Figure 1. Compactability of unmanned and conventional maintenance sections.

2.3.2. Thickness

In accordance with the T0912-2019 in the JTG 3450-2019, to obtain the thickness of the core sample, a vernier caliper was used to measure the height between the surface and bottom along four symmetrical directions across the circumference of the core sample. Then, the average value of the four positions was calculated. The average thickness and qualification rate of the core samples are depicted in Figure 2. As shown in Figure 2a, it is evident that, in comparison with the design values, the average thickness of the unmanned maintenance section, particularly in Section 4, is lower than the design value, while the average thickness of other maintenance sections is higher than the design value. As shown in Figure 2b, regarding qualification rates, Sections 2 and 6 under the unmanned maintenance sections exhibit relatively low rates, at 80% and 83.3%, respectively. Conversely, under the conventional maintenance mode, the average thickness values for each section surpass the design values, and the qualification rates are all 100%. This implies that the thickness evaluation results for the conventional maintenance sections are slightly superior to those for the unmanned maintenance sections. This is attributed to the fact that some sections maintained by unmanned methods exhibit uneven thicknesses, primarily concentrated at the starting and ending points of construction, with a thinner thickness. Consequently, this non-uniformity results in the overall average thickness of the section being lower than the design value.

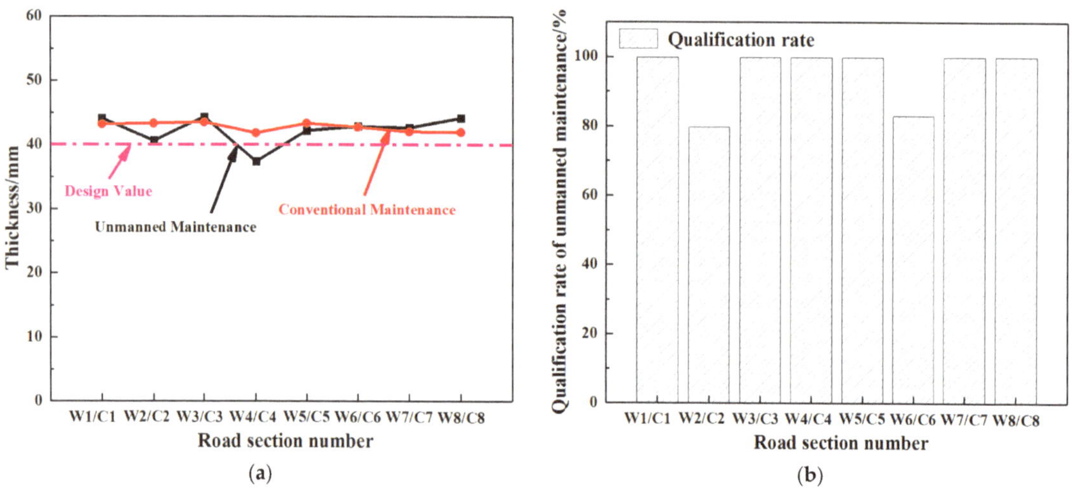

Figure 2. Thickness and qualification rate of unmanned and conventional maintenance sections. (**a**) Thickness of two maintenance method sections. (**b**) Qualification rate of unmanned maintenance sections.

2.3.3. Surface Smoothness

In accordance with the T0934-2008 in the JTG 3450-2019, the vehicle-mounted laser profiler was used to measure the International Roughness Index (*IRI*) of the main lane and passing lane. The speed of the vehicle was kept at 50–80 km/h. Once the vehicle exited the testing section, the testing personnel ceased data collection and recording. At the same time, in accordance with the Highway Performance Assessment Standards (JTG 5210-2018) [22], the Riding Quality Index (*RQI*) was calculated by Equation (2).

$$RQI = \frac{100}{1 + \alpha_O e^{\alpha_1 IRI}} \tag{2}$$

where α_O (0.026) and α_1 (0.65) are the parameters.

The results of *IRI* and *RQI* of unmanned maintenance sections are shown in Table 3. The test results indicate that the overall average *IRI* for the main lane construction section is 1.23 m/km, with an *RQI* average of 94.50. For the pass lane construction section, the overall average *IRI* is 1.30 m/km, with an *RQI* average of 94.26. The smoothness index is assessed as excellent in both lanes.

Table 3. IRI and RQI of unmanned maintenance sections.

Road Section Number	Main Lane		Pass Lane	
	IRI	*RQI*	*IRI*	*RQI*
W1	1.35	94.13	1.11	94.93
W2	1.23	94.53	1.39	93.98
W3	1.13	94.86	1.42	93.85
W4	1.30	94.31	1.39	93.96
W5	1.31	94.25	1.47	93.66
W6	1.19	94.68	1.40	93.93
W7	0.98	95.32	1.12	94.90
W8	1.40	93.93	1.13	94.87

In addition, in accordance with the Inspection and Evaluation Quality Standards for Highway Maintenance Engineering Section 1 Civil Engineering (JTG 5220-2020) [23], the qualification rate of the *IRI* for the section (the ratio of the number of points in each section meeting the specification requirements to the total number of test points) was calculated, as shown in Figure 3. It can be observed that the qualification rates for unmanned maintenance sections and conventional maintenance sections are quite close, but both show cases where the qualification rate for the *IRI* does not reach 100%. This is mainly due to the lack of qualification of the *IRI* in the starting and ending areas of the sections. Thus, it can be concluded that the quality control of smoothness at the starting and ending sections of construction sections is a weak link in unmanned maintenance. Additionally, the multiple repairs conducted in localized areas of the main lane before starting the unmanned maintenance have a negative impact on road smoothness, so the qualification rate for the smoothness of the main lane is lower than the pass lane.

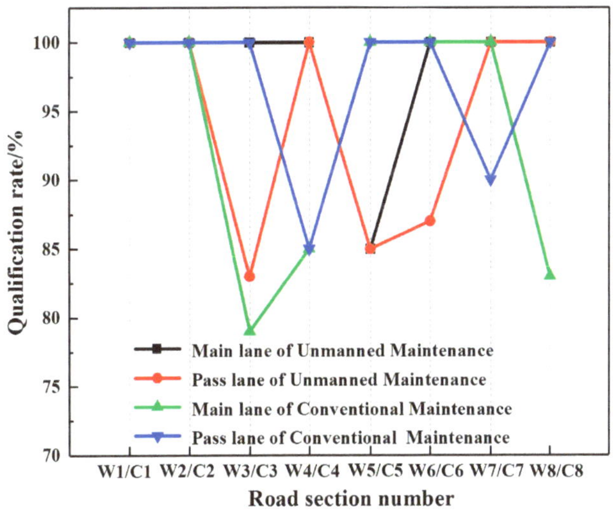

Figure 3. Qualification rate of the *IRI* of unmanned and conventional maintenance sections.

2.3.4. Permeability Coefficient and Constructure Depth

The permeability of asphalt pavement reflects an indirect indicator of the gradation composition of asphalt mixtures on the road surface. It is also a crucial parameter for the water stability of asphalt pavements. In accordance with the T 0971-2019 in JTG 3450-2019, the permeability coefficient of the maintenance sections was measured. In this test, a pavement permeability tester was used. The instrument had a volume of 600 mL, with calibration lines at 100 mL and 500 mL. Firstly, water was poured into the graduated cylinder up to the 100 mL calibration lines, and then the stopwatch was immediately started to begin timing. When the time reached 3 min, the volume of water was recorded, and the test was stopped simultaneously. If the time elapsed was less than 3 min and the volume of water had already dropped to 500 mL, the time was immediately recorded, and the test was stopped simultaneously. The permeability coefficient C_w can be calculated by Equation (3):

$$C_w = \frac{V_2 - V_1}{t_2 - t_1} \times 60 \tag{3}$$

where V_1 is the volume of water for the first time; V_2 is the volume of water for the second time; t_1 is the time for the first time; and t_2 is the time for the second time.

The construction depth of the road surface is a significant indicator of surface roughness, primarily used to assess the macro-roughness, drainage performance, and skid resistance of the road surface. In accordance with the T 0966-2008 in JTG 3450-2019, the vehicle-mounted laser profilometer was used to obtain the pavement constructure depth. The results of the permeability coefficient and construction depth measurements for the unmanned and conventional maintenance sections are shown in Figure 4. The detection data indicate that both the unmanned maintenance section and conventional maintenance section exhibit satisfactory results for the two parameters. The permeability coefficient and construction depth compliance rates reach 100%, which aligns with the design requirements, indicating that both maintenance methods effectively meet the specified standards. This suggests that intelligent unmanned maintenance technology yields comparable results to conventional manual maintenance methods in terms of these critical parameters, affirming its suitability for use in road construction projects.

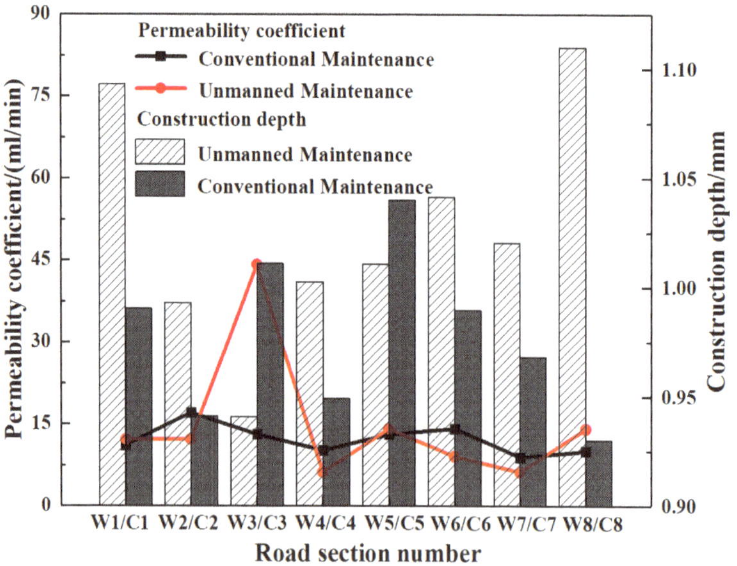

Figure 4. Permeability coefficient and construction depth of conventional maintenance sections.

2.3.5. Performance Comparison of Unmanned Maintenance between Daytime and Nighttime

In this study, Sections 1–6 are designated as daytime maintenance sections, while Sections 7 and 8 are nighttime maintenance sections. To further compare the level of unmanned maintenance under different working conditions during the day and at night, the detection data from Sections 1–6 are averaged to represent the quality assessment results of daytime unmanned maintenance, while the detection data from Sections 7 and 8 are averaged to represent the quality assessment results of nighttime unmanned maintenance for comparative analysis, as shown in Table 4. It can be observed that the thickness, construction depth, and permeability coefficient of the nighttime all meet the specification requirements and show little difference compared to the daytime. The compaction index of the nighttime construction sections is slightly lower than that of the daytime construction. However, the coefficient of variation of the *IRI* calculated from the original data per hundred meters shows that the coefficient of variation (0.12) during nighttime is higher compared to that during daytime (0.07), indicating greater difficulty in controlling the smoothness during nighttime maintenance. Nighttime maintenance is more susceptible to environmental factors such as temperature and wind speed, which may lead to a rapid decrease in the temperature of the asphalt mixture, thereby affecting compaction and smoothness indices. Further optimization of maintenance processes during nighttime can be carried out to make them more suitable for nighttime environments.

Table 4. Comparison of road quality between unmanned maintenance at daytime and nighttime.

Type	Daytime	Nighttime	Standard Value
Compactability	96.8	95.5	/
Thickness	41.7	43.3	>40 mm
Smoothness	94.5	94.6	/
Permeability coefficient	16.0	10.0	<80 mL/min
Construction depth	1.06	1.12	>0.55 mm
Coefficient of variation for *IRI*	0.07	0.12	/

3. Fuzzy Comprehensive Evaluation of Intelligent Unmanned Maintenance Technology

3.1. Establish Evaluation Index System

The evaluation index system is shown in Figure 5. The road quality, safety, technology application, and socio-economic benefits were selected as the criteria layer evaluation indicators to evaluate unmanned maintenance technology. This is because road quality is an important indicator for measuring the success of unmanned maintenance construction technology. High-quality roads ensure stable and durable facilities post-construction, thereby reducing maintenance costs and the need for subsequent repair work. Furthermore, the safety of the operators and the working environment during the road construction process are equally important indicators. Evaluating safety can consider indicators such as accident rates and the effectiveness of safety management measures to ensure the safety of personnel and equipment during maintenance. Additionally, unmanned maintenance technology typically relies on the support of information technology, such as sensors, drones, and artificial intelligence. Assessing the application of information technology can reflect technological proficiency and innovation, which are significant for enhancing construction efficiency and reducing costs. Lastly, the application of unmanned maintenance technology not only concerns its technical performance but also encompasses its impact on socio-economic factors. Evaluating socio-economic benefits can consider indicators such as project costs, construction periods, and the efficiency of human resource utilization to determine the technology's contribution to overall socio-economic benefits.

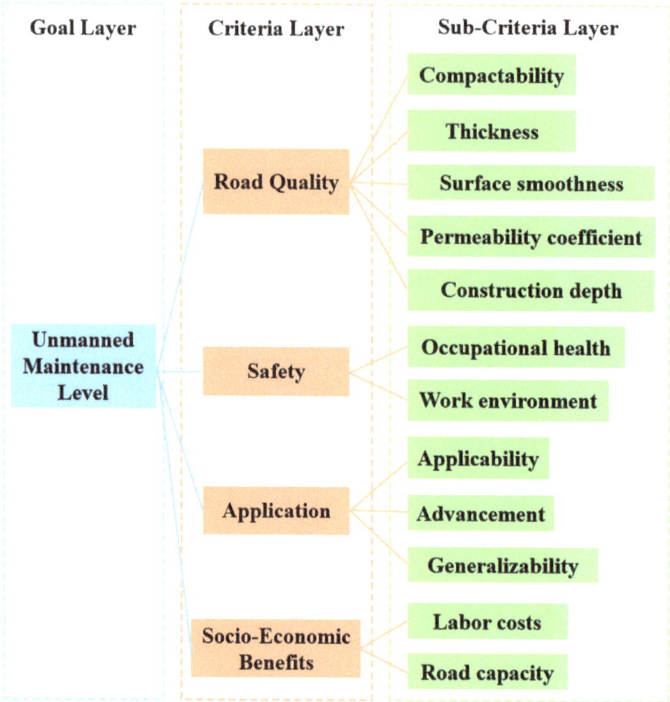

Figure 5. Evaluation index system.

In addition, the reason for selecting compactability, thickness, surface smoothness, permeability coefficient, and construction depth as secondary evaluation indicators for road quality is that these indicators directly reflect the quality of road construction. Indicators such as compactability, thickness, and surface smoothness can measure the structural stability and smoothness of the road surface, while permeability coefficient and construction depth relate to the durability and anti-permeability ability of the road surface. The reason for selecting workplace safety and occupational health safety as secondary evaluation indicators for safety is that workplace safety and occupational health safety are critical factors in ensuring the life safety and health of construction workers. Workplace safety includes safety management and environmental protection at the construction site, while occupational health safety focuses on various health risks faced by construction workers during work. The reason for selecting applicability, advancement, and generalizability as secondary evaluation indicators for technology application is that these indicators can comprehensively evaluate the actual effects and promotion potential of technology in road construction. Applicability considers whether technology can adapt to different construction environments and needs, advancement focuses on whether the technology is at the industry forefront, and generalizability considers whether the technology has the potential for promotion and application in different regions and projects. The reason for selecting labor cost and road capacity as secondary evaluation indicators for socio-economic benefits is that these indicators involve the impact of road construction on the socio-economic aspects. Labor cost directly affects construction costs and efficiency, while road traffic capacity relates to the impact of road construction on traffic flow and travel efficiency, thereby affecting the realization of socio-economic benefits.

3.2. Evaluation Method

In engineering practice, engineers frequently encounter decision-making challenges. The objects under evaluation are influenced by multiple factors and exhibit fuzzy characteristics. Fuzzy refers to concepts with unclear and ambiguous boundaries, such as the relative beauty or ugliness of decorative materials [24]. Hence, there is a need for an evaluation method to consider all relevant factors and address the ambiguity of boundaries. The commonly used evaluation methods include the analytic hierarchy process (AHP), fuzzy comprehensive evaluation method (FCE), Principal Component Analysis, Grey Comprehensive Evaluation Method, etc. The FCE method is one of the most fundamental approaches in fuzzy mathematics, allowing for the evaluation of multiple indicators of an object's state at a comprehensive evaluation level. It is based on the theory of membership degree, which can convert qualitative indicators into quantitative indicators and comprehensively evaluate the membership degree levels of the evaluated objects based on multiple parameters and dimensions. The fuzzy rule base of fuzzy comprehensive evaluation is a fundamental component that governs the decision-making process in detail [25]. It comprises a set of rules that define the relationship between input variables and output assessments within a fuzzy logic system. These rules are typically formulated based on expert knowledge or empirical data. Each rule specifies how input variables, which may have uncertain or imprecise values, contribute to the overall evaluation outcome. By employing fuzzy logic techniques, the rule base accommodates the inherent ambiguity and uncertainty in real-world scenarios, allowing for a more flexible and nuanced decision-making process. Furthermore, in the practical application process of the evaluation system, this study combines the fuzzy comprehensive evaluation method with AHP [26], decomposing factors related to decision-making into objectives and criteria, and conducting qualitative and quantitative analyses on this basis, allowing experts to provide important levels for pairwise comparison of indicators, thereby ensuring higher reliability.

3.2.1. Fuzzy Comprehensive Evaluation

The establishment process of the FCE model includes six main steps: defining the evaluation indicator set, establishing the set of evaluation grades, constructing the membership functions, determining the fuzzy matrix, and forming the fuzzy evaluation vector. The specific steps are shown as follows:

The first step is to establish an evaluation indicator set, which is a collection of various factors that constitute the assessment objects. For the evaluation of maintenance effectiveness, a comprehensive judgment is required from multiple aspects such as road quality, implementation safety, the degree of information technology application, generated socio-economic benefits, etc. All these factors form the set of the evaluation indicator system, namely the evaluation factor set, denoted as $U = \{U_1, U_2, \cdots, U_n\}$. Here, $U_i = (i = 1, 2, \cdots, n)$ represents an evaluation factor, and n is the number of individual factors at the same level.

The second step involves establishing the set of evaluation grades. This is necessary due to the different evaluation values associated with each indicator, often resulting in varying levels such as excellent, good, fair, poor, and so forth, for assessing maintenance effectiveness. It is denoted as $V = \{V_1, V_2, \cdots, V_m\}$. Here, V_j ($j = 1, 2, \cdots$, m) represents the evaluation level standards, providing different grades for each indicator.

The third step is to construct membership functions. After constructing the set of evaluation factors, it is necessary to quantitatively characterize these factors. This involves determining the degree of membership of each factor to the set of evaluation grades, thereby obtaining the fuzzy matrix.

Firstly, the fuzzy evaluation matrix $R_i = (r_{i1}, r_{i2}, \cdots, r_{im})$ is constructed, where R_i is the fuzzy matrix for the ith indicator in the evaluation factors corresponding to each evaluation grade V_i, V_2, $\cdots$, V_m. For the safety, information technology application, and socio-economic benefit indicators, the membership degrees are calculated using the expert scoring method. The same evaluation grade is used for each indicator. Finally, the evaluation data from 20 experts are taken to calculate the membership degrees. The calculation equation for r_{ij} is as follows [26]:

$$r_{ij} = \frac{c_{ij}}{\sum_{j=1}^{m} c_{ij}} \quad (4)$$

where c_{ij} represents the number of occurrences where the ith factor U_i receives the jth comment V_j; r_{ij} (i = 1, 2, $\cdots$, n; j = 1, 2, $\cdots$, m; n is the number of evaluation factors, m is the number of evaluation grades) represents the proportion of occurrences of the indicator factor U_i receiving the comment V_j. For the road quality indicators, membership degrees are calculated using testing data.

The fourth step is to construct a fuzzy matrix based on the membership degree subset, as shown in Equation (5) [27].

$$R = \begin{pmatrix} R_1 \\ \vdots \\ R_n \end{pmatrix} = \begin{pmatrix} r_{11} & \cdots & r_{1m} \\ \vdots & \ddots & \vdots \\ r_{n1} & \cdots & r_{nm} \end{pmatrix} \quad (5)$$

The fifth step is to form the fuzzy evaluation vector, which can be calculated by multiplying the weight vector W by the fuzzy matrix R, as shown in Equation (6) [27].

$$B = W \cdot R = (w_1,\ w_2, \ldots, w_n) \begin{bmatrix} r_{11} & \cdots & r_{1m} \\ \vdots & \ddots & \vdots \\ r_{n1} & \cdots & r_{1m} \end{bmatrix} \quad (6)$$

3.2.2. Analytic Hierarchy Process

To obtain the fuzzy evaluation vector, the weight vector W should be obtained. In general, the roles played by various evaluation indicators in a comprehensive evaluation are not the same. The results of the comprehensive evaluation depend not only on the assessments of each factor but also largely on the impact of each factor on the overall evaluation. This requires determining a weight allocation among the evaluation indicators, represented as a fuzzy vector $W = (W_1, W_2, \cdots, W_n)$ on set $U = \{U_1, U_2, \cdots, U_n\}$, where Wi is the weight of the ith evaluation indicator. In this study, the AHP is applied to quantitatively analyze the importance of each indicator and determine the weights of each indicator. The weight calculation can be carried out according to the following steps:

(1) Establishing hierarchical structure model: This step is used to construct a hierarchical organizational model for the evaluation of the application effectiveness of intelligent unmanned maintenance technology, as shown in Figure 5.

(2) Constructing judgment matrix: This step involves constructing the judgment matrices to obtain the relative importance of different evaluation indicators. For each pair of evaluation indicators to be compared, a scale is used to express the relative importance between them. Five basic scales (1, 3, 5, 7, and 9) of absolute numbers can be used, representing equal importance, moderate importance, strong importance, very strong importance, and extreme importance, respectively. The numbers between these scales (2, 4, 6, and 8) express intermediate importance. This method decomposes the evaluation objectives into multiple levels, assessing the relative importance of different factors within each level of the objectives. This study establishes a judgment matrix by inviting experts to quantitatively score the importance of each indicator.

(3) Calculate the maximum eigenvector (λ_{max}) and corresponding eigenvector (α): These parameters can be calculated by Equation (7) [28]:

$$\lambda_{max} = \frac{1}{n} \sum_{i=1}^{n} \frac{(P\alpha)_i}{n\alpha_i} \quad (7)$$

where n is the order of the judgment matrix.

(4) Consistency ratio check. In this step, the consistency index (CI) and consistency ratio (CR) are calculated to conduct a consistency check on the judgment matrix, aiming to enhance the reliability of the AHP, as shown in Equations (8) and (9) [26].

$$CR = \frac{CI}{RI} \quad (8)$$

$$CI = \frac{\lambda_{max} - n}{n - 1} \quad (9)$$

where RI is a random consistency index. Generally, if the CR is less than an accepted threshold (typically 0.1), the judgment matrix is considered consistent; otherwise, a reevaluation of the relative importance of pairwise comparisons is necessary.

Finally, the judgment matrices are used to calculate the weight vectors. Finally, obtain the weight values (W_i) and the weight vector $W = (W_1, W_2, \cdots, W_n)$, where $\sum_{i=1}^{n} w_i = 1$.

4. Results and Discussion

4.1. Determination of Evaluation Factors Set and Evaluation Grade Set

As shown in Figure 6, in this study, four criteria layer evaluation indicators, including road quality, safety, application, and socio-economic benefits, were selected to evaluate unmanned maintenance technology. These evaluation indicators were labeled from U_1 to U_5, respectively. Four grades, namely, excellent (V_1), good (V_2), normal (V_3), and poor (V_4), were used to set up the evaluation set. To further quantify the application effect of unmanned maintenance technology, specific scores are assigned to each evaluation grade, resulting in $P = \{P_1, P_2, P_3, P_4\} = \{100, 80, 60, 40\}$. V_1 signifies a very high level of achievement for each indicator, with an excellent application effect of unmanned maintenance technology. V_2 represents a relatively good level of achievement for each indicator, with a good application effect of unmanned maintenance technology. V_3 denotes a normal level of achievement for each indicator, with an average application effect of unmanned maintenance technology, requiring improvement. V_4 indicates a poor application effect of the technology.

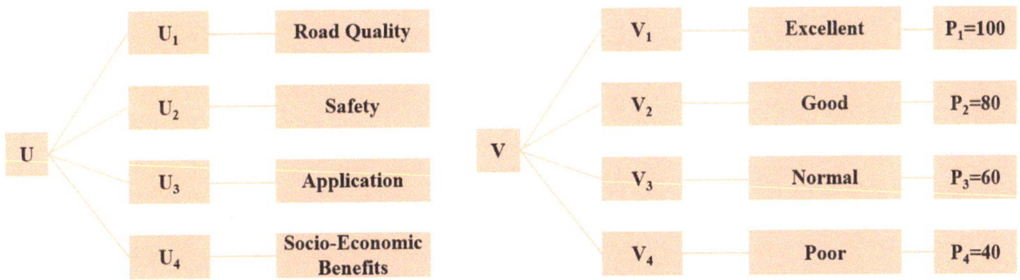

Figure 6. Set of evaluation indicators and evaluation grades.

4.2. AHP for Weight Determination

A panel of 10 experts and 10 on-site maintenance engineering technicians, all involved in the project, are invited to form an evaluation group. Based on the constructed evaluation index system, experts and technicians conduct pairwise comparisons of the indicators at the same level and with the indicators of the previous level. They assign values representing the relative importance, resulting in judgment matrices. The weights for the single-level ranking of the criteria layer and sub-criteria layer evaluation factors are then calculated, and consistency checks are performed. According to the judgment matrix, the weight coefficients for each indicator are obtained, as shown in Table 5.

Table 5. Weight of evaluation factors.

Criteria Layer Evaluation Indicators	Weight	Sub-Criteria Layer Evaluation Indicators	Weight
Road quality	0.510	Compactability (N_1) Thickness (N_2) Surface smoothness (N_3) Permeability coefficient (N_4) Construction depth (N_5)	0.485 0.227 0.143 0.089 0.057
Safety	0.330	Occupational health (N_6) Workplace (N_7)	0.25 0.75
Application	0.100	Applicability (N_8) Advancement (N_9) Generalizability (N_{10})	0.623 0.239 0.137
Socio-economic benefits	0.059	Labor costs (N_{11}) Road capacity (N_{12})	0.667 0.333

The weight set for the criteria layer is {0.51, 0.330, 0.100, 0.059}, indicating that the importance of the evaluation indicators is road quality (U_1) > safety (U_2) > application (U_3) > socio-economic benefits (U_4). This further emphasizes that road quality is the most crucial factor in evaluating the effectiveness of intelligent unmanned maintenance technology, ensuring that the technologies and measures applied in maintenance projects meet the required standards. Therefore, strict requirements should be imposed on the implementation process of the target project, with the quality of the road repeatedly checked. During the implementation process, there should be ample communication and discussion between management personnel and maintenance technicians to continuously optimize and improve, ensuring that the applied unmanned construction and maintenance technology meets the standards for maintenance construction. Safety (U_2) in maintenance projects holds the second-largest weight among the criteria layer evaluation indicators, indicating that highway maintenance projects prioritize safety. Managers should ensure safety through technical measures to guarantee the safety of project implementation.

From Table 5, for road quality (U_1), three indicators, including compactability, thickness, and surface smoothness, have relatively large weights, all exceeding 0.140. This suggests that when applying unmanned maintenance technology, attention should be focused on paving and compaction operations, strictly adhering to process requirements to ensure that all test indicators meet standard requirements. In the criteria layer indicator safety (U_2), the proportion of workplace safety is the largest, indicating that the safety of engineering facilities and the surrounding environment is a key focus during the implementation of highway maintenance projects. For the application of intelligent technology, the highest weight is assigned to applicability, highlighting the significant potential this technology holds for future utilization within the domain of road maintenance. In terms of socio-economic benefits, it is evident that labor costs carry significant weight, reflecting the substantial advantage of unmanned construction in reducing labor costs compared to traditional manual maintenance construction methods.

4.3. Fuzzy Comprehensive Evaluation

The membership degrees of the criteria layer evaluation indicators to different evaluation grades are illustrated in Figure 7. It can be observed that both safety and socio-economic benefits have membership degrees greater than 0.8, reaching the excellent grade (V_1). Especially for socio-economic benefits, there are no instances of the normal grade (V_3) and poor grade (V_4), and they have the highest membership degree (0.883) on the excellent grade among all evaluation indicators. It can be concluded that the socio-economic benefits generated by the application of unmanned maintenance technology are quite favorable. In addition, road quality has a membership degree of 0.785 in the excellent grade (V_1), which is the smallest among the four indicators. Moreover, the proportion of cases categorized as poor is relatively large, highlighting the need for further refinement and improvement.

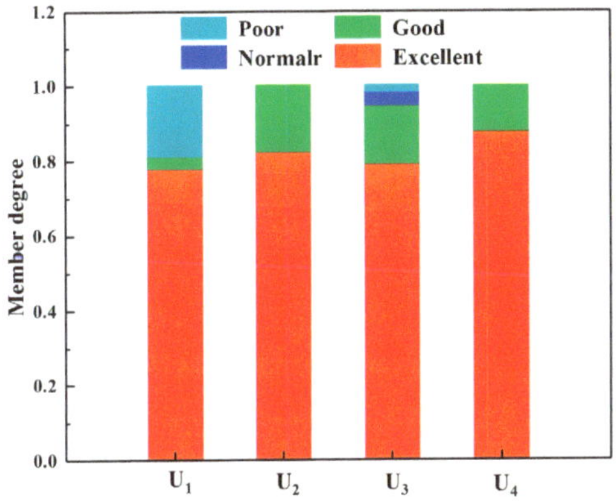

Figure 7. Member degrees of criteria layer evaluation indicators.

The membership degrees of the sub-criteria layer evaluation indicators to different evaluation grades are illustrated in Figure 8. It can be observed that compactability (N_1), thickness (N_2), surface smoothness (N_3), and generalizability (N_{10}) all exhibit a poor grade (V_4). Compactability and thickness are identified as weak indicators, and the permeability coefficient (N_4) and construction depth (N_5) belong to the excellent grade with a membership degree of 1, aligning with the analysis results of the inspection data. In addition, generalizability (N_{10}) has the lowest membership degree in the excellent grade (V_1), standing at only 0.5, indicating a need for significant optimization and improvement. Smoothness (N_3), occupational health and safety (N_7), and labor cost (N_{11}) show relatively high membership degrees for the excellent grade (V_1), indicating that unmanned maintenance technology for highway pavements performs well in terms of labor cost and occupational health and safety aspects.

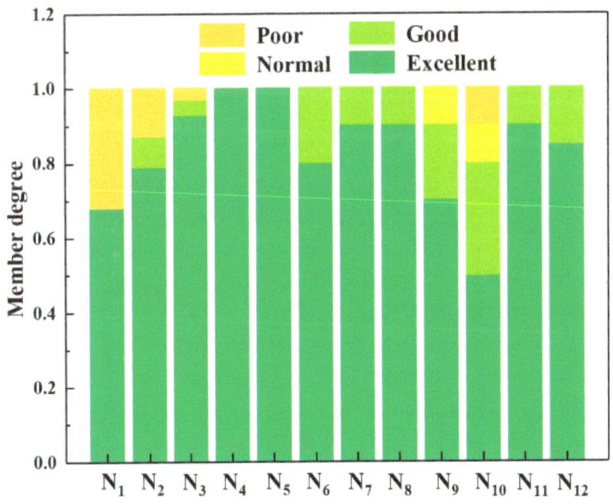

Figure 8. Member degrees of sub-criteria layer evaluation indicators.

Finally, the comprehensive evaluation vectors of all the involved criteria layer evaluation indicators are calculated, as shown in Equations (10)–(13), giving the comprehensive evaluation vectors of all the involved asphalt mastics:

$$U_1 = W_1 \times R_1 = (0.785,\ 0.026,\ 0,\ 0.189) \tag{10}$$

$$U_2 = W_2 \times R_2 = (0.825,\ 0.175,\ 0,\ 0) \tag{11}$$

$$U_3 = W_3 \times R_3 = (0.797,\ 0.151,\ 0.038,\ 0.014) \tag{12}$$

$$U_4 = W_4 \times R_4 = (0.883,\ 0.117,\ 0,\ 0) \tag{13}$$

So, the fuzzy matrix can be obtained, as follows:

$$R = \begin{pmatrix} 0.785 & 0.026 & 0 & 0.189 \\ 0.825 & 0.175 & 0 & 0 \\ 0.797 & 0.151 & 0.038 & 0.014 \\ 0.883 & 0.117 & 0 & 0 \end{pmatrix} \tag{14}$$

Combined with the weight vector $\mathrm{W} = (0.51, 0.33, 0.1, 0.059)$ for the criteria layer indicators, the fuzzy vector B can be obtained:

$$\mathrm{B} = \mathrm{W} \times \mathrm{R} = (0.805,\ 0.093,\ 0.004,\ 0.098) \tag{15}$$

It can be observed that the membership degree of unmanned maintenance technology in the excellent grade is the highest, reaching 0.805. In other words, the final comprehensive evaluation result of unmanned maintenance technology is rated as excellent. To further quantify the application effect of unmanned maintenance technology, each grade level is assigned to specific scores, namely, four evaluation grades (V_1, V_2, V_3 and V_4) with assigned scores of 100, 80, 60, and 40, respectively. In this case, the comprehensive score F is calculated by Equation (16):

$$\mathrm{F} = \mathrm{B} \times \mathrm{P}^{\mathrm{T}} \tag{16}$$

The quantified value F for the overall evaluation of the application effectiveness of intelligent unmanned maintenance technology is 92.10. This indicates that the application's effectiveness is very good.

5. Conclusions

In this study, intelligent unmanned road construction technology is utilized to explore its potential in road maintenance engineering. The engineering technical indicators of asphalt pavement after unmanned maintenance are investigated and compared. Meanwhile, using the analytic hierarchy process (AHP) and fuzzy evaluation method, an evaluation index model that considered road quality, safety, technology application, and socio-economic benefits was established to assess the application effectiveness of intelligent unmanned maintenance technology. The main conclusions can be drawn as follows:

(1) The road quality of the unmanned maintenance method is inferior to traditional manual maintenance methods, especially in terms of the compactability and surface smoothness at the starting and ending points of the maintenance section, which need improvement. However, the quality of the unmanned maintenance method still meets specification requirements.

(2) The weight set of the four types of criteria layer evaluation indicators suggests that the importance of the evaluation indicators follows the order of road quality (U_1) > safety (U_2) > application (U_3) > socio-economic benefits (U_4). Consequently, stringent

requirements must be placed on the implementation process of the target project, with continual monitoring of road quality.

(3) Regarding socio-economic benefits, labor costs hold considerable weight, underscoring the significant advantage of unmanned construction in reducing labor costs compared to traditional manual maintenance construction methods.

(4) Both safety and socio-economic benefits exhibit membership degrees exceeding 0.8, attaining the excellent grade (V_1). Furthermore, road quality attains a membership degree of 0.785 in the excellent grade (V_1), the lowest among the four indicators. Additionally, the proportion of road quality classified as poor is relatively high, emphasizing the necessity for further refinement and enhancement.

(5) The membership degree of unmanned maintenance technology in the excellent grade is the highest, reaching 0.805, and the quantified value for the overall evaluation of the application effectiveness of unmanned maintenance technology is 92.10. This means that the final comprehensive evaluation result of unmanned maintenance technology is rated as excellent.

The integrated technology used in this study enables the precise edge-rolling of road boundaries and the unmanned compaction of standardized road sections based on satellite positioning. However, in scenarios such as tunnels or under bridges where satellite signals are obstructed, the lack of high-precision location information leads to increased trajectory errors. In the future, a multi-sensor fusion positioning method can be used to address the localization issues of unmanned equipment in the absence of satellite signals, thus improving adaptability to different scenarios. In addition, it is necessary to align with the industry's development stages, thus optimizing and enhancing the established index system and evaluation methods.

Author Contributions: Conceptualization, J.W.; Data curation, J.C.; Funding acquisition, G.H.; Investigation, G.S. and R.Z.; Methodology, J.W.; Supervision, J.W.; Validation, G.H. and J.C.; Visualization, H.W.; Writing—original draft, G.H.; Writing—review and editing, G.S., R.Z., J.C. and H.W. All authors have read and agreed to the published version of the manuscript.

Funding: This research was funded by Zhejiang Provincial Department of Transportation Science and Technology Plan Project under Grant number 2021058.

Data Availability Statement: All data, models, and codes generated or used in this study are included in the submitted manuscript.

Conflicts of Interest: Authors Gensheng Hu, Gongzuo Shi and Jianfeng Chen were employed by the company Zhejiang Communication Investment Expressway Operation Management Co., Ltd. Authors Runhua Zhang, Haichang Wang and Junzhe Wang were employed by the company Testing Technology (Beijing) Co., Ltd. of China Academy of Transportation Sciences. The remaining authors declare that the research was conducted in the absence of any commercial or financial relationships that could be construed as a potential conflict of interest.

References

1. Mujtaba, H.; Khalid, U.; ur Rehman, Z.; Farooq, K. Recycling of reclaimed subbase materials in flexible pavement design. *Road Mater. Pavement Des.* **2021**, *23*, 2713–2732. [CrossRef]
2. Ijaz, N.; Ye, W.; Rehman, Z.U.; Ijaz, Z.; Junaid, M.F. New binary paper/wood industry waste blend for solidification/stabilisation of problematic soil subgrade: Macro-micro study. *Road Mater. Pavement Des.* **2023**, *24*, 1215–1232. [CrossRef]
3. Fu, C.; Liu, K.; Liu, Q.; Xu, P.; Dai, D.; Tong, J. Exploring directional energy conversion behavior of electromagnetic-based multifunctional asphalt pavement. *Energy* **2023**, *268*, 126573. [CrossRef]
4. Wu, W.; Cavalli, M.C. Differing perspectives on the use of high-content SBS polymer-modified bitumen. *Constr. Build. Mater.* **2024**, *411*, 134433. [CrossRef]
5. Qureshi, W.S.; Power, D.; Ullah, I.; Mulry, B.; Feighan, K.; McKeever, S.; O'Sullivan, D. Deep learning framework for intelligent pavement condition rating: A direct classification approach for regional and local roads. *Autom. Constr.* **2023**, *153*, 104945. [CrossRef]
6. Elseicy, A.; Alonso-Díaz, A.; Solla, M.; Rasol, M.; Santos-Assunçao, S. Combined use of GPR and other NDTs for road pavement assessment: An Overview. *Remote Sens.* **2022**, *14*, 4336. [CrossRef]

7. Zhang, J.; Zhu, Z.; Liu, H.; Zuo, J.; Ke, Y.; Philbin, S.P.; Zhou, Z.; Feng, Y.; Ni, Q. System framework for digital monitoring of the construction of asphalt concrete pavement based on IOT, BeiDou navigation system, and 5G technology. *Buildings* **2024**, *411*, 503. [CrossRef]
8. Liang, J.; Guan, Z. Design of intelligent automatic leveling control system based on ultrasonic distance measuring technology. In Proceedings of the 2007 IEEE International Conference on Automation and Logistics, Shandong, China, 18–21 August 2007; Volumes 1–6, pp. 1620–1624.
9. Qi, Y.; Ma, W.; Zhang, J. Construction deviation control method of transverse super large inclined asphalt pavement. *Arch. Civ. Eng.* **2023**, *69*, 165–176.
10. Chen, B.; Yu, X.; Dong, F.; Zheng, C.; Ding, G.; Wu, W. Compaction quality evaluation of asphalt pavement based on intelligent compaction technology. *J. Constr. Eng. Manag.* **2021**, *147*, 04021099. [CrossRef]
11. Lu, L.; Dai, F.; Zaniewski, J.P. Automatic roller path tracking and mapping for pavement compaction using infrared thermography. *Comput. Aided Civ. Infrastruct. Eng.* **2021**, *36*, 1416–1434. [CrossRef]
12. Dunston, P.S.; Monty, J. Practices for Seamless Transmission of Design Data from Design Phase to Construction Equipment Operation: A Synthesis Study, No. FHWA/IN/JTRP-2008/08. In Joint Transportation Research Program, Indiana Department of Transportation and Purdue University, Indiana. 2009. Available online: https://docs.lib.purdue.edu/cgi/viewcontent.cgi?article=2672&context=jtrp (accessed on 1 May 2009).
13. Snyder, M. Stringless Paving, No. HWA-HIF-18-011, Federal Highway Administration, United States. 2019. Available online: https://www.fhwa.dot.gov/pavement/pubs/hif18011.pdf (accessed on 1 January 2019).
14. Peyret, F.; Jurasz, J.; Carrel, A.; Zekri, E.; Gorham, B. The computer integrated road construction project. *Autom. Constr.* **2000**, *9*, 447–461. [CrossRef]
15. Ye, W.; Wang, T.; Zhou, C. Application of millimeter 3D automatic paving system in highway engineering construction. In Proceedings of the 20th COTA International Conference of Transportation Professionals, Xi'an, China, 14–16 August 2020; pp. 1604–1615.
16. Bian, Y.; Fang, X.; Yang, M.; Zhao, Z. Automatic rolling control for unmanned vibratory roller based on fuzzy algorithm. *J. Tongji Univ.* **2017**, *45*, 1830–1838.
17. Xie, R. Research on intelligent construction technology of asphalt pavement complete equipment. *Value Eng.* **2021**, *40*, 129–131.
18. Wang, J.; Yang, J.; Zhang, R.; Xie, W.; Wu, G. Research and practice of operating expressway maintenance based on unmanned automatic driving technology. In Proceedings of the CICTP 2023, Beijing, China, 14–17 June 2023; pp. 310–319.
19. *JTG F40-2004*; Technical Specifications for Construction of Highway Asphalt Pavements. Research Institute of Highway Ministry of Transport: Beijing, China, 2004.
20. *JTG 3450-2019*; Field Test Methods of Highway Subgrade and Pavement. China Communications Press: Beijing, China, 2019.
21. *JTG E20-2011*; Standard Test Methods of Bitumen and Bituminous Mixtures for Highway Engineering. Ministry of Transport of the People's Republic of China: Beijing, China, 2011.
22. *JTG 5210-2018*; Highway Performance Assessment Standards. Ministry of Transport of the People's Republic of China: Beijing, China, 2018.
23. *JTG 5220-2020*; Inspection and Evaluation Quality Standards for Highway Maintenance Engineering Section 1 Civil Engineering. Research Institute of Highway Ministry of Transport: Beijing, China, 2020.
24. Zadeh, L.A. Fuzzy sets. *Inf. Control* **1965**, *9*, 338–353. [CrossRef]
25. Xu, X.; Yu, F.; Pedrycz, W.; Du, X. Multi-source fuzzy comprehensive evaluation. *Appl. Soft Comput.* **2023**, *135*, 110042. [CrossRef]
26. Fu, C.; Liu, K.; Zhi, W.; Liu, P.; Oeser, M. Comprehensive evaluation of thermally conductive functional layer on snow-melting performance for electric heating bridge system. *Cold Reg. Sci. Technol.* **2023**, *219*, 104099. [CrossRef]
27. Jing, H.; Yao, P.; Song, L.; Zhang, J.; Zhao, Y.; Zhang, Z. Analysis on optimization decision of pavement maintenance mode based on analytic hierarchy process. *Dyna* **2021**, *96*, 605–611. [CrossRef]
28. Zhang, Z.; Liu, P.; Rochlani, M.; Falla, G.C.; Leischner, S.; Oeser, M. A fuzzy comprehensive evaluation method to select the optimal mineral filler for optimization of the performance of asphalt mastics. *Constr. Build. Mater.* **2021**, *298*, 123847. [CrossRef]

buildings

Article

Nano-Scale and Macro-Scale Characterizations of the Effects of Recycled Plastics on Asphalt Binder Properties

Ahmad Al-Hosainat [1], Munir D. Nazzal [2,*], Savas Kaya [3] and Toufiq Reza [4]

[1] Terracon Consultants, Inc., Columbus, OH 43230, USA; ahmad.al-hosainat@terracon.com
[2] Center for Smart, Sustainable & Resilient Infrastructure (CSSRI), Department of Civil and Architectural Engineering and Construction Management, University of Cincinnati, Cincinnati, OH 45221, USA
[3] School of Electrical Engineering and Computer Science, Ohio University, Athens, OH 45701, USA
[4] Florida Institute of Technology, Melbourne, FL 32901, USA
* Correspondence: munir.nazzal@uc.edu or nazzalmd@ucmail.uc.edu; Tel.: +1-513-556-3736

Abstract: This paper summarizes the results of one of the first comprehensive laboratory studies that was conducted to evaluate the effects of adding different contents of recycled polyethylene terephthalate (rPETE) as a modifier to an asphalt binder on the rheological and mechanical properties of the modified binder as well as on the agglomeration behavior between the rPETE and asphalt binder at a multiscale level. The high-temperature and low-temperature performances of the modified binder were investigated at the macro-scale and compared with those of the unmodified binder using dynamic shear rheometer (DSR) and bending-beam rheometer (BBR) rheological tests, as well as asphalt binder cracking device (ABCD) testing. The nano-scale evaluation of the binder properties, including the surface roughness, bonding energy, and reduced modulus, was accomplished using atomic force microscopy (AFM). The results indicated that the addition of rPETE enhanced the high- and intermediate-temperature rheological properties of the modified PG 64-22 binder. The low-temperature rheological properties and resistance to cracking decreased slightly with increasing rPETE content in the asphalt binder. However, this reduction was not remarkable when adding 4%, 8%, and 10% rPETE contents. The asphalt binder modified with 4% rPETE had a low-temperature grade of −22, similar to that of the unmodified binder, indicating that 4% rPETE can be added to the binder to improve its high- and intermediate-temperature properties without reducing its resistance to low-temperature damage. The AFM tapping-mode results indicated that the inclusion of rPETE in the asphalt binder improved the stiffness properties of the modified binder as compared with those of the control asphalt binder. In addition, the rPETE-modified binders showed rougher surfaces than the control binder. The addition of rPETE to the binder increased the values of the reduced modulus and bonding energy compared with those of the control binder.

Keywords: polyethylene terephthalate; rheological properties; agglomeration; surface roughness; reduced modulus; bonding energy; atomic force microscope

Citation: Al-Hosainat, A.; Nazzal, M.D.; Kaya, S.; Reza, T. Nano-Scale and Macro-Scale Characterizations of the Effects of Recycled Plastics on Asphalt Binder Properties. *Buildings* **2024**, *14*, 642. https://doi.org/10.3390/buildings14030642

Academic Editors: Songtao Lv and Xinghai Peng

Received: 30 December 2023
Revised: 22 February 2024
Accepted: 26 February 2024
Published: 29 February 2024

1. Introduction

In the last two decades, the need for using plastics in industry or recycling waste plastics has become very important owing to the harmful effects of these materials on the environment and the pollution they produce. The pollution risk of waste plastics comes from the fact that they are, unlike some other waste materials, non-biodegradable for a relatively long time that can be more than 100 years. There are three methods for disposing of plastic materials, including using landfills, using incinerators to burn them, which also represents a very severe environmental concern because of very large numbers and amounts of volatiles that are generated from using this method [1], or recycling. The interest in recycling plastics is at a high level, and many methods for recycling plastics have been developed to mitigate the environmental risks for not using the right procedures to dispose of these materials. One of the alternatives to recycling plastic materials is to try to mix them or reuse them in industry and the production of other construction, medical, agricultural, or chemical industrial materials. Enhancing the rheological properties of the asphalt binder that is used in highway construction is a suitable alternative that can mediate the mentioned problems by enhancing the binder's rheological and mechanical properties and adding some additives or modifiers, such as polymers, to the binder. These modifiers can add more stiffness to the binder or improve its flexibility. To evaluate the performance of the asphalt binder and understand the interactions between the modified binder and its additives, the evaluation of the rheological properties of the modified binder and the intermolecular forces between the binder and the additives is essential to determine the potential benefits for adding any modifier to the asphalt binder. In the past two decades, several additives, such as fibers, polymers, and plastics, have been used in previous studies to enhance the binder's performance [2–7]. The use of polymers is a common practice in the pavement industry, as polymers improve the high- and intermediate-temperature properties of the asphalt binder without reducing its resistance to low-temperature cracking; the most used types of polymers for modifying asphalt binders include rubber, styrene–butadiene–styrene (SBS), styrene–butadiene rubber (SBR), Elvaloy®, and polyethylene [8,9]. Fibers and Sasobit® are also being used currently in the pavement industry to improve the rheology of the asphalt binder at high temperatures [5,10].

Previous studies have evaluated the effects of using recycled PETE as a modifier of the asphalt binder on the rheological properties of the modified binder [11–19]. Polyethylene terephthalate, commonly abbreviated as PET, PETE, or the obsolete PETP or PET-P, is the most common thermoplastic polymer resin of the polyester family and is used in fibers for clothing (>60%) and containers for liquids and foods (30%). PETE is a clear, strong, and lightweight plastic that is widely used for packaging foods and beverages. PETE consists of polymerized units of the monomer ethylene terephthalate, with repeating ($C_{10}H_8O_4$) units [20]. PETE is commonly recycled and has the number "1" as its resin identification code (RIC) [21]. It is worth noting that the procedure for extracting PETE from recycled plastics is complex and requires processing the recycled plastics at high temperatures and high pressures [22,23]. It was indicated that the optimal processing of the waste polyethylene (WPE)-modified binder can be conducted at a shear rate of 3750 rpm for 1.5 h at 150 °C to obtain the improved storage stability of the binder. It was also indicated that the thermal stability of modified bitumen was higher than that of the unmodified binder. However, it was generally independent of the processing procedure [24]. PETE-modified asphalt binder showed improved penetration, softening points, and temperature susceptibility values, which would result in better performance at high temperatures [25]. The rotational viscosity, G*, and rutting parameter ($G^*/\sin\delta$) of the asphalt binder were improved by increasing the content of PE in the binder, while no improvement in the phase angle results was indicated when increasing the PE content [26]. Almeida e Silva et al. indicated that PETE-modified binder exhibited an improved consistency, an elastic response, and lower oxidation levels, which indicates an improved resistance to aging [27]. Lower penetration values and higher softening point values contribute to the PETE-modified asphalt binder as compared with the unmodified one. It was also found that although the rutting parameter for the binder was improved by the addition of PETE, it could not be considered as being storage stable at high temperatures [28].

In addition, measuring the interaction between the modified asphalt binder and its additives can provide a better understanding of its performance and its interaction with different materials. Nazzal and coworkers indicated from atomic force microscopy (AFM) imaging results that the width of the "bee-like" structures (revealing dispersed black and white successions of elongated structures within a relatively flat matrix) within the binder was reduced by adding Sasobit, whereas no significant effect on these structures was found for the other warm-mix asphalts (WMAs) used in the study [29]. In another study by Abu Qtaish and coworkers [30], it was indicated based on AFM-testing results that although the aging of the binders increased their reduced modulus values, it also reduced their bonding energies. It was also indicated from AFM, Fourier-transform infrared (FTIR) spectroscopy, and gel permeation chromatography (GPC) testing results that less or equal aging will occur in WMA binders over pavement's service life as compared to HMA binders. The effects of the reclaimed asphalt pavement (RAP) source's content on the cracking resistance of asphalt mixtures containing high RAP contents was also evaluated using micro-scale and macro-scale tests [31]. Based on AFM testing results, it was indicated that the microstructures obtained for RAP binders were different from those of virgin asphalt binders, and significantly lower adhesions were obtained for the two RAP binders from different sources as compared with those of the virgin asphalt binders. Using polymer additives was the most significant factor that affected the binder's AFM adhesive forces. In addition, the adhesive bonds between the aggregates and binder inside the mixture had more a pronounced effect on the indirect tensile strength than the cohesive bonds inside the binder, as stress concentrates at the aggregate–binder interface, which leads to an increased tendency for cracking to occur in that area [32].

In summary, there is increasing interest currently in using recycled PETE as a modifier of the asphalt binder to enhance its physical and rheological properties, especially the rutting performance of the modified binder [11,13,17], and improving the properties of asphalt mixtures and their resistances to rutting and fatigue cracking [12,19]. Also, there is increased potential in the current state-of-the-art uses of recycled PETE in the asphalt binder and asphalt mixtures from an environmental perspective to help in recycling waste plastic materials and reduce the environmental concerns related to the increasing quantities of these waste materials and the conventional ways for disposing of them [14,15,18]. This leads to introducing the use of rPETE in the asphalt pavement industry as an effective modifier that helps to improve the properties of asphalt pavement materials for sustainable and clean asphalt production [15].

Despite several research efforts that have been made to characterize and simulate the performance of the asphalt binder and understand the interaction of the binder with its additives or with mineral aggregates using AFM testing, to date, no research has investigated the interaction between rPETE as an additive of the asphalt binder and the binder itself using AFM. Therefore, very limited information is available in the literature regarding the interaction between these two materials at the micro-scale and the effects of the rPETE addition on the mechanical properties of the modified binder. Thus, research is needed to evaluate these properties and compare them with the results of macro-scale tests. This paper presents one of the first research studies to discuss the addition of rPETE as a modifier for the asphalt binder and the simulation of the interaction between rPETE and the binder at the micro-scale using AFM. This study also evaluates the rheological and mechanical properties of the rPETE-modified asphalt binder at the macro-scale using Superpave® rheological tests, which allow for the understanding of the properties of the rPETE-modified binder and its interaction with the rPETE modifier at both the micro-scale and macro-scale. It is worth noting that the rPETE material was extracted in the laboratory by performing complex reactions at high temperatures and high pressures because recycled plastics cannot be directly used as a modifier for the asphalt binder and, therefore, processing for extraction is required for having a rPETE material without any impurities.

2. Objectives

The first objective of this paper is to evaluate the diffusion and agglomeration of the PETE material in the asphalt binder using AFM. The second objective is to evaluate the rheological and mechanical properties of the PETE-modified binder using laboratory rheological tests.

3. Materials and Methods

3.1. Materials

3.1.1. Asphalt Binder

The asphalt binder from Barrett Paving, which is frequently used in highway construction, was selected for evaluation in this study. It is the unmodified binder having a performance grade of PG 64-22. Table 1 presents the properties of the selected asphalt binder, as obtained from the results of laboratory rheological testing conducted in this study.

Table 1. Properties of asphalt binder selected in this study.

Property	Specification	Sample Results
Original (Unaged) Binder		
Dynamic Shear, 10 rad/s; $G^*/\sin\delta$, kPa	≥1.00 at 64 °C	1.3993 at 64 °C
Pressure-Aging Vessel Residue		
Dynamic Shear, 10 rad/s; $G^* \cdot \sin\delta$, kPa, 28 °C	≤5000 at 28 °C	1659.5 at 28 °C
Bending-Beam Creep Slope, *m-value*, −12 °C	≥0.300 at −12 °C	0.303 At −12 °C
Cracking Device PG Grade, °C	N/A	−22.9

3.1.2. Hydrothermal Recycling of PETE

End-of-life PETE bottles were hydrothermally recycled to evaluate them as asphalt binders. Commercial PETE bottles were first cut to a 2 cm × 2 cm size. A 1.5 gal Parr batch stirred reactor manufactured by Parr Instrument Company (Moline, IL, USA) was used to recycle the PETE. Details about the hydrothermal method can be found elsewhere [22,23]. In short, 200 g of PETE and 2 L of deionized (DI) water were poured into the reactor. The reactor was sealed and heated at 5 K/min from room temperature to 260 °C, and the isothermal condition was kept for 30 min. After the end of the reaction time, the reactor was cooled by passing cool water until the reactor temperature reached the ambient temperature. The recycled PETE (rPETE) was then filtered from the process liquid using vacuum filtration. The rPETE was dried in a drying oven overnight and stored in a Ziplock® bag until further characterization and application. The yield of the rPETE was 80.1 ± 0.5%. Table 2 presents the basic properties of the selected rPETE material, as obtained from laboratory testing conducted in this study.

Table 2. Typical basic properties of rPETE material.

Property	Value
Specific Gravity at 25 °C	1.38
Average Molecular Weight (g/mol)	228.199
Average Melting Point (°C)	260
Glass-Transition Temperature (°C)	67–81
Average Modulus of Elasticity (GPa)	3.25
Average Yield Tensile Strength (MPa)	73.7

3.2. Mixing Procedure

Five different contents of rPETE by the weight of the asphalt binder were selected for modifying the binder in this study (0%, 4%, 8%, 10%, and 12%). Each content of rPETE was mixed with the asphalt binder by preheating the liquid binder at 165 °C for 10 min; rPETE was then gradually added to the binder while mixing at a slow velocity of 500 rpm for 5 min using a mechanical mixer. The binder–rPETE mix was then mixed at a rate of 2200 rpm for 60 min at 165 °C until the mix was visually homogeneous [24,25,28]. After mixing, the modified binder was re-stored in a metal can at room temperature and covered to prevent any contamination or loss of properties.

3.3. Rheological Testing

The evaluation of the high-temperature, intermediate-temperature, and low-temperature rheological properties of the control and rPETE-modified asphalt binders was conducted using Superpave rheological tests. The rolling thin-film oven (RTFO), and pressure-aging vessel (PAV) were used to simulate the short-term and long-term aging of the binder, respectively. The RTFO device was used to apply the short-term aging of the asphalt binders in the laboratory according to the AASHTO T 240 standard [33]. In addition, the PAV device was used to apply the long-term aging of the asphalt binders in the laboratory to the RTFO short-term binder samples according to the AASHTO R 28 standard [34].

3.3.1. Dynamic Shear Rheometer (DSR) Testing

DSR equipment was used to test the rheological properties of the control and rPETE-modified asphalt binders before and after aging according to the AASHTO T 315 standard [35] to evaluate the resistances of these binders to rutting and fatigue at high and intermediate temperatures, as well as repeated load conditions. In addition, the DSR device was used to determine the high-temperature performance grades for these binders.

In the DSR test, binder samples with a diameter of 25 mm and a thickness of 1 mm were prepared from unaged and RTFO-aged asphalt binders and tested at the high temperature, starting from 46 °C and incrementing by 6 °C to the failure temperature, at which the rutting parameter is less than the standard value. Also, binder samples with a diameter of 8 mm and a PAV residue thickness of 2 mm were tested at temperatures ranging from 16 °C to 42 °C in increments of 6 °C for each test temperature.

During the test, the binder sample was sandwiched between top and bottom plates, and the test was performed in the strain-control mode. The top plate oscillated at 10 rad/s (1.59 Hz) in a sinusoidal waveform on top of the fixed bottom plate. The DSR measured the maximum applied stress, the resulting maximum strain, and the time lag between them. The values of the complex modulus (G^*) and phase angle (δ) were then calculated automatically by the software.

3.3.2. Bending-Beam Rheometer (BBR) Testing

A BBR device was used to test the low-temperature rheological properties (stiffness and creep) of the control and rPETE-modified asphalt binders after PAV aging according to the AASHTO T 313 standard [36]. In addition, BBR was used to determine the low-temperature performance grades for these binders according to the AASHTO PP 42 standard [37].

In the BBR test, a beam with a size of 6.25 × 12.5 × 127 mm was used as a mold for the binder sample. The binder sample was simply supported at two points at a distance of 102 mm in a controlled-temperature fluid bath. The samples were tested at two temperatures (−6 °C and −12 °C). The sample was then loaded at the midpoint with a 100 g load at a force of 0.98 N. The deflection of the loaded sample was measured at 8, 15, 30, 60, 120, and 240 s. Beam stiffness, often called "creep stiffness", was calculated for these times. The BBR test was conducted on two beam samples.

3.3.3. Asphalt Binder Cracking Device (ABCD) Testing

ABCD testing was conducted on the control and rPETE-modified asphalt binders according to the AASHTO TP 92-14 standard [38] after PAV aging in the laboratory to evaluate their resistances to low-temperature thermally induced cracking. The cracking temperature is a parameter describing the lowest temperature that the binder can tolerate before thermally induced cracking develops.

3.4. AFM Nano-Scale Testing

AFM is a non-destructive effective imaging tool that was developed by Binning et al. [39] for imaging and characterizing a variety of surfaces at atomic or molecular levels [40]. AFM has several advantages over conventional microscopy techniques [41]. Different forces can be measured using AFM, such as, mechanical, friction, contact, van der Waals, capillary, chemical bonding, magnetic, and electrostatic [42]. In this study, an Agilent 5500 LS atomic force microscope manufactured by Agilent (Santa Clara, CA, USA) was used to test the adhesive/cohesive forces and agglomeration properties of the prepared control and rPETE-modified asphalt binder samples. PicoView 1.20 computer software was used in the AFM operation steps and data extraction, including controlling the AFM stage, setting the scanning parameters, displaying the force–distance (force spectroscopy) curves, and converting the scanning data to images.

3.4.1. Sample Preparation

The procedure of the AFM sample preparation was conducted by pouring the liquid asphalt binders on a glass microslide. The same binders used in the macro-scale rheological testing were used for preparing the AFM samples to ensure reliability in the macro- and micro-scale testing results. The method of the sample preparation used in this study is a modified version of the method in [43], as explained in [44]. This method was used in this study to evaluate the adhesive and cohesive forces as well as the healing properties of the control and rPETE-modified binders. In this procedure, a predetermined amount of asphalt binder (96 mg) was placed on a glass microscope slide at a specific desired thickness (0.4 ± 0.02 mm). Controlling the thickness of asphalt binder was achieved by placing the predetermined amount of the binder between four strips of a high-temperature-resistance tape; the liquid binder sample was then heated at 154 °C and allowed to spread in the predefined area (1.6 × 1.5 cm^2). Figure 1 shows a typical asphalt binder sample used in AFM testing.

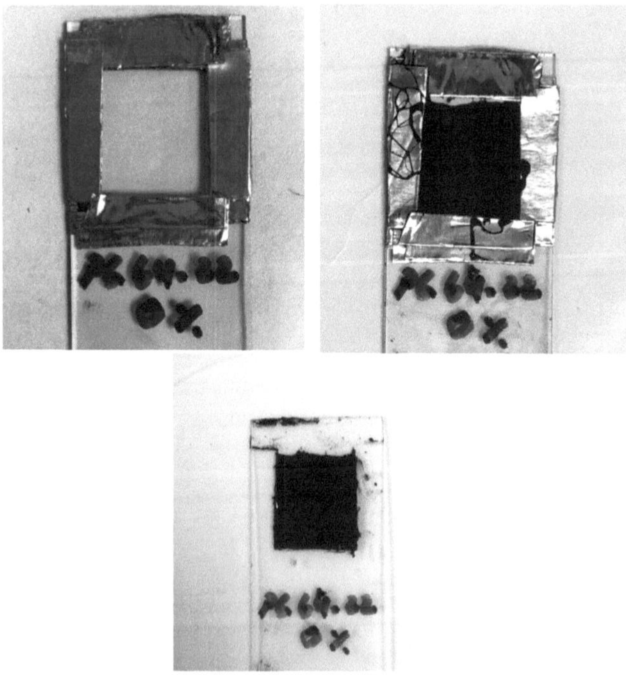

Figure 1. AFM sample preparation steps.

3.4.2. AFM Techniques

The tapping mode was used in obtaining surface images of the samples. The tapping mode is a dynamic AFM mode that is referred to as an intermittent contact mode and an AC mode. In this mode, a piezoelectric element connected to the tip holder assembly allows the tip to oscillate at its resonance frequency. Surface topographical and phase images were obtained for the binder samples using the AFM tapping mode to evaluate the microstructure, viscoelastic properties, and mechanical properties of the binder. In addition, the surface roughness and flatness of the binder samples were evaluated by taking images in the tapping mode before measuring the cohesive and adhesive forces in the binder using the AFM contact mode.

The force spectroscopy AFM technique was used at a constant temperature of 25 °C to evaluate the interaction between the rPETE and asphalt binder in the rPETE-modified binders by measuring the adhesive/cohesive forces. In this mode, the cantilever-tip assembly approaches and retracts from the sample surface by moving a half-cycle downward and upward in the *z*-direction, respectively. Figure 2 presents a typical force–distance curve generated from a single indentation experiment. The curve consists of two steps: the approaching step and the retracting step. In the first step, as the tip approaches closer to the surface, a slight increase in the force between the tip and the sample starts to appear; then, a remarkable increase in cantilever deflections occurs when the tip contacts the surface. The tip then continues to penetrate the sample until reaching the preselected indentation depth. In the second step, the tip starts retracting from the sample surface until it completely separates when it overcomes the adhesion force between itself and the asphalt sample surface.

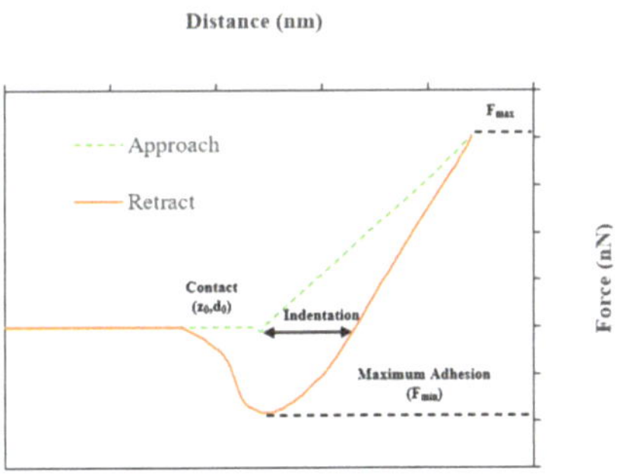

Figure 2. Typical force–distance curve generated from AFM force spectroscopy mode testing.

The reduced elastic modulus of the binder samples as well as the bonding energy (the total energy required for tip separation) were calculated by analyzing the force–distance curves resulting from the force spectroscopy indentation. The calculation of the reduced elastic modulus ($E_{reduced}$) was conducted using Equation (1), which was proposed by Fischer-Cripps [45] and is based on Sneddon's modification of the Hertzian model for a stiff-tip indentation in a flat, soft sample.

$$E_{reduced} = \frac{\pi}{2} \frac{F}{\delta^2 \tan \alpha} \quad (1)$$

where

F: measured force (nN); δ: indentation depth (μm) = $z - d$; α: half-opening angle of the tip (degrees); d: cantilever deflection (μm); and z: piezo-driver displacement.

Equation (2) was also used to estimate the total bonding energy needed for tip separation ($E_{bonding}$) by calculating the area of the negative forces under the force–distance curve in the retraction region [46].

$$E_{bonding} = \int_{z_0}^{z_1} F dz \approx \frac{\Delta z}{2N} \sum_{i=1}^{N} [F(z_{i+1}) + F(z_i)] \quad (2)$$

4. Results and Analysis

4.1. Results of Rheological Testing

4.1.1. Results of DSR Testing

Rutting is one type of distress that occurs in pavement at high temperatures owing to heavy traffic loads. The parameter $G^*/\text{Sin}\delta$ corresponds to the ability for the binder to resist rutting. The rutting parameter is a stress-controlled parameter. Maximizing the value of $G^*/\text{Sin}\delta$ for the asphalt binder must be achieved to minimize each cycle's dissipated energy and reduce the distress potential. The Superpave system assigned the limits for the rutting parameter to minimum values of 1 kPa for the unaged binder and 2.2 kPa for the RTFO-aged binder. Figure 3 presents the plot for the rutting parameter versus the testing temperature for the unmodified binder and rPETE-modified binders obtained from DSR testing of unaged binder samples. As the test temperature increased, the rutting parameter value decreased exponentially at a high rate at low temperatures and at a lower rate at higher temperatures. The rutting parameter remarkably increased when adding the rPETE modifier to the asphalt binder, indicating an improved rutting resistance for the rPETE-modified binder. In addition, the value of the rutting parameter increased with increasing content of rPETE inside the binder up to 10% and then started to decrease when 12% rPETE content was added to the binder, indicating an excessive amount of rPETE, which resulted in lower resistance to permanent deformation (G^*) and higher values of the phase angle. This suggests that adding 10% rPETE content can be considered as the best in improving the rutting resistance of the asphalt binder.

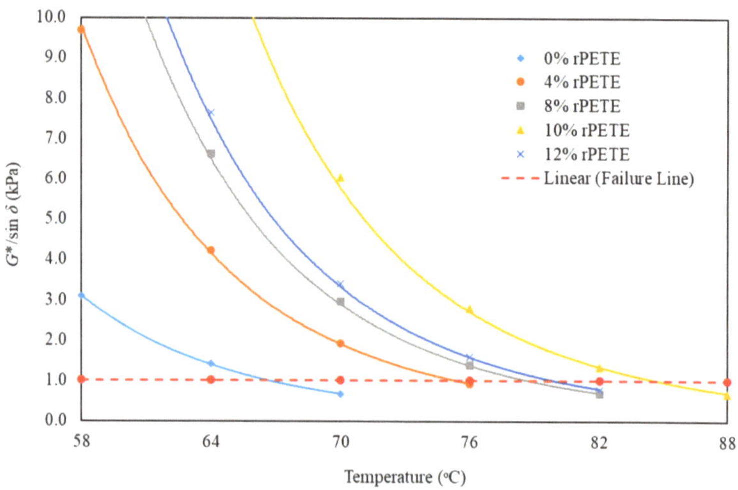

Figure 3. Results for rutting parameter versus temperature for unaged asphalt binders.

Fatigue cracking in thin pavements is considered as a strain-controlled parameter, and it can be minimized by decreasing the dissipated energy per load cycle. Table 3 presents the results of the failure temperature for the control binder and rPETE-modified binder with different rPETE contents, as obtained from DSR testing of unaged samples of these binders. As seen in Table 3, the addition of the rPETE modifier to the asphalt binder raised the failure temperature compared with that of the unmodified (control) asphalt binder. The addition of 10% rPETE to the binder resulted in the best enhancement effect on the unaged binder's rheology, with a resulting failure temperature of 84.53 °C, which enables the PG high-temperature grade of the unaged binder to be raised from 64 to 82 when adding 10% rPETE. Adding 8% of the rPETE modifier to the asphalt binder resulted in increasing the failure temperatures to values above 76 °C, which enables the PG grade to be raised to 76 instead of 64. The binders containing 4% or 12% of the rPETE modifier exhibited the lowest failure temperatures among the rPETE-modified binders, but they were still higher than the value obtained for the unmodified binder, and the PG high-temperature grade for the 4% rPETE binder or 12% rPETE binder could be raised from 64 to 70.

Table 3. Results for DSR testing on control and rPETE-modified unaged binders.

rPETE Content (%)	Continuous High-Temperature PG Grade (°C)
0%	66.70
4%	75.27
8%	78.73
10%	84.53
12%	79.83

Figure 4 presents the plot of the fatigue parameter versus the test temperature for the unmodified binder and rPETE-modified binders obtained from DSR testing of PAV-aged binder samples. As test temperature increased, the $G^* \cdot \text{Sin}\delta$ value decreased exponentially at a higher rate at low temperatures and at a lower rate at high temperatures. The fatigue parameter increased with increasing content of the rPETE inside the binder up to 10% and then started to decrease when a 12% rPETE content was added to the binder, indicating an excessive amount of rPETE, which resulted in lower resistance to permanent deformation (G*) and higher values of the phase angle.

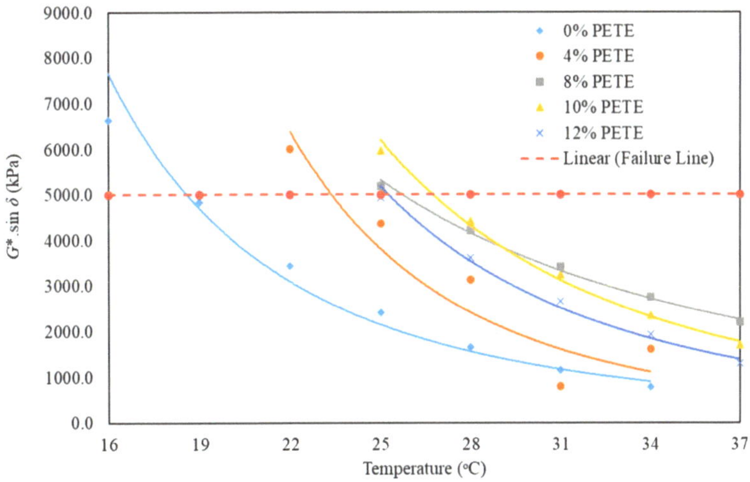

Figure 4. Results of fatigue parameter versus temperature for PAV-aged asphalt binders.

4.1.2. Results of BBR Testing

Table 4 presents the results of the BBR testing for the PAV-aged samples obtained from the control and rPETE-modified binders. The values of the creep stiffness (S_{60}) and *m-value* at 60 s were obtained for each binder at −6 °C and −12 °C to calculate the continuous lower grade. As seen in the table and both figures, the creep stiffness of the asphalt binder increased exponentially with increasing rPETE content at both testing temperatures. However, the values were still lower than the 300 MPa Superpave limit. On the other hand, the m-value decreased exponentially with increasing rPETE content at both temperatures, which indicates that adding higher contents of rPETE to the asphalt binder will have a negative effect on its stiffness. All the rPETE-modified binders except the 4% rPETE-modified binder failed at the −12 °C testing temperature because the *m-value* for these binders were lower than the lower limit (0.3) at that temperature. Therefore, the continuous lower-temperature grade was obtained for each binder to determine the failure temperature for each binder (Table 4). The binder with the 4% rPETE content had the coldest temperature among all the rPETE-modified binders, with a performance grade value of −22 °C. This suggests that adding this content of rPETE to the asphalt binder can result in a slightly lower resistance to low-temperature failure compared to the control binder. Asphalt binders modified with 8% or 10% rPETE contents showed almost similar lower grade values (failure temperatures), with a slightly higher value for the 10% rPETE-modified binder. Adding 12% rPETE to the binder had a remarkably negative impact on the low-temperature properties of the modified binder because it had a very warm failure temperature compared with that of the control binder. Table 4 also presents the results of the ΔT_c parameter, which corresponds to the relaxation properties of the asphalt binder resulting from non-load-related cracking or aging-related embrittlement distresses in the asphalt pavement. The ΔT_c parameter basically defines the relationship between the stiffness and the relaxation of the asphalt binder. Positive ΔT_c values indicate a stiffness-controlled asphalt binder, whereas negative ΔT_c values indicate an m-value-controlled asphalt binder [47]. Equation (3) was used to calculate the values of the ΔT_c parameter for the control and rPETE-modified asphalt binders using the results of the stiffness and m-value obtained from BBR testing.

$$\Delta T_c = T_{c,S} - T_{c,m} \tag{3}$$

where:

$T_{c,S}$: The critical low temperature of the asphalt binder at S_{60} is exactly equal to the specification value of 300 MPa;
$T_{c,m}$: The critical low temperature of the asphalt binder at the *m-value* is exactly equal to the specification value of 0.3.

Table 4. Results of BBR testing on control and rPETE-modified PAV-aged binders.

rPETE Content (%)	T (°C)	S_{60} (MPa)	*m-Value*	Pass/Fail (*m-Value* ≥ 0.3 and $S \le 300$ MPa)	Stiffness Continuous Grade (°C)	*m-Value* Continuous Grade (°C)	ΔT_c (°C)
0	−6	68.7	0.36	Pass	−30.60	−22.28	−10.32
	−12	163.7	0.303	Pass			
4	−6	79.4	0.337	Pass	−30.75	−22.1	−8.65
	−12	169.1	0.301	Pass			
8	−6	81.6	0.328	Pass	−30.73	−19.86	−10.87
	−12	172.4	0.284	Fail			
10	−6	84.1	0.327	Pass	−28.48	−19.64	−8.84
	−12	187.9	0.283	Fail			
12	−6	85.1	0.314	Pass	−26.85	−17.57	−8.46
	−12	204.7	0.26	Fail			

As seen in Table 4, the value of the ΔT_c parameter was negative for all the asphalt binders tested using BBR; this indicates that all the binders are m-value-controlled asphalt binders.

4.1.3. Results of ABCD Testing

Figure 5 presents the results of the cracking temperatures obtained from ABCD testing conducted on the control and rPETE-modified binders. As seen in Figure 5, the addition of the rPETE modifier to the asphalt binder did not lead to an increase in resistance to low temperatures for the modified binder, and the cracking temperature increased with increasing rPETE content inside the binder. These results agree with the results obtained from the BBR test. Table 5 presents the cracking temperatures and low-temperature PG grades obtained from the ABCD tests. It is noted that even though the PG grade decreased with increasing rPETE content inside the binder, the noted failure temperature was colder than that obtained from BBR testing. This may be attributed to the limitation of the BBR to evaluate the inclusion of the rPETE inside the binder because it measures the stiffness of the binder without evaluating the fracture (cracking) performance. ABCD tests can be used to investigate the fracture performance of the asphalt binder at low temperatures, which makes them more suitable for obtaining binder properties at these low temperatures. It was reported in [47] that the stiffness and relaxation parameters alone do not fully characterize the cracking resistance of asphalt binders even though they are related to the cracking resistance of binders to some extent. In addition, the failure properties of the strain tolerance of complex and modified asphalt binders cannot be fundamentally and fully understood with measurements in the linear viscoelastic (LVE) range without inducing damage to the binder sample. Therefore, a new parameter, ΔT_f, (the binder failure index) was proposed in the new approach to combine the temperature at the critical stiffness (T_c at S = 300 mPa), as obtained from BBR testing, and the ABCD cracking temperature, T_{cr}, as shown in Equation (4). "A positive ΔT_f suggests that a binder used in a surface layer similarly aged and subjected to ABCD cooling rates would crack at a temperature below the temperature corresponding to 300 MPa ($T_c(S)$) and vice-versa" [48].

$$\Delta T_f = T_c(S) - T_{cr} \tag{4}$$

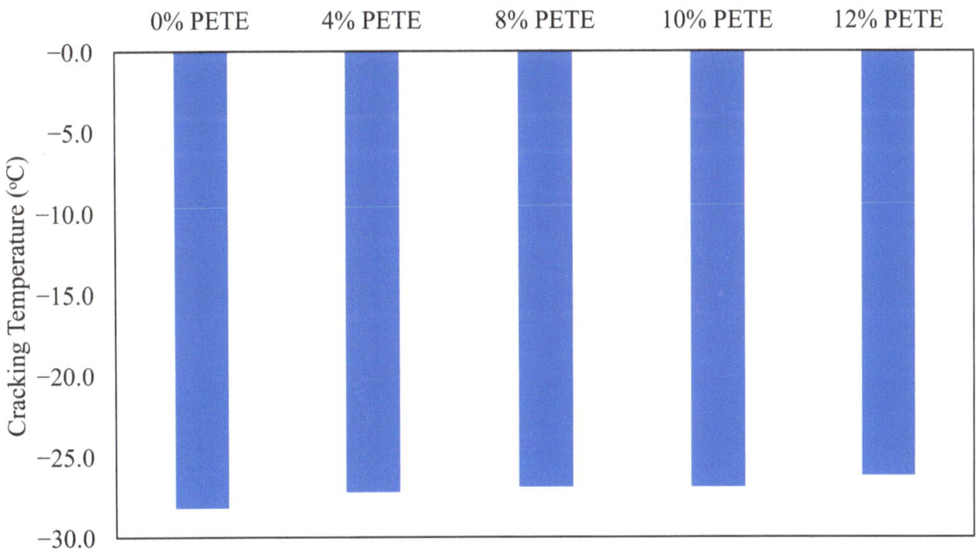

Figure 5. Results for cracking temperature versus rPETE content for PAV-aged asphalt binders.

Table 5. Results for ABCD testing on control and rPETE-modified PAV-aged binders.

rPETE Content (%)	ABCD Cracking Temperature (°C)	ABCD PG Grade (°C)
0	−28.2	−22.9
4	−27.2	−22.1
8	−26.9	−21.8
10	−26.9	−21.8
12	−26.2	−21.3

Table 6 presents the results of the binder failure indices for the unmodified and rPETE-modified binders, as calculated from the results of the BBR and ABCD testing. As seen in Table 6, the asphalt binders modified with higher contents of rPETE (10% and 12%) showed higher (better) ΔT_f values than the control binder and binders modified with lower rPETE contents.

Table 6. Results of binder failure indices calculated from BBR and ABCD testing results.

rPETE Content (%)	BBR Temperature at Critical Stiffness (S = 300 MPa) (°C)	ABCD Cracking Temperature (°C)	Binder Failure Index (°C)
0	−30.60	−28.20	−2.40
4	−30.75	−27.20	−3.55
8	−30.73	−26.90	−3.83
10	−28.48	−26.90	−1.58
12	−26.85	−26.20	−0.65

4.1.4. Summary of Asphalt Binder Performance-Grading Tests

Based on the results of the DSR, BBR, and ABCD tests discussed in previous sections, the performance grades for the unmodified and rPETE-modified binders could be determined. A summary of the rheological testing results and the corresponding PG grades is shown in Table 7. It can be seen in Table 7 that the addition of rPETE to the asphalt binder resulted in increasing the high-temperature PG grade compared with that of the control binder and decreasing the low-temperature maximum PG grade by one unit (−6 °C), except for the binder modified with a 4% rPETE content, which showed a minimal negative effect on the low-temperature PG grade of the modified binder that had a value similar to that of the control binder (−22 °C). These results for the 4% rPETE content can be considered as being better than the improvement resulting by adding other additives to the asphalt binder, such as fibers, Sasobit or other waxes, or some nano-materials. The improvement in the asphalt binder's rheology by adding 4% rPETE can also be compared with the improvement resulting from adding polymers to the binder, especially by adding the SBS polymer, which improves the high- and intermediate-temperature performances of the binder without reducing the low-temperature performance.

Table 7. Asphalt binder performance grades for unmodified binder and rPETE-modified binders.

Parameter	Control Binder	4% rPETE	8% rPETE	10% rPETE	12% rPETE	Criteria
DSR Temperature for Unaged Binder (°C)	64	70	76	82	76	$G^*/\sin\delta \geq 1.0$ kPa
DSR Temperature for PAV Residue (°C)	25	28	34	31	31	$G^* \cdot \sin\delta \leq 5000$ kPa

Table 7. *Cont.*

Parameter	Control Binder	4% rPETE	8% rPETE	10% rPETE	12% rPETE	Criteria
DSR PG Continuous Grade (°C)	66.70	75.27	78.73	84.53	79.83	
BBR PG Continuous Grade (°C)	−22.28	−22.1	−19.86	−19.64	−17.57	$S \leq 300$ MPa *m-value* ≥ 0.3
BBR PG Grade for PAV Residue (°C)	−22	−22	−16	−16	−16	
ABCD PG Continuous Grade (°C)	−22.9	−22.1	−21.8	−21.8	−21.3	
PG Grade	PG 64-22	PG 70-22	PG 76-16	PG 82-16	PG 70-16	

4.2. Results for AFM Nano-Scale Testing

4.2.1. Results for AFM Tapping Mode

Figures 6–10 present the topographical and phase images of the PG 64-22 control and rPETE-modified binders. It can be seen that some nonhomogeneity exists within the asphalt matrix, as represented by the revealed dispersed black and white successions of elongated structures within the relatively flat matrix. These black and white successions are called "bee-like" structures and have been identified in many previous studies [49–54]. They were attributed, in many studies, to the asphaltene content in the asphalt binder [50,55]. The topographical and phase images for the rPETE-modified asphalt binder samples indicate that the inclusion of rPETE in the asphalt binder reduced the size of the bee-like structures to appear in long chains with a much smaller width than those in the control binder. This may be attributed to the obstruction in the movement of the asphaltene molecule chains owing to the presence of resin thermoplastic polymer molecules of rPETE. As seen from the phase images of the control asphalt binder (Figure 6) and the rPETE-modified binder (Figures 7–10), an increase in the phase contrast between the dispersed domains and the flat matrix was noted with the addition of rPETE. Based on that, stiffer dispersed domains will result after adding rPETE, which may result in improving the stiffness properties of the rPETE-modified asphalt as compared with those of the control asphalt binder.

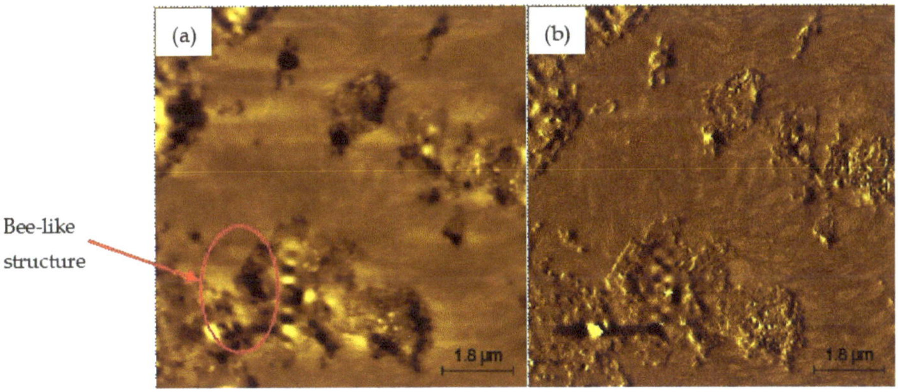

Figure 6. AFM images of PG 64-22 control binder: (**a**) topographical image; (**b**) phase image.

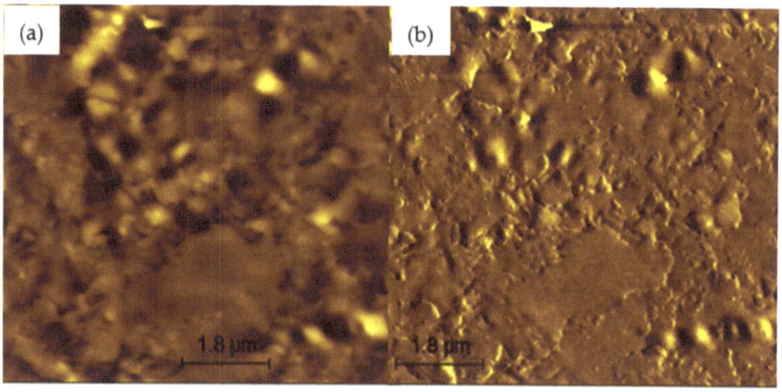

Figure 7. AFM images of 4% rPETE-modified binder: (**a**) topographical image; (**b**) phase image.

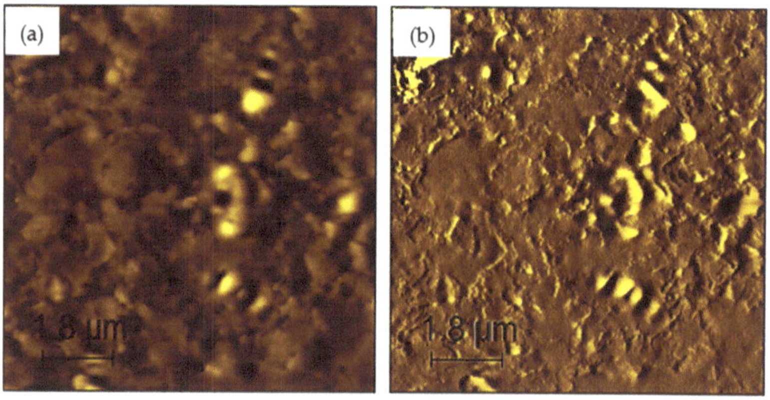

Figure 8. AFM images of 8% rPETE-modified binder: (**a**) topographical image; (**b**) phase image.

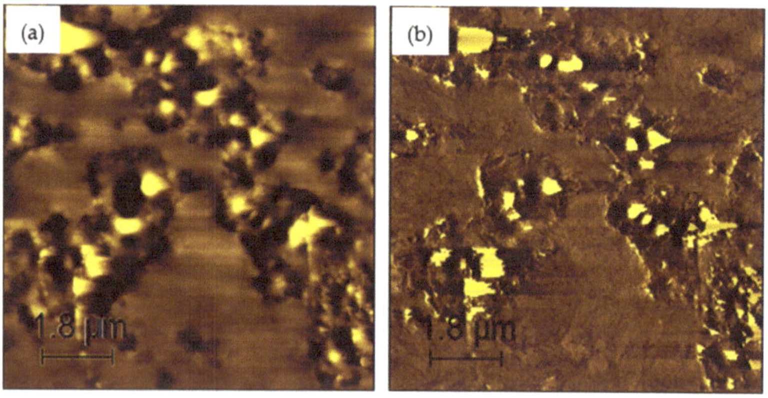

Figure 9. AFM images of 10% rPETE-modified binder: (**a**) topographical image; (**b**) phase image.

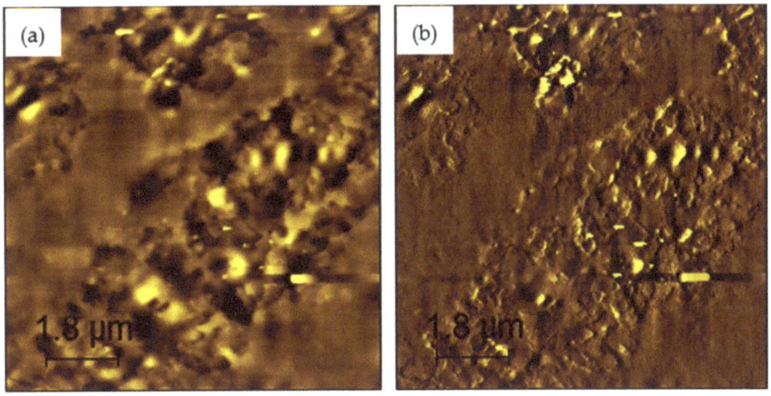

Figure 10. AFM images of 12% rPETE-modified binder: (**a**) topographical image; (**b**) phase image.

The surface roughness of the control and rPETE-modified binders was also evaluated by conducting a roughness analysis on the obtained topographical images of each binder sample using WSxM version 5.0 software [56]. In this analysis, the average roughness was estimated by calculating the absolute mean of the difference between the average height and the height of each single point of the sample. Figure 11 presents the average roughness values for the PG 64-22 control and rPETE-modified binders. As seen in Figure 11, the rPETE-modified binder samples had much higher roughness values compared with those of the control binder sample, and the roughness value increased with increasing rPETE content. This may be attributed to the increased number of bee-like structures in the modified binders, which have long chains with a smaller width.

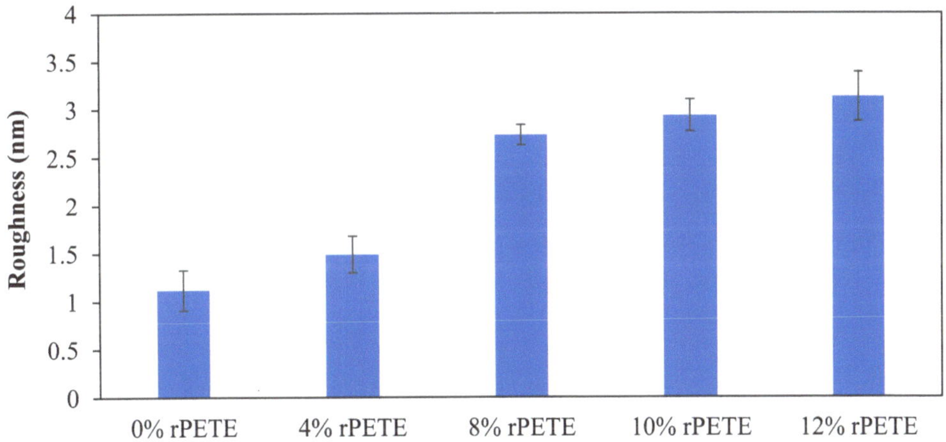

Figure 11. Results of surface roughness for control and rPETE-modified binders.

4.2.2. Results of AFM Force Spectroscopy Mode

Figure 12 presents typical force–distance curves for the different binder types considered in this study. Figure 13 presents the results of the reduced modulus ($E_{reduced}$) for the PG 64-22 control and rPETE-modified binders. The reduced modulus values for the rPETE-modified binders were remarkably higher than that of the control binder; this may be attributed to the short-term aging resulting from the mixing of the rPETE with the binder at a relatively high temperature. Also, this may be attributed to the properties of the rPETE material, which improved the stiffness of the binder. The binder with the 8% rPETE content showed the highest reduced modulus value among all the binders, followed by the binder with 4% rPETE. This means an excessive level of stiffness is achieved by the binder, especially when adding 8% rPETE. The asphalt binders modified with 10% or 12% rPETE contents showed higher values of the reduced modulus than the control binder. However, these values were considered as being reasonable because the modified binders did not exhibit higher levels of stiffness that might affect their adhesion properties.

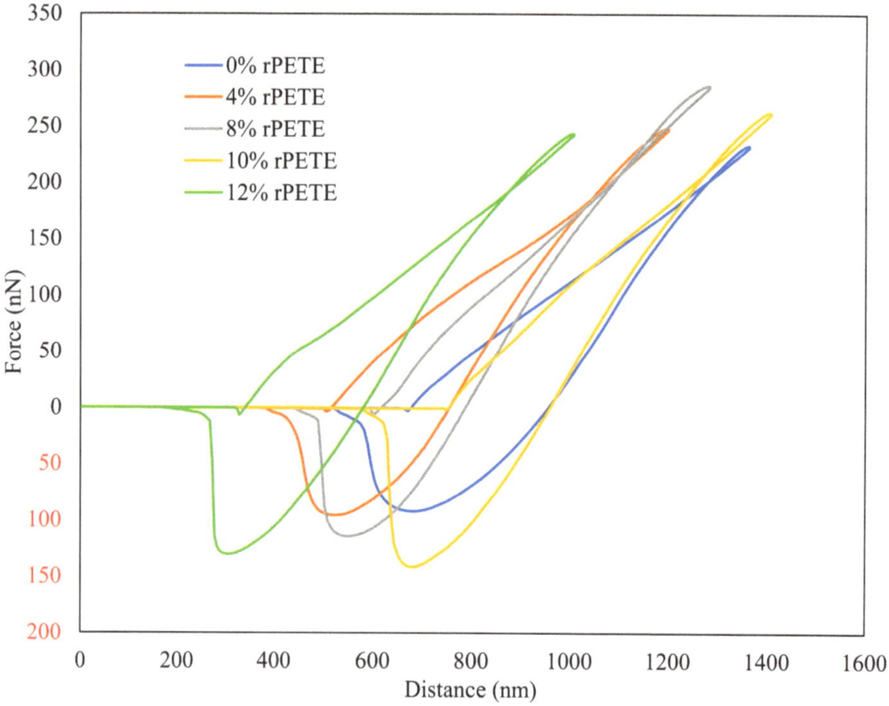

Figure 12. Typical force–distance curves for control and rPETE-modified asphalt binders. Red numbers represent negative values.

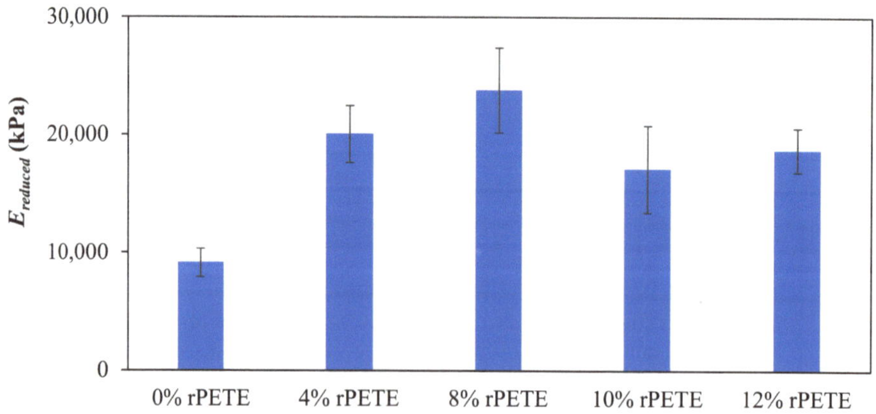

Figure 13. Results of reduced modulus for control and rPETE-modified binders.

Figure 14 presents the results of the bonding energy for the PG 64-22 control and rPETE-modified binders. In general, the addition of rPETE to the asphalt binder resulted in raising the value of its bonding energy compared with that of the control binder. This improvement in the bonding energy increased with the rPETE content increasing. However, the binder with the 10% rPETE content exhibited the highest bonding energy. This indicates improved adhesion properties between the binder and the rPETE material because the AFM tip needed more energy to pull out from the sample surface after retraction, especially when the 10% rPETE content was used. Increasing the rPETE content to more than 10% will have an adverse effect, as an excessive amount of rPETE will reduce the adhesion between the tip and the binder surface, resulting in lower bonding energy values. When combining the results obtained for the reduced modulus and the bonding energy for this binder, it can be seen that adding this content to the binder resulted in excessive values of stiffness, as represented by the reduced modulus and reduced adhesion between the binder and the rPETE. Finally, the content of 10% rPETE was selected to represent the best performance in regard to the adhesive properties, stiffness, and roughness.

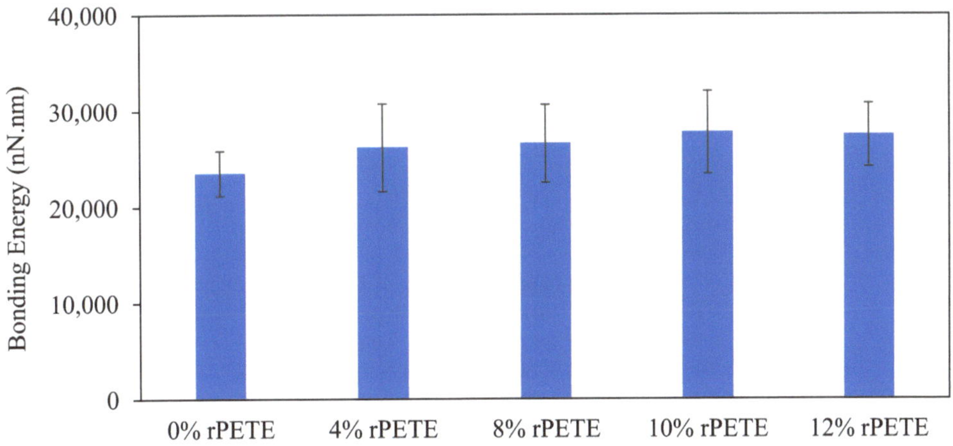

Figure 14. Results for bonding energy for control and rPETE-modified binders.

4.2.3. Statistical Analysis of the Results for AFM Testing

The statistical evaluation of the results for the micro-scale AFM testing and the significance of the effects of the rPETE addition on the performances of the modified binder and the binder–rPETE adhesion was conducted by applying analysis of variance (ANOVA) and post-ANOVA least square means (LSMs) using Statistical Package for the Social Sciences (SPSS)®. A linear completely random design (CRD) model was used in these analyses.

The results of the ANOVA on the results of the AFM bonding energy, reduced modulus, and average roughness parameters for the unaged and RTFO-aged control and rPETE-modified binders are presented in Table 8. It is noted that the p-values in Table 8 were less than 0.05 at the 95% confidence level, which indicates a statistically significant effect for the rPETE content on the results of the AFM parameters. Table 9 presents the results for the ranking of the asphalt binders based on the results of the bonding energy, reduced modulus, and roughness, as determined by the post-ANOVA LSM analysis. In Table 9, a descending-order listing was given to the groups, with the letter "A" contributing to the highest mean, followed by the other letters in the appropriate order. As seen in Table 9, the 10% rPETE content had the highest value of the bonding energy compared to the other binders, followed by the 12% rPETE content. On the other hand, the 8% rPETE content had the highest value of the reduced modulus compared to the other binders, followed by the 4% PETE content. The 12% rPETE content had the highest roughness value compared to the other binders, followed by the 10% rPETE content. However, the control binder with no rPETE addition had statistically the lowest values of the bonding energy, reduced modulus, and average roughness among all the binders.

Table 8. Results of ANOVA on AFM-testing of unmodified and rPETE-modified asphalt binders.

Type 3 Tests of Fixed Effects					
Effect	**Parameter**	**Num DF**	**Den DF**	**F-Value**	***p*-Value**
rPETE Content	Bonding Energy	4	50	2.392	0.064
	$E_{reduced}$	4	45	29.185	<0.0001
	Roughness	4	14	67.734	<0.0001

Table 9. Results of post-ANOVA analysis on AFM testing of unmodified and rPETE-modified asphalt binders.

rPETE Content	Bonding Energy	Letter Group	rPETE Content	$E_{reduced}$	Letter Group	rPETE Content	Roughness	Letter Group
10%	27,749.7	A	8%	23,799.2	A	12%	3.127	A
12%	27,483.2	A	4%	20,067.1	AB	10%	2.932	A
8%	26,609.3	AB	12%	18,707.5	B	8%	2.733	A
4%	26,211.2	B	10%	17,104.7	B	4%	1.489	B
0%	23,513.5	C	0%	9022.6	C	0%	1.119	B

5. Conclusions

This paper documents the results of a research study that included the preparation of rPETE-modified asphalt binders to evaluate the effects of adding rPETE in different contents on the high-temperature, intermediate-temperature, and low-temperature performances of the modified binders, using rheological macro-scale testing, as well as on the interaction and bonding properties between the asphalt binder and rPETE at the nano-scale, using AFM testing, and to compare them with those of the unmodified binder. The following conclusions can be drawn based on the results of this study:

1. The addition of the rPETE enhanced the high- and intermediate-temperature rheological properties and slightly reduced the low-temperature rheological properties of the modified PG 64-22 binder;
2. The results from the AFM testing indicated that the inclusion of rPETE in the asphalt binder improved the stiffness properties of the modified binder as compared with those of the control asphalt binder. In addition, the rPETE-modified binders showed rougher surfaces and remarkably higher values of the reduced modulus than the control binder;
3. As seen from the results of the AFM bonding energy, the rPETE-modified binders showed higher bonding energy values compared with those of the control binder. This improvement in the bonding energy increased with increasing rPETE content. The addition of the 10% rPETE content resulted in the highest bonding energy among all the binders;
4. The mixture with the 10% rPETE content resulted in the best high- and intermediate-temperature binder performances. Although the 4% rPETE binder showed improved high- and intermediate-temperature properties without reducing the low-temperature grade of the modified binder, 4% can be considered as the optimal content of rPETE as a modifier for the asphalt binder.

Author Contributions: Conceptualization, M.D.N.; Methodology, A.A.-H., M.D.N. and S.K.; Formal analysis, A.A.-H. and S.K.; Investigation, M.D.N. and A.A-H..; Data curation, A.A.-H. and T.R.; Writing—original draft, A.A.-H. and M.D.N.; Writing—review & editing, A.A.-H. and M.D.N.; Supervision, M.D.N. All authors have read and agreed to the published version of the manuscript.

Funding: This research received no external funding.

Data Availability Statement: All the data, models, and code generated or used during the study appear in the submitted article.

Conflicts of Interest: Author Ahmad Al-Hosainat was employed by the company Terracon Consultants, Inc. The remaining authors declare that the research was conducted in the absence of any commercial or financial relationships that could be construed as a potential conflict of interest.

References

1. Shahnawaz, M.; Sangale, M.K.; Ade, A.B. Plastic Waste Disposal and Reuse of Plastic Waste. In *Bioremediation Technology for Plastic Waste*; Springer: Singapore, 2019. [CrossRef]
2. Jamshidi, A.; Hamzah, M.O.; Aman, M.Y. Effects of Sasobit® content on the rheological characteristics of unaged and aged asphalt binders at high and intermediate temperatures. *Mater. Res.* **2012**, *15*, 628–638. [CrossRef]
3. Kök, B.V.; Yılmaz, M.; Akpolat, M. Evaluation of the conventional and rheological properties of SBS+Sasobit modified binder. *Constr. Build. Mater.* **2014**, *63*, 174–179. [CrossRef]
4. Tarefder, R.A.; Yousefi, S.S. Rheological Examination of Aging in Polymer-Modified Asphalt. *J. Mater. Civ. Eng.* **2016**, *28*, 04015112. [CrossRef]
5. Ghuzlan, K.A.; Al Assi, M.O. Sasobit-Modified Asphalt Binder Rheology. *J. Mater. Civ. Eng.* **2017**, *29*, 04017142. [CrossRef]
6. Al-Omari, A.; Taamneh, M.; Khasawneh, M.A.; Al-Hosainat, A. Effect of crumb tire rubber, microcrystalline synthetic wax, and nano silica on asphalt rheology. *Road Mater. Pavement Des.* **2020**, *21*, 757–779. [CrossRef]
7. Abuaddous, M.; Taamneh, M.M.; Rabab'ah, S.R. The potential use of recycled polyethylene terephthalate (RPET) plastic waste in asphalt binder. *Int. J. Pavement Res. Technol.* **2021**, *14*, 579–587. [CrossRef]
8. Becker, Y.; Mendez, M.P.; Rodriguez, Y. Polymer modified asphalt. *Vis. Tecnol.* **2001**, *9*, 39–50. [CrossRef]
9. Behnood, A.; Gharehveran, M.M. Morphology, rheology, and physical properties of polymer-modified asphalt binders. *Eur. Polym. J.* **2019**, *112*, 766–791. [CrossRef]
10. Ma, F.; Jin, Y.; Fu, Z.; Dai, J.; Zhang, P.; Zhang, C.; Wen, Y. Influencing factors and evaluation methods of reinforcement effect of fiber-modified asphalt binder. *Polym. Compos.* **2022**, *43*, 8986–8999. [CrossRef]
11. White, G.; Reid, G. Recycled waste plastic for extending and modifying asphalt binders. In Proceedings of the 8th Symposium on Pavement Surface Characteristics (SURF 2018), Brisbane, QLD, Australia, 2–4 April 2018; pp. 2–4.
12. Ahmad, A.F.; Razali, A.R.; Razelan, I.S.M. Utilization of polyethylene terephthalate (PET) in asphalt pavement: A review. In *IOP Conference Series: Materials Science and Engineering*; IOP Publishing: Bristol, UK, 2018; Volume 203, p. 012004.
13. Khedaywi, T.; Haddad, M.; Bataineh, H. Effect of waste plastic polyethylene terephthalate on properties of asphalt cement. *Innov. Infrastruct. Solut.* **2023**, *8*, 232. [CrossRef]
14. Chen, S. An Evaluation of Waste Plastic in Asphalt Pavement towards a Circular Economy. Ph.D. Thesis, University of Missouri–Columbia, Columbia, MO, USA, 2020.
15. Wu, S.; Montalvo, L. Repurposing waste plastics into cleaner asphalt pavement materials: A critical literature review. *J. Clean. Prod.* **2021**, *280*, 124355. [CrossRef]
16. Vorobyev, D.; Borisenko, Y.; Shvachev, D.; Rudak, S. Effective Polymer-Modified Bitumen Based on PET Waste. In Proceedings of the FORM 2022: Construction the Formation of Living Environment, Moscow, Russia, 20–22 April 2022; Springer International Publishing: Cham, Switzerland, 2022; pp. 417–427.
17. Rotimi, A. Reuse of Waste Plastic as an Additive in Asphalt Concrete. *ATBU J. Sci. Technol. Educ.* **2022**, *10*, 49–59.
18. Leng, Z.; Padhan, R.K.; Sreeram, A. Production of a sustainable paving material through chemical recycling of waste PET into crumb rubber modified asphalt. *J. Clean. Prod.* **2018**, *180*, 682–688. [CrossRef]
19. Majka, T.M.; Ostrowski, K.A.; Piechaczek, M. Research on the Development of a Way to Modify Asphalt Mixtures with PET Recyclates. *Materials* **2023**, *16*, 6258. [CrossRef]
20. Kawai, F.; Kawabata, T.; Oda, M. Current State and Perspectives Related to the Polyethylene Terephthalate Hydrolases Available for Biorecycling. *ACS Sustain. Chem. Eng.* **2020**, *8*, 8894–8908. [CrossRef]
21. Bhanderi, K.K.; Joshi, J.R.; Patel, J.V. Recycling of polyethylene terephthalate (PET or PETE) plastics—An alternative to obtain value added products: A review. *J. Indian Chem. Soc.* **2023**, *100*, 100843. [CrossRef]
22. Reza, M.T.; Wirth, B.; Lüder, U.; Werner, M. Behavior of selected hydrolyzed and dehydrated products during hydrothermal carbonization of biomass. *Bioresour. Technol.* **2014**, *169*, 352–361. [CrossRef] [PubMed]
23. Saha, N.; McGaughy, K.; Reza, M.T. Elucidating hydrochar morphology and oxygen functionality change with hydrothermal treatment temperature ranging from subcritical to supercritical conditions. *J. Anal. Appl. Pyrolysis* **2020**, *152*, 104965. [CrossRef]

24. Fang, C.; Liu, P.; Yu, R.; Liu, X. Preparation process to affect stability in waste polyethylene-modified bitumen. *Constr. Build. Mater.* **2014**, *54*, 320–325. [CrossRef]
25. Mohammed, D.T.; Hussein, Z.H. Evaluation of pyrolisis PET utilization in asphalt binder. *Evaluation* **2014**, *3*, 114–121.
26. Ghuzlan, K.A.; Al-Khateeb, G.G.; Qasem, Y. Rheological Properties of Polyethylene-Modified Asphalt Binder. *Athens J. Technol. Eng.* **2015**, *2*, 75–88. [CrossRef]
27. Silva, J.d.A.A.e.; Lucena, L.C.d.F.L.; Rodrigues, J.K.G.; Carvalho, M.W.; Costa, D.B. Use of Micronized Polyethylene Terephthalate (Pet) Waste in Asphalt Binder. *Pet. Sci. Technol.* **2015**, *33*, 1508–1515. [CrossRef]
28. Ameri, M.; Nasr, D. Properties of asphalt modified with devulcanized polyethylene terephthalate. *Pet. Sci. Technol.* **2016**, *34*, 1424–1430. [CrossRef]
29. Nazzal, M.D.; Abu-Qtaish, L.; Kaya, S.; Powers, D. Using Atomic Force Microscopy to Evaluate the Nanostructure and Nanomechanics of Warm Mix Asphalt. *J. Mater. Civ. Eng.* **2015**, *27*, 04015005. [CrossRef]
30. Abu Qtaish, L.; Nazzal, M.D.; Abbas, A.; Kaya, S.; Akinbowale, S.; Arefin, M.S.; Kim, S.-S. Micromechanical and Chemical Characterization of Foamed Warm-Mix Asphalt Aging. *J. Mater. Civ. Eng.* **2018**, *30*, 04018213. [CrossRef]
31. Obaid, A.; Nazzal, M.D.; Abu Qtaish, L.; Kim, S.S.; Abbas, A.; Arefin, M.; Quasem, T. Effect of RAP Source on Cracking Resistance of Asphalt Mixtures with High RAP Contents. *J. Mater. Civ. Eng.* **2019**, *31*, 04019213. [CrossRef]
32. Nazzal, M.D.; Abu Qtaish, L.; Al-Hosainat, A.; Abu Talha, S.; Kaya, S.; Abbas, A.R. Evaluation of Moisture Damage in Asphalt Mixtures at Macro- and Nanoscales. *J. Mater. Civ. Eng.* **2021**, *33*, 04021369. [CrossRef]
33. *AASHTO T 240*; Standard Method of Test for Effect of Heat and Air on a Moving Film of Asphalt Binder (Rolling Thin-Film Oven Test). American Association of State Highway and Transportation Officials: Washington, DC, USA, 2023.
34. *AASHTO T R28*; Standard Practice for Accelerated Aging of Asphalt Binder Using a Pressurized Aging Vessel (PAV). American Association of State Highway Transportation Officials: Washington, DC, USA, 2023.
35. *AASHTO T 315-22*; Standard Method of Test for Determining the Rheological Properties of Asphalt Binder Using a Dynamic Shear Rheometer (DSR). American Association of State Highway and Transportation Officials: Washington, DC, USA, 2022.
36. *AASHTO T 313-22*; Standard Method of Test for Determining the Flexural Creep Stiffness of Asphalt Binder Using the Bending Beam Rheometer (BBR). American Association of State Highway and Transportation Officials: Washington, DC, USA, 2022.
37. *AASHTO T PP42*; Standard Practice for Determination of Low-Temperature Performance Grade (PG) of Asphalt Binders. American Association of State Highway and Transportation Officials: Washington, DC, USA, 2007.
38. *AASHTO TP 92–14*; Standard Method of Test for Determining the Cracking Temperature of Asphalt Binder Using the Asphalt Binder Cracking Device (ABCD). Standard by American Association of State and Highway Transportation Officials: Washington, DC, USA, 2014.
39. Binnig, G.; Quate, C.F.; Gerber, C. Atomic Force Microscope. *Phys. Rev. Lett.* **1986**, *56*, 930–933. [CrossRef]
40. Das, P.K.; Baaj, H.; Tighe, S.; Kringos, N. Atomic force microscopy to investigate asphalt binders: A state-of-the-art review. *Road Mater. Pavement Des.* **2016**, *17*, 693–718. [CrossRef]
41. Blanchard, C.R. Atomic Force Microscopy. *Chem. Educ.* **1996**, *1*, 1–8. [CrossRef]
42. Nazzal, M.D.; Qtaish, L.A. *The Use of Atomic Force Microscopy to Evaluate Warm Mix Asphalt (No. FHWA/OH-2012/19)*; Ohio Department of Transportation: Columbus, OH, USA, 2013.
43. Tarefder, R.A.; Zaman, A.M. Nanoscale Evaluation of Moisture Damage in Polymer Modified Asphalts. *J. Mater. Civ. Eng.* **2010**, *22*, 714–725. [CrossRef]
44. AbuQtaish, L.; Nazzal, M.D.; Kaya, S.; Kim, S.-S.; Abbas, A.; Abu Hassan, Y. AFM-Based Approach to Study Blending between RAP and Virgin Asphalt Binders. *J. Mater. Civ. Eng.* **2018**, *30*, 04017300. [CrossRef]
45. Fischer-Cripps, A.C. Critical review of analysis and interpretation of nanoindentation test data. *Surf. Coat. Technol.* **2006**, *200*, 4153–4165. [CrossRef]
46. Pauli, T.; Grimes, W.; Cookman, A.; Huang, S.-C. Adherence Energy of Asphalt Thin Films Measured by Force-Displacement Atomic Force Microscopy. *J. Mater. Civ. Eng.* **2014**, *26*, 04014089. [CrossRef]
47. Asphalt Institute Technical Advisory Committee. *Use of the Delta Tc Parameter to Characterize Asphalt Binder Behavior*; Asphalt Institute: Lexington, KY, USA, 2019; ISBN 978-1-934154-77-9.
48. Elwardany, M.; Planche, J.-P.; King, G. Universal and practical approach to evaluate asphalt binder resistance to thermally-induced surface damage. *Constr. Build. Mater.* **2020**, *255*, 119331. [CrossRef]
49. Loeber, L.; Sutton, O.; Morel, J.; Valleton, J.-M.; Muller, G. New direct observations of asphalts and asphalt binders by scanning electron microscopy and atomic force microscopy. *J. Microsc.* **1996**, *182*, 32–39. [CrossRef]
50. Jäger, A.; Lackner, R.; Eisenmenger-Sittner, C.; Blab, R. Identification of four material phases in bitumen by atomic force microscopy. *Road Mater. Pavement Des.* **2004**, *5* (Suppl. 1), 9–24. [CrossRef]
51. Masson, J.-F.; Leblond, V.; Margeson, J. Bitumen morphologies by phase-detection atomic force microscopy. *J. Microsc.* **2006**, *221*, 17–29. [CrossRef] [PubMed]
52. Kringos, N.; Schmets, A.; Scarpas, A.; Pauli, T. Towards an understanding of the self-healing capacity of asphaltic mixtures. *Heron* **2011**, *56*, 45–74.
53. Nazzal, M.D.; Kaya, S.; Gunay, T.; Ahmedzade, P. Fundamental Characterization of Asphalt Clay Nanocomposites. *J. Nanomech. Micromech.* **2013**, *3*, 1–8. [CrossRef]

54. Holcombe, E.W.; Nazzal, M.D.; Mogawer, W.; Austerman, A.J.; Kaya, S. Evaluating Asphalt Binders Prepared with Different Processes to Meet the Same Performance Grade: Use of Atomic Force Microscope. *Transp. Res. Rec.* **2017**, *2632*, 99–109. [CrossRef]
55. Loeber, L.; Muller, G.; Morel, J.; Sutton, O. Bitumen in colloid science: A chemical, structural and rheological approach. *Fuel* **1998**, *77*, 1443–1450. [CrossRef]
56. Horcas, I.; Fernández, R.; Gómez-Rodríguez, J.M.; Colchero, J.; Gómez-Herrero, J.; Baro, A.M. WSXM: A software for scanning probe microscopy and a tool for nanotechnology. *Rev. Sci. Instrum.* **2007**, *78*, 013705. [CrossRef] [PubMed]

buildings

MDPI

Article

Performance Evaluation of a Multifunctional Road Marking Coating for Tunnels Based on Nano SiO_2 and TiO_2 Modifications

Xiujie Quan [1], Liang Yang [2], Hui Li [1], Yan Chen [1] and Shuang Shi [1,*]

1 School of Transportation, Southeast University, Nanjing 211189, China; 230228856@seu.edu.cn (X.Q.); huili94@zju.edu.cn (H.L.); 220223120@seu.edu.cn (Y.C.)
2 Suqian Highway Business Development Center, Suqian 223800, China; a987shi@yahoo.com.cn
* Correspondence: shishuang@seu.edu.cn

Abstract: Multifunctional road marking coatings with the functions of high-temperature stability, degradation of exhaust gas, and self-cleaning are of great significance for the safe operation and environmental protection of tunnels. This article uses active acrylic resin and an organosilicon hydrophobic agent as the base material, selects expanded vermiculite and glass microspheres as insulation fillers, and uses ammonium polyphosphate, pentaerythritol, melamine, and aluminum hydroxide as high-thermal-stability systems to prepare a two-component road marking coating base material. Then, nano SiO_2 and modified nano TiO_2 are added as modifiers to prepare a multifunctional road marking coating for tunnels. The physical and chemical properties of multifunctional road marking coatings are evaluating based on laboratory tests including thermogravimetry and derivative thermogravimetry, differential scanning calorimetry, infrared spectroscopy, scanning electron microscopy, exhaust degradation, and contact angle tests. The results indicate that the developed multifunctional road marking coating effectively reduces the thermal conductivity of the carbon layer through physical changes in the flame retardant system and the heat resistance formed by the high breaking bond energy of nano SiO_2 during the combustion process. It forms a ceramic-like structure of titanium pyrophosphate with nano TiO_2 that is beneficial for improving flame retardancy without generating harmful volatile gases and has good flame retardant properties. N–V co-doping reduces the bandgap of TiO_2, broadens the absorption range of visible light by nano TiO_2, improves the catalytic efficiency of visible light, and achieves the degradation efficiency of the four harmful components NO_x, HC, CO, and CO_2 in automotive exhaust by 23.4%, 8.3%, 2.5%, and 2.9%, respectively. The solid–liquid phase separation in the multifunctional road marking coating in the tunnel causes the formation and accumulation of nano SiO_2 and TiO_2 particles on the coating surface, resulting in a microstructure similar to the "micro–nano micro-convex" on the lotus leaf surface and making a water droplet contact angle of 134.2° on the coating surface.

Keywords: nano SiO_2; nano TiO_2; road marking; coating; performance

Citation: Quan, X.; Yang, L.; Li, H.; Chen, Y.; Shi, S. Performance Evaluation of a Multifunctional Road Marking Coating for Tunnels Based on Nano SiO_2 and TiO_2 Modifications. *Buildings* **2024**, *14*, 459. https://doi.org/10.3390/buildings14020459

Academic Editor: Abderrahim Boudenne

Received: 9 January 2024
Revised: 2 February 2024
Accepted: 2 February 2024
Published: 7 February 2024

1. Introduction

Road markings, as an important component of traffic safety facilities, provide drivers with guidance and management information as a carrier. They are a type of safety facility that transmits legal information to drivers and pedestrians through graphics, symbols, lines, and text to control, warn, and guide traffic. Reasonable setting of road markings can improve transportation efficiency. In addition, with the rapid development of highway construction and urban three-dimensional transportation systems in China, many long tunnels have emerged. The tunnel environment is relatively closed, and the number of vehicles has increased year by year, resulting in a sharp increase in tunnel fire accidents [1]. The road markings inside the tunnel can provide traffic safety information for extreme situations such as fires and can protect the lives of disaster victims and rescue personnel after accidents occur. At the same time, the ventilation inside the tunnel is not smooth, making it difficult for car exhaust fumes to be discharged promptly, resulting in a significantly high concentration of exhaust gas in the tunnel. The exhaust gas contains harmful substances such as NO_x, HC, CO, CO_2, seriously endangering the health of drivers and passengers. It is necessary to degrade the exhaust gas in the tunnel. In addition, the dust and stains accumulated on the road surface inside the tunnel for a long time are difficult to clean promptly, breeding harmful microorganisms and bacteria and affecting the visibility of road markings. Therefore, multifunctional road marking coating with the functions of high-temperature stability, degradation of exhaust gas, and self-cleaning is of great significance for the safe operation and environmental protection of tunnels.

In recent years, an increasing number of experimental studies have been conducted on shield tunnel linings damaged by fire [2–6]. As road tunnels are in a semi-enclosed state, in the event of a fire, the smoke and heat generated by the fire cannot be dissipated within a short period, resulting in situations such as dense smoke, low visibility, and high temperature. All these are extremely unfavorable for escape and rescue [7–11]. Kim et al. [12] developed a fire-resistant coating consisting of cement, priming ash, and polypropylene fibers with sufficient strength to meet service load safety requirements while providing fire resistance, and the coating material was validated through fire testing using the RABT tunnel fire scenario. Alhawat et al. [13] studied the temperature and spalling behavior of tunnel linings by producing large tunnel lining prototypes using fireproof concrete with bulk fly ash and nano SiO_2 (HVFNS). Researchers have extensively investigated the refractory properties of bulk fly ash (HVFA) concrete and mortar using nano SiO_2 [14,15]. The results showed that the addition of nano SiO_2 to cement pastes significantly improved the thermal stability of the cement system. Nano SiO_2-added samples show less strength loss after 500 °C compared to samples with silica powder [16]. Pereyra et al. [17] chose nanostructured self-curing inorganic silicates (high silica-to-alkali ratio) as the film-forming material, and the results showed that the non-polluting coating system has excellent corrosion resistance and good high-temperature stability. Zinc phytate (ZnPA) chelated geopolymer and in situ polymerized chitosan oligosaccharide (COS)/DOPO were explored as a challenging and exciting flame-retardant organic–inorganic hybrid coating using the sol–gel method, and the results showed that ZnPA addition (0.5 wt%) enhanced the flame-retardant properties of geopolymer coatings [18]. Ng et al. [19] highlighted the thermal decomposition and fire-protective behavior of two polymer-based coatings for concrete, which creates a steep temperature gradient between the coating surface and the coating–concrete interface. Abniki et al. [20] prepared a novel flame-retardant coating material using a modification of epoxy resin with phenylboronic acid in the presence of tetra-n-butyl ammonium bromide catalyst during reaction at 80–120 °C.

Nano TiO_2, as a photocatalyst, has attracted widespread attention due to its excellent photochemical properties, good stability, low cost, and non-toxic properties. It has been widely applied and studied in areas such as air purification and sewage treatment. To further improve the absorption rate of visible light by nano TiO_2, metal ion doping and non-metal ion doping modification were commonly used to modified nano TiO_2. Metal ion modification of nano TiO_2 can cause lattice defects in TiO_2 and improve the crystallinity of the catalyst, introducing doping energy levels into the bandgap of TiO_2, extending the absorption band edge. The method of metal non-metal co-doping has been a hot research topic in recent years. The use of dual elements to dope and modify TiO_2 aims to improve the photocatalytic activity of TiO_2 through different mechanisms of synergistic effects between co-doped ions. Due to the synergistic effect of ions, not only does it improve the photocatalytic activity, but it also extends the TiO_2 absorption spectrum to the visible light region, improving its photocatalytic ability under visible light and making the catalytic effect more obvious. The composite-modified nano TiO_2^{3+}, N, and WO_3 containing Fe have good exhaust gas degradation efficiency under special lighting conditions in highway tunnels, which can promote the improvement of the driving environment in highway tunnels [21]. Because the light sources in highway tunnels do not generate ultraviolet light, nano TiO_2 cannot be used for photocatalysis. Therefore, researchers used modified nano TiO_2 to improve its photocatalytic performance in the visible light range [22–24]. Zhang et al. [25] designed an innovative photocatalytic microsurface mixture (PMM) reinforced with polypropylene (PP) fibers and nano TiO_2 to purify severe air pollution in traffic-intensive urban areas and tunnels. Xia et al. [26] [27] investigated TiO_2 aqueous coatings using photocatalytic oxidation to degrade pollutants under visible and ultraviolet (UV) light. To improve the efficiency and durability of semiconductor photocatalysts for degrading automobile exhaust fumes, Liu et al. [28] designed a highly active degradation microcell by cross-linking nano TiO_2, wherein the finely crumbled rubber surface was coated with a coupling agent.

The hydrophobicity of the coating surface is related to its surface wettability, which is mainly influenced by the surface roughness and compound composition. Improving the surface roughness and reducing the surface energy can significantly enhance the surface hydrophobicity. Hydrophobic surfaces not only have self-cleaning properties but also have waterproof, anti-corrosion, anti-adhesion, and anti-UV properties. The surface of lotus leaves exhibits excellent hydrophobicity and self-cleaning capabilities, mainly due to the synergistic effect of their dual micro rough structure and low-surface-energy plant wax. According to lotus leaves and rice leaves, it is proposed that micro/nano second-order rough structures are the fundamental reason for hydrophobicity. Based on this, various methods for constructing rough structures have been developed to achieve biomimetic hydrophobic self-cleaning methods such as the electrochemical method, etching method, vapor deposition method, self-assembly method, sol–gel method, etc. However, these methods often require expensive equipment and complex processes to be implemented, and they cannot be produced on a large scale, resulting in significant limitations. The nanoparticle filling method, which has emerged in recent years, is a relatively simple method for preparing superhydrophobic surfaces and is expected to overcome the aforementioned difficulties. Lu et al. [29] prepared epoxy-modified silicone resin by blending method and hydrophobically modified TiO_2 using a polytetrafluoroethylene (PTFE) coating method to provide raw material support for self-cleaning pavement coatings. Jiang et al. [30] modified the surface wettability of TiO_2 using electron-beam interradiation-initiated graft polymerization using 3-methacryloyloxypropyltrimethoxysilane (MPS) and dodecafluoroheptyl methacrylate (DFHMA) with methacrylic acid methacrylate as the modifier. Through fluoride modification, TiO_2 reached a superhydrophobic state and enhanced anti-icing properties [31–33]. Superhydrophobic modification of TiO_2 can also occur through fluorinated deposition of fluoride plasmas produced by thermal decomposition of polytetrafluoroethylene in nitrogen. Due to the grafting of groups such as $\backslash CF_3$, $\backslash CF_2$, and CH_2CF_2 on the surface of the treated TiO_2, the contact angle exceeds 150° and a superhydrophobic state is achieved [34]. Xia et al. [35] used surface-modified TiO_2 with stearic acid under ethanol conditions to obtain hydrophobic TiO_2.

Based on the above literature analysis, this study plans to use a two-component road marking material based on active acrylic resin and organosilicon hydrophobic agent as a base material, selects expanded vermiculite and glass microspheres as insulation fillers, and selects ammonium polyphosphate, pentaerythritol, melamine, and aluminum hydroxide as flame-retardant materials, followed by the addition of nano SiO_2 and modified nano TiO_2 as modifiers to prepare a multifunctional road marking coating for tunnels. Tests such as thermogravimetry and derivative thermogravimetry (TG-DTG), differential scanning calorimetry (DSC), infrared spectroscopy (FTIR) analysis, scanning electron microscopy (SEM), the exhaust gas degradation test, and contact angle measurement were conducted to obtain various physical and chemical properties of multi-functional road marking coating in tunnels.

2. Preparation of Materials and Test Methods

2.1. Preparation of Materials

A two-component road marking material was fabricated using active acrylic resin and an organosilicon hydrophobicity agent was used as a base material. Expanded vermiculite and glass microspheres were selected as insulation fillers. Ammonium polyphosphate, pentaerythritol, melamine, and aluminum hydroxide were selected as flame-retardant materials. Then, nano SiO_2 and modified nano TiO_2 were added as modifiers. Finally, a multifunctional road marking coating for tunnels was prepared using high-speed stirring for 30 min. Among them, the mass content of each component in units of 100 g is shown in Table 1, and the prepared marking coating material was added to a blender to mix and prepare a white semi-viscous liquid as shown in Figure 1.

Table 1. Raw Material Ratio.

Ingredient	Content (%)	Provenience	Ingredient	Content (%)	Provenience
Polyacrylic Resin Component A	19.73	Xuchuan Chemical (Suzhou) Co., Ltd., Suzhou, China	Ammonium Polyphosphate	3.99	Sinopharm Chemical Reagent Co., Ltd., Shanghai, China
Curing agent component B	15.51	Xuchuan Chemical (Suzhou) Co., Ltd., Suzhou, China	Pentaerythritol	3.46	Sinopharm Chemical Reagent Co., Ltd., Shanghai, China
Aluminum Hydroxide	7.24	Zibo Aotai new material technology CO., Ltd., Zibo, China	Melamine	3.24	Sinopharm Chemical Reagent Co., Ltd., Shanghai, China
Expanded Vermiculite	2.16	Nantong Xingchen Synthetic Material, Co., Ltd., Nantong, China	Reagent	7.82	Jiangsu Baoying Asphalt Technology Co., Ltd., Baoying, China
Modified nano TiO_2	10.46	Jiangsu Chuangwei Transportation Technology Development Co., Ltd., Nanjing, China	Silicone Water Repellent	13.94	Nantong Xingchen Synthetic Material, Co., Ltd., Nantong, China
Glass Beads	3.48	Jiangsu Chuangwei Transportation Technology Development Co., Ltd., Nanjing, China	Nano SiO_2	8.97	Nantong Xingchen Synthetic Material, Co., Ltd., Nantong, China

Figure 1. Modified nano TiO_2 and nano TiO_2 co-emulsions.

Modified nano TiO_2, which was nitrogen–vanadium co-doped nano TiO_2, was chosen for use in the sol–gel method, with the optimal doping amount of nitrogen and vanadium chosen for monodoping selected as the amount for co-doping; thus, the doping concentration of nitrogen was 3.0% and vanadium was 1.0%, the calcination temperature was 400 °C, and the N–V co-doped nano TiO_2 was prepared as follows:

(1) A certain amount of urea (N doping concentration of 3.0%) was added to 30 mL of anhydrous ethanol at room temperature with continuous stirring, and 17 mL of butyl titanate was added slowly with stirring for 30 min to obtain a yellow uniform and transparent solution A.

(2) Measures of 28 mL of anhydrous ethanol and 20 mL of glacial acetic acid were mixed at room temperature; a certain amount of ammonium metavanadate (V doping amount of 1.0%) was added to 8 mL of distilled water in a water bath, stirring and heating until the ammonium metavanadate was completely dissolved in the deionized water; the two solutions were then fully mixed to form solution B, and the resulting solution was placed in a dispensing funnel for spare.

(3) Under the condition of a constant temperature of 30 °C, solution B was added to solution A at a constant rate (1–2 drops per second) with continuous stirring to produce a yellow sol.

(4) The stabilized sol obtained was aged at room temperature for 7 days to form a gel.

(5) The gel was dried at 80 °C for 24 h, milled, and calcined at 400 °C for 2.5 h to obtain N–V modified nano TiO_2 and expressed as TiO_2 + N3.0% + V1.0%.

The specific steps are shown in Figure 2.

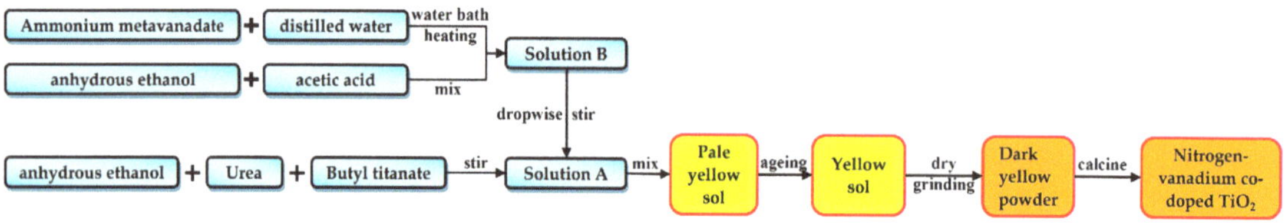

Figure 2. Preparation of N–V co-doped nano TiO_2.

2.2. *Test Methods*

2.2.1. Flame Retardant Tests

TG-DTG, DSC, FTIR, and SEM tests were conducted to evaluate the flame retardancy of multifunctional road marking coating in tunnels.

(1) TG-DTG tests

The TG-DTG test aims to make the sample under the control of a certain temperature program and observe the change process of the mass of the sample with temperature or time. The TA910S synchronous thermal analyzer was used when testing. A measure of 6.5 ± 0.5 mg of the sample was weighed and placed in an Al_2O_3 crucible. Under the conditions of N_2 carrier gas, a flow rate of 20 mL·min^{-1}, and a pyrolysis temperature of 30–800 °C, it was placed in a microbalance of the synchronous thermal analyzer for testing. The heating rate was 5, 10, 15, 20, and 25 K·min^{-1}.

(2) DSC tests:

The DSC test can characterize the physical or chemical change process by measuring the change in its thermodynamic properties. It is a thermal analysis method to measure the relationship between the power difference of the input sample and the reference substance and the temperature under the condition of programmed temperature control. The information recorded during the experiment was the heat difference between the sample and the reference sample while maintaining the same temperature. A TA-Q20 differential scanning calorimeter was used to test multifunctional road marking coating in tunnels. Measures of 5–10 mg of samples were taken, with a testing temperature of 30–800 °C, a heating rate of 10 °C/min, and a N_2 flow rate of 40 mL/min.

(3) FTIR tests

The FTIR test involves the use of the interferometer interference frequency modulation working principle. The light is emitted by the light source through the Michelson interferometer into interference light, and then the interference light irradiates the sample; then, the receiver receives the interference light with sample information, and Fourier transform is used to obtain the sample spectrum. In this study, an IRAffinity-1SWL Fourier transform infrared spectrometer was employed to test the multifunctional road marking coating in the tunnel, and transmission technology was used for testing, with a resolution of 4 cm^{-1}.

(4) SEM tests

SEM is a method of grating scanning on the surface of the sample after focusing the electron beam emitted by the electron gun. It observes and analyzes the composition, morphology and structure of the sample surface by detecting the signal generated by the electron acting on the sample. In this study, the microstructure of the carbon layer of the burned sample was obtained using a JEM-2100 (HR) field emission transmission electron microscope.

2.2.2. Exhaust Degradation Tests

In this study, TiO_2 was modified to purify automobile exhaust fumes under tunnel illumination conditions. Due to the influence of tunnel traffic, the TiO_2 degradation effect was tested in the tunnel, and the safety of testers could not be guaranteed. Therefore, it was difficult to directly test the effect of photocatalytic materials on automobile exhaust degradation in the tunnel. Therefore, the test was carried out indoors according to the actual situation of the tunnel, and the basic conditions in the tunnel were poor lighting conditions and poor ventilation. According to these basic requirements, purification efficiency test equipment was made.

The test equipment for automobile exhaust gas purification was shown in Figure 3. The device was mainly composed of the following parts: an automobile exhaust analyzer, reaction box, ordinary incandescent lamp, ultraviolet lamp, filter dryer, gasoline engine, sample table, and negative pressure hole (vacuum) reaction box. The negative-pressure hole reaction box was the core part of the test equipment, the external dimensions were length (680 mm) × width (530 mm) × height (530 mm), and the volume of the whole reaction chamber was 162.5 L. The high-precision automobile exhaust analyzer has automatic zero adjustment, cleaning, temperature and pressure compensation and other functions, fully in line with the accuracy requirements of this test. An ultraviolet lamp was used in this experiment to study the degradation effect of the modified sample on the tail gas under ultraviolet light. Incandescent lamps were used to study the degradation effect of modified samples on exhaust gas under simulated tunnel lighting conditions. A gasoline engine was used to create automobile exhaust fumes (NO_x, HC, CO, and CO_2). A vacuum pump was used to give the reaction box negative pressure and thus make it easy to exhaust gas into.

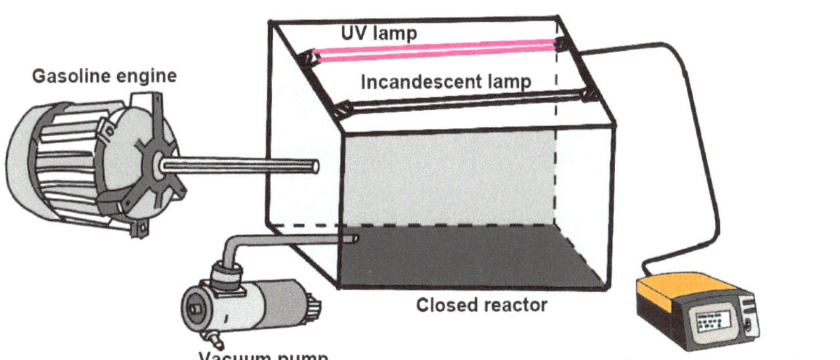

Figure 3. Experimental apparatus schematic diagram for photocatalytic activity test of TiO_2 samples to vehicle exhaust.

The test steps were as follows: (1) start the exhaust gas analyzer: firstly, carry out the leakage test, prepare for the test, and then connect the detection probe of the exhaust gas analyzer and exhaust gas collection with the degradation device and the gasoline engine exhaust pipe; then, carry out exhaust gas collection with the degradation device intake valve. (2) Put in the specimen: put the photocatalyst into the test chamber. To make full contact between the sample and the light source, the modified nano TiO_2 photocatalytic material was evenly spread on the sample table, and the reaction box was sealed and ready for testing. (3) Exclude air: turn on the vacuum pump so that the reaction box is in a semi-vacuum state. (4) Pass the air into the car exhaust gas: the exhaust gas used for the test comes from the exhaust gas of the small gasoline engine when it is working; the gasoline engine is first started, and when it is in stable operation, the gas is then discharged into the reaction box. (4) Record the results of the specimen: turn on the ordinary incandescent lamp or ultraviolet light source, record the initial concentration of each gas, and exit the test state, and then start the test mode once every 10 min to collect the data until 60 min to end a group of tests.

2.2.3. Self-Cleaning Tests

(1) Self-cleaning performance tests

The self-cleaning performance test was conducted according to GB/T31815-2015 [36], using the collection of dust landing on leaves, the preparation of rainwater effluent to simulate the precipitation environment indoors, and the preparation of test boards coated with multifunctional road marking coating in tunnels. A test board without multifunctional coating was selected as the comparison board, and the multifunctional coating was applied to two test boards. Then, the prepared rainwater effluent was uniformly sprayed on the surfaces of the other two test boards, and the specimen boards were placed in a ventilated condition to dry naturally for 2 h. The test was conducted once a day according to the above steps five consecutive times. The comparison board and the two tested test boards were compared to observe the color depth and density of the sewage stains on the coating surface and visually assess the sewage stain rating of the test boards.

(2) Water droplet contact angle tests

In addition, a water droplet contact angle tester was used to test the contact angle of the surface of multifunctional road marking coating in tunnels. The prepared tunnel multifunctional road marking coating samples were tested in the standard environment using the hanging drop method, with a drop of deionized water on the surface of the coating and the use of image analysis technology to measure the size of the contact angle between the water droplets and the surface of the tunnel multifunctional road marking coating specimens; each sample was measured three times, taking the average value for the final measurement results. According to the size of the contact angle, used to analyze the wetting of the water droplets and the surface of the multifunctional road marking coating in the tunnel, when the water droplets and solid contact angle are present for the complete wetting phenomenon, $\theta < 90^\circ$ indicates the hydrophilic wetting phenomenon; otherwise, it is the hydrophobic wetting phenomenon.

3. Results and Discussion

3.1. Evaluation of Thermal Stability

Tunnels constitute a semi-enclosed environment; the high-temperature stability of the material is related to its service life. To explore the effect of modified titanium dioxide and silica addition on the thermal stability of road marking, and to clarify its high-temperature stability performance, thermogravimetry analysis–differential scanning calorimetry–Fourier transform infrared spectroscopy (TG-DSC-FTIR) coupling technology was used to compare and analyze the thermomechanical properties of road marking base material and modified multifunctional marking.

3.1.1. TG-DTG Results Analysis

The thermal decomposition process of multifunctional road marking coatings is relatively complex. Its substrate is composed of active acrylic resin, an organosilicon hydrophobic agent, ammonium polyphosphate (APP), pentaerythritol (PER), melamine (MEL), and aluminum hydroxide ($Al(OH)_3$). Modified nano titanium dioxide and silicon dioxide are used as additives to endow road marking coatings with multi-functionality. TG and DTG curves and typical decomposition parameters are used to analyze and study the thermal stability performance of multifunctional road marking coatings. The results are shown in Figure 4.

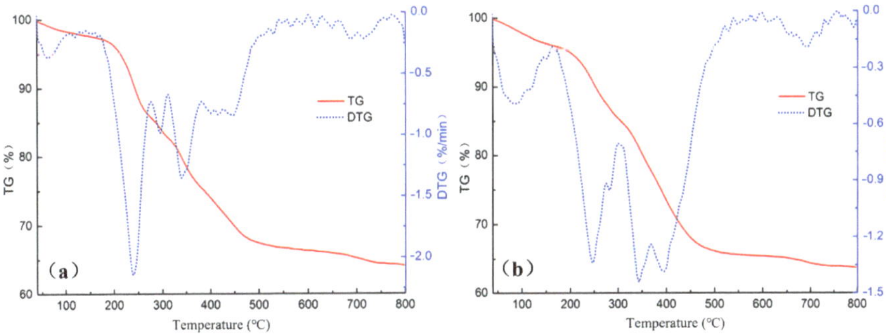

Figure 4. Thermogravimetric and micro commercial thermogravimetric profiles of multifunctional road marking coatings: (**a**) base material; (**b**) modified.

As can be seen from Figure 4a, when the temperature is lower than 143.1 °C, the road marking base material loses very little weight, about 2.3%, and the specimen shows a strong heat absorption peak at 64.3 °C, which is mainly due to a small amount of water being released from the tunnel road marking; the most important weight loss process of the specimen occurs in the second stage of combustion (143.1–275.8 °C), and the weight loss of the stage is about 11.7%. The weight loss in this stage is about 11.7%, and the maximum heat loss rate is −2.3%/min, mainly because in this stage APP is decomposed slowly by heat, NH_3 begins to escape, and a small amount of PE begins to be oxidized to liberate water and carbon dioxide; at this stage, $Al(OH)_3$ in the coating also begins to decompose by absorbing heat, and it releases crystalline water, which evaporates at a high temperature.

The maximum thermal weight loss rates of the coatings in the third and fourth stages were −1.1%/min and −1.5%/min, respectively, and the temperature intervals in which the two stages were located were the temperature bands in which the PER and MEL in the coatings were fully reacted, with the PER starting to decompose at 270 °C, completing its decomposition at 345 °C and reaching the maximum value of its weight loss rate at 315 °C, with the MEL starting to decompose at 296 °C, completing its decomposition at 370 °C, with its weight loss rate reaching the maximum value of its weight loss rate at 365 °C. The decomposition of MEL starts at 296 °C and finishes at 370 °C, and its weight loss reaches its maximum value at 365 °C. APP as a carbonization promoter will participate in these two stages throughout, with a large amount of APP decomposition to produce NH_3, water and phosphoric acid, and then dehydration reaction to produce pyrophosphoric acid, followed by dehydration to carbon formation reaction with PER to produce a carbon layer in the molten state; at the same time, MEL is also subjected to heat to react and release NH_3, so as to produce a multi-hollow carbon layer inside the carbon layer in the molten state: 310–325 °C for the $Al(OH)_3$ dehydration reaction stage as in Equation (1), which mainly occurs in the third stage, and 350–400 °C for the formation of Al_2O_3 reaction as in Equation (2), which mainly occurs in the fourth stage; $Al(OH)_3$ releases water from crystallization, and a large amount of heat is absorbed in the process of high-temperature evaporation, which promotes the high-temperature stability of the coating, and since the maximum rate of heat weight loss of both PER and MEL occurs in the fourth stage, the maximum rate of heat weight loss in this stage is greater than the maximum heat weight loss rate of the third stage. The maximum thermal weight loss rate in this stage is greater than that in the third stage because the maximum weight loss rate of both PER and MEL occurs in the fourth stage, and the weight loss of the coating in the fifth stage mainly comes from the remaining APP and the complete decomposition of $Al(OH)_3$.

$$Al(OH)_3 \rightarrow AlOOH + H_2O \uparrow$$

$$2AlOOH \rightarrow Al_2O_3 + H_2O \uparrow$$

Comparing with Figure 4b, it can be seen that the first stage of the multifunctional coating has a small weight loss interval from 40 to 164.6 °C, with the weight loss being about 4.0%, and the starting thermal decomposition temperature of 95 °C is lower than that of the road marking coating base material (121 °C) and the temperature interval is larger than that of the road marking coating base material, which is mainly due to the evaporation of the water residual in the coating as well as thermal decomposition of a part of the substance and the organosilicon hydrophobic agent. When the temperature continues to rise, the multifunctional coating begins to melt and decompose and then enters the second stage, wherein the weight loss reaches 10.3%, and the thermal stability of the multifunctional coating is improved in this stage, which is not only due to the modified nano TiO_2 having excellent thermal stability but also because part of the pyrophosphoric acid decomposed by APP reacts with the modified nano TiO_2 to generate titanium pyrophosphate, which can effectively improve the thermal stability of the system.

The maximum heat loss decomposition temperature in the third stage is 392 °C, and this temperature is the peak of the decomposition temperature of multifunctional coatings, which is 154.3 °C higher than the decomposition temperature of road marking coating base material, which may be due to the addition of nano SiO_2, silicone water-repellent agent and its formation of Si-O bonding energy being as high as 4225 kJ/mol, which enhances the heat-resistant properties of multifunctional coatings. The fourth stage is the largest multifunctional coating weight loss stage; the weight loss in this stage is about 12.3% and the maximum thermal weight loss rate is $-1.5\%/\text{min}$; the thermal weight loss in this stage is mainly due to the further thermal decomposition of APP into phosphoric acid, metaphosphoric acid, etc., and the decomposition of $Al(OH)_3$ to release water from crystallization as well as the organosilicon hydrophobic agent in the decomposition of organic carbonaceous materials at high temperatures to generate carbon dioxide (CO_2) and water. In the fifth stage of multifunctional coating, the weight loss of the coating is less than 1.4%, and the maximum heat weight loss rate is $-0.2\%/\text{min}$, respectively. During this stage, the weight loss of the coating mainly comes from the decomposition of residual APP and PER.

3.1.2. DSC Results Analysis

To study the thermal effect and thermochemical reaction process of multifunctional road marking coatings in tunnels at high temperatures and to further analyze the thermodynamic behavior of multifunctional road marking coatings in tunnels, the marking coatings were analyzed using the DSC test, and since the TG-DTG test found that the marking coatings were accompanied by a change in mass during the thermal decomposition process, the TG curves were further analyzed by placing the TG curves into DSC curves; the results are shown in Figure 5 and Table 2.

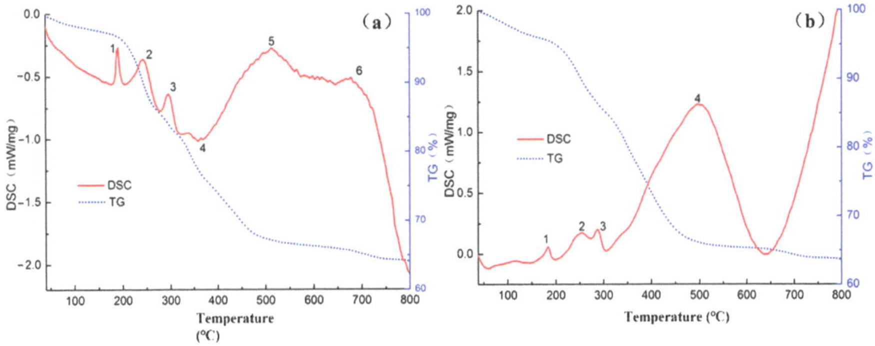

Figure 5. DSC curves of multifunctional road marking coatings: (**a**) base material; (**b**) modified.

Table 2. DSC analysis results of multifunctional road marking coatings: (a) base material; (b) modified.

Materials	Characteristic Peak	Starting Temperature (°C)	Peak Temperature (°C)	Termination Temperature (°C)
(a) Road marking coating base material	1	173	187	208
	2	206	240	278
	3	278	297	316
	4	339	361	420
	5	420	510	572
	6	641	676	716
(b) Multifunctional marking paint	1	143	180	202
	2	202	260	275
	3	275	305	310
	4	310	635	640

From Figure 5a and Table 2(a), it can be seen that the DSC curves of the road marking coating base material have five obvious heat absorption peaks (Peak 1, Peak 2, Peak 3, Peak 5, and Peak 6) and one obvious exothermic peak (Peak 4), in which the heat absorption of Peak 5 is the largest and that of Peak 1, Peak 2 and Peak 3 is smaller. Combined with the results of thermogravimetric analysis, it can be seen that Peak 1 should be the release of the combined water in the coating, generating water vapor generated by the heat absorption. Peak 2 corresponds to the heat absorption peaks of APP and $Al(OH)_3$; at this time, APP began to absorb heat to release ammonia, water and phosphoric acid, etc., and the decomposition of aluminum hydroxide will also begin to absorb heat to release the water from crystallization and evaporation at high temperatures. Peak 3 corresponds to the peaks in absorption of the heat of the PER and the MEL as well as the APP continuing to absorb heat decomposition, and the generated phosphoric acid reaching about 300 °C began to further dehydration to generate pyrophosphoric acid and polyphosphoric acid; due to Peak 4 for the exothermic peak, it can be speculated that in the road marking coating, the base material of the various groups recombines into a new molecular chain segment with the release of heat. Peak 5 of the heat absorption is mainly related to the decomposition of a large number of Al $(OH)_3$ in the coating to produce Al_2O_3 and water; the small heat absorption peak in Peak 6 is produced when the remaining APP is decomposed.

From Figure 5b and Table 2(b), it can be seen that the DSC curve of the multifunctional road marking coating has four obvious exothermic peaks with the increase in temperature, in which it can be seen that Peak 4 absorbs the largest amount of heat and Peak 3 absorbs the smallest amount of heat; comparing with Figure 5a, it can be seen that, with the incorporation of modified nano TiO_2 and nano SiO_2, it changes the thermal decomposition process of the coating, and there is no obvious exothermic peak, and they improved its high-temperature stability performance overall and did not produce new molecular groups.

Combined with the results of TG analysis, it can be seen that the first three peaks of the multifunctional marking coatings are similar to the base material, mainly due to the evaporation of the combined water in the coating as well as the decomposition of the substrate system's APP, PER and MEL synergistic reaction of heat absorption, and Peak 4 at 635 °C demonstrated a strong heat absorption peak; the stage of absorption of heat is larger, indicating that this stage of the multifunctional marking coatings' thermal stability has the largest contribution, which is mainly due to the multifunctional marking coatings' complex composition system, with PER generating charcoal melting at the same time; the $Al(OH)_3$ decomposition reaction is also more active, and APP's further reaction with modified nano TiO_2 to generate titanium pyrophosphate, etc., absorbs a large amount of heat, with organosilicon water repellent agent with a Si-O bonding energy of 4225 kJ/mol, much greater than the bond energy of the C-C (345 kJ/mol) and C-O bond energy (351 kJ/mol); thus, the breakage of Si-O bond will also absorb heat.

3.1.3. Analysis of Combustion Volatiles FTIR Results

In order to study whether toxic gaseous volatiles are generated during the combustion process of multifunctional road marking coatings in tunnels at high temperatures, real-time tracking and detection of the combustion volatiles of multifunctional road marking coatings in tunnels were carried out using FTIR analysis technology; the results are shown in Figure 6.

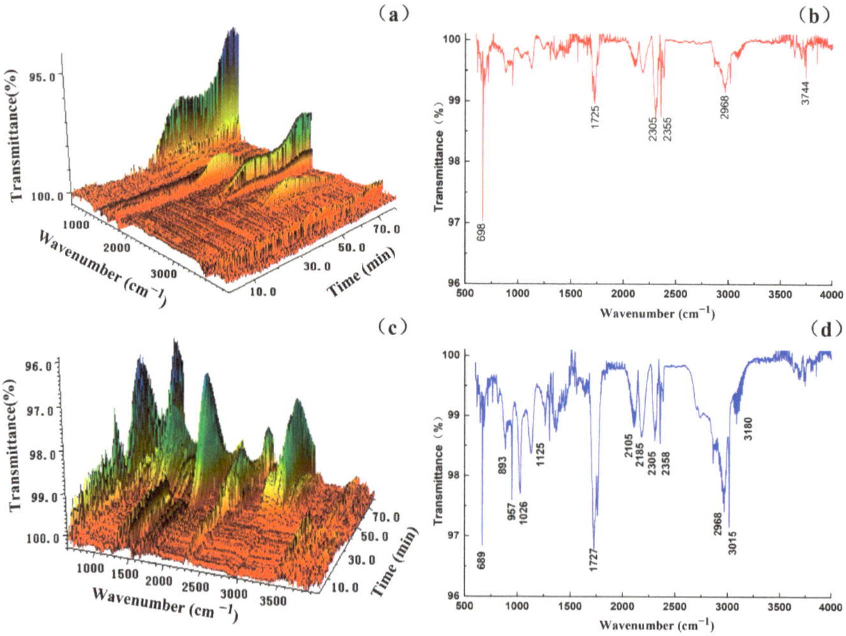

Figure 6. D FTIR spectra of volatiles during combustion of multifunctional road marking coatings: (**a**) base material; (**c**) modified; (**b**) 500 °C FTIR spectra; (**d**) 530 °C FTIR spectra.

From Figure 6a, it can be seen that the road marking coating base material 3D FTIR spectra in the absorption band are weak; only at higher temperatures are there weak peaks, which indicates that the base material combustion process of volatile substances will not produce harmful gases; furthermore, Figure 6c shows that the tunnel multifunctional road marking coatings at 400 °C or below show better thermal stability and almost no volatile substance precipitation, and above, 400 °C there are weak CO_2, H_2O and other characteristic absorption peaks, indicating that adding modified nano TiO_2 and SiO_2 to optimize the thermal stability almost does not produce harmful gases during the combustion process. After 400 °C, there are weak CO_2, H_2O, and other characteristic absorption peaks, indicating that the multifunctional road marking coatings with added modified nano TiO_2 and SiO_2 are effective in optimizing the thermal stability during the combustion process and also produce limited amounts of harmful gases.

Considering that the characteristic peaks of volatiles are the most obvious after combustion of the base and modified materials at 500 °C and 530 °C, these two temperatures are chosen for recording the FTIR spectra, and the results are shown in Figure 6b,d. From Figure 6b, it can be seen that the continuous peak appeared at 4000~3500 cm^{-1}, which is mainly related to the release of gaseous H_2O, which is due to the evaporation of the combined water in the multifunctional road marking paint in the tunnel due to the heat and the water generated by the decomposition of $Al(OH)_3$. The peaks at 2305 cm^{-1} and 2355 cm^{-1} are the characteristic absorption peaks of antisymmetric telescopic vibration of CO_2, which indicates that a certain amount of CO_2 is generated by the combustion of multifunctional road marking coatings in tunnels at high temperatures, and it may be caused by the oxidative decomposition of the base material (a combination of deacidification reaction of carboxylic functional groups and hydroxyl side chain breakage), and the peaks near 2968 cm^{-1} are the absorption peaks of methylene symmetric and antisymmetric telescopic vibration caused by the telescopic vibration of the carbonyl group (C=O) at 1725 cm^{-1} for the absorption peak of the telescopic vibration of the carbonyl group (C=O); when the carbonyl group undergoes an oxidation reaction, carboxylate groups are generated and heat is released. Therefore, the volatiles released from the base material at high-temperature combustion are mainly CO_2, H_2O, and trace amounts of gases containing methylene and carboxyl groups.

Comparing with Figure 6b, it can be seen from Figure 6d and Table 3 that the characteristic peaks of gaseous H_2O appeared at 4000~3500 cm^{-1}, the characteristic absorption peaks due to antisymmetric telescopic vibration of CO_2 appeared at 2305 cm^{-1} and 2358 cm^{-1}, the characteristic peaks due to methylene symmetry and antisymmetric telescopic vibration appeared near 2968 cm^{-1}, with characteristic peaks due to methylene symmetry and antisymmetric telescopic vibration near 2968 cm^{-1}, and with the telescopic vibration absorption peak of the carbonyl group (C=O) at 1727 cm^{-1}. In addition to this, other weak characteristic peaks also appear; the presence of NH_3 is indicated by the presence of organosilicon hydrophobic agents, nano SiO_2, and modified nano TiO_2 in the coating, which affect the combustion process of the coating. A stretching vibration absorption peak of -NH- appeared at 3500–3100 cm^{-1}, and multiple characteristic absorption peaks (893 cm^{-1}, 957 cm^{-1}, 1026 cm^{-1}) appeared near 1000 cm^{-1}, indicating the presence of NH_3. The characteristic absorption peaks at 2185 cm^{-1} and 2105 cm^{-1} were caused by the fracture between C=O and C-O-C to produce CO. The reaction of carbonyl (C=O) functional groups removing CO to produce CO_2 is endothermic. Therefore, the volatiles released by the combustion of multifunctional coatings at high temperatures are mainly CO_2, H_2O, and NH_3, as well as trace amounts of gases containing methylene and carboxyl groups, CO, etc.

3.1.4. Micro-Morphological Analysis of the Charcoal Layer after Combustion

In order to study the protection effect of multifunctional road marking coatings on the base material after high-temperature combustion in tunnels, SEM (scanning electron microscopy) was used to observe the microscopic morphology of the charcoal level; the results are shown in Figure 7.

Figure 7a shows the surface structure of the carbon layer of the road marking coating base material, from which it can be observed that there is a laminar structure, but it is not obvious, and the carbon layer appears to be particularly weak and the laminar structure mostly contains small holes, the reason for which may be the inert gases released by the heat of the coating at high temperatures, which dissipate the laminar structure or break it. Figure 7b shows the fault structure of the carbon layer, from which many irregular honeycomb porous foam structures can be seen, indicating that the cross-section quality of the carbon layer meets the requirements. Therefore, from the consideration of the carbon layer cross-section and surface directions, it can be concluded that the road marking paint base material has excellent thermal stability.

Table 3. Infrared Absorption Frequency Characteristics.

Bond-Type	Type of Vibration	Frequency (cm^{-1})
-NH-	Stretching vibration	3500~3300
O=C=O	Asymmetric stretching vibration	2305, 2358
-C-H-	Symmetric stretching vibration	3300~3000
-C-H	Out-of-plane bending vibration	689
-O-	In-plane bending vibration	1125
-OH	Stretching vibration	4000~3500
-NH	Out-of-plane bending vibration	893, 957
-NH	Deformation vibration	1026
C=O	Stretching vibration	2185
C=O	Conjugate	1740~1700
C-O-C	Stretching vibration	2105
$-CH_2-$	Stretching vibration	2968

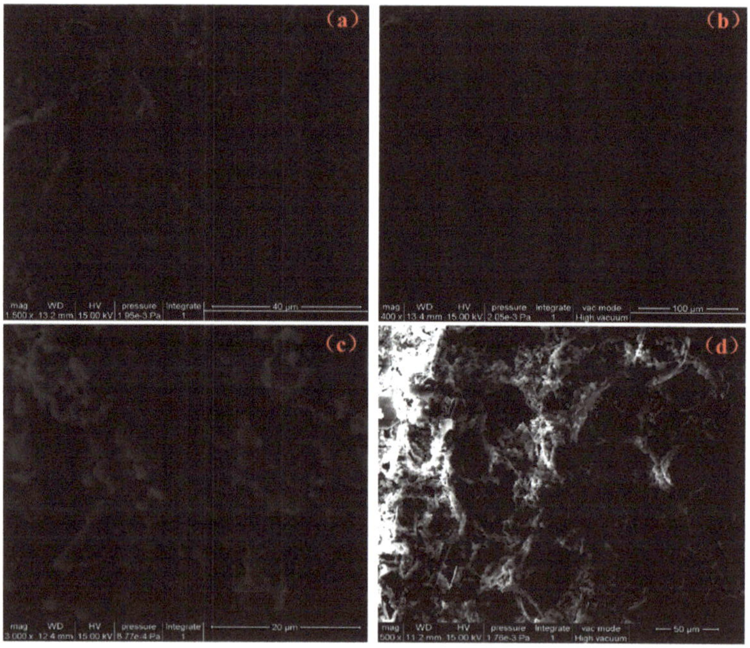

Figure 7. Microscopic morphology of multifunctional road marking coatings: (**a**) surface of the base material; (**b**) fault of the base material; (**c**) surface after modification; (**d**) fault after modification.

Figure 7c shows the surface structure of the charcoal layer of the multifunctional road marking coating in the tunnel. The surface bubble structure of the charcoal layer is obvious, and the charcoal product is relatively intact in structure, with almost no structural damage and good strength, which is conducive to preventing the escape of inert gases and to improving the thermal stability of the coating. Figure 7d shows the fault structure of the carbon layer of multifunctional road marking paint. From the figure, it can be observed that the carbon fault had an irregular honeycomb porous foam structure; the honeycomb structure was obvious and uniform, and the surface of the carbon layer formed a kind of white material. Considering this as well as the results of the thermodynamic analysis, it could be argued that this is because of the tunnel multifunctional road marking paint containing modified nano TiO_2, ammonium polyphosphate, and other ingredients. The modified nano TiO_2 reacts with ammonium polyphosphate at a high temperature to form more stable TiP_2O_7 (titanium pyrophosphate), and its reaction formula is as follows:

$$2TiO_2 + (NH_4)_4P_4O_{12} \rightarrow 2TiP_2O_7 + 4NH_3 + 2H_2O$$

TiP_2O_7 is a white ceramic-like porous substance, which plays an enhanced role in the skeleton structure of the carbon layer, and the carbon layer does not easily collapse under high temperatures, so the rate of heat loss of the carbon layer decreases significantly, and the antioxidant performance of the carbon layer is obviously improved, which indicates that the addition of modified nano TiO_2 and SiO_2 can enhance its high-temperature stability to a certain extent and optimize its durability. And the more the substance is generated, the more uniform the distribution, the longer the fire resistance limit of the multifunctional road marking coating, and the better the high temperature stability performance, and the TiP_2O_7 formation process generates a protective gas, NH_3, that can block the contact between the base material and oxygen. At the same time, the hydrophobic nano SiO_2 in the tunnel multifunctional road marking coatings demonstrates better dispersion, and the coating at high temperatures generated by the charcoal layer containing SiO_2 is also more uniform, which can effectively inhibit the oxygen and heat to the coating of the internal propagation, which improves the tunnel multifunctional road marking coatings' flame-retardant effect. In summary, it can be seen that the surface and cross-section of the modified carbon layer is not only structurally more complete but also generates TiP_2O_7 ceramic substances conducive to the stability of the coating at high temperatures, which greatly improves the high-temperature resistance of the coating and effectively extends its service life.

4. Assessment of Tail Gas Degradation

According to the theory of semiconductor solid-state physics, there is a certain band gap in the energy band of TiO_2, which is between the valence band and conduction band. When the photon energy irradiated on the surface of TiO_2 is equal to or greater than the band gap width (Eg), the electrical balance between the valence band and conduction band will be lost. An electron (e^-) in the valence band has an excited transition to the conduction band, the conduction band will have an additional electron (e^-), and the original valence band will form a hole (h^+) and, eventually, an electron–hole pair. The electron–hole pair has a strong oxidation-reduction ability, but the life span is very short, and it will migrate to the surface of TiO_2. During the migration process, (h^+) and (e^-) in the electron–hole pair will undergo two changes: first, some electrons (e^-) and holes (h^+) will recombine and release heat energy to the outside, making the electron–hole pair lose activity and reducing the photocatalytic efficiency of TiO_2; second, some electrons (e^-) and holes (h^+) are separated and react with O_2 and H_2O adsorbed on the surface of TiO_2 crystal to form oxidizing agents such as hydroxyl radical (OH), superoxide ion radical ($\cdot O^{2-}$) and $\cdot HO_2$ radical.

Studies have shown that when the PH value is less than 4, O_2 adsorbated on the surface of TiO_2 crystal will not only undergo reduction reaction to form hydroxyl radical but will also react with electron (e^-) to form superoxide ion free radical ($\cdot O^{2-}$).

$$O_2 + e^- \rightarrow O^{2-}$$

$$H_2O + \cdot O^{2-} \rightarrow \cdot OOH + OH^-$$

$$2\cdot OOH \rightarrow O_2 + H_2O_2$$

$$\cdot OOH + H_2O + e^- \rightarrow H_2O_2 + OH^-$$

$$H_2O_2 + e^- \rightarrow \cdot OH + OH^-$$

$$H_2O_2 + \cdot O^{2-} \rightarrow \cdot OH + OH^-$$

When the PH value is greater than 10, holes (h^+) can oxidize OH^- and H_2O adsorbated on the surface of TiO_2 crystal to form hydroxyl radical ($\cdot OH$).

$$OH^- + h^+ \rightarrow \cdot OH$$

$$H_2O + h^+ \rightarrow \cdot OH + H^+$$

It can be seen from the above chemical reaction formula that the phase-separated electron–hole pair reacts to generate the hydroxyl radical ($\cdot OH$) and superoxide ion ($\cdot O^{2-}$) with strong oxidation during the migration process. These strong oxidants can gradually oxidize organic pollutants and some inorganic pollutants into harmless inorganic small molecules such as CO_2 and H_2O, so as to achieve the purpose of environmental governance. Therefore, it can be applied in the field of automobile exhaust treatment, and the process of purifying automobile exhaust is shown in Figure 8.

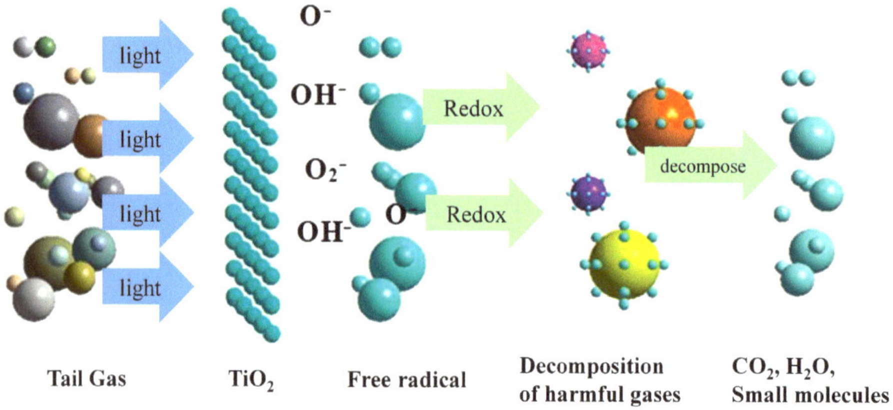

Figure 8. Schematic diagram of photocatalytic degradation of automobile exhaust fumes.

To investigate the degradation law of multi-functional road marking coating on four harmful components, NO_x, HC, CO, and CO_2, in automobile exhaust fumes in the tunnel, this study presents the degradation law of NO_x, HC, CO, and CO_2 by taking the change in concentration of harmful components, degradation efficiency η (%) and degradation rate k as the evaluation indexes, respectively. Among them, the concentration test of harmful components was measured by photocatalyst purification of automobile exhaust test equipment, and the degradation efficiency η (%) was calculated as:

$$\eta = \frac{C_0 - C}{C_0} \times 100\% \quad (1)$$

where C_0 is the initial concentration of the reaction of the harmful components and C is the concentration of the harmful components after the end of the reaction. The difference between C_0 and C is the change in the volume concentration of the degraded harmful gases, and the larger the value is, the more obvious the degradation effect of nano TiO_2 is on this kind of gas.

Then, the photocatalytic degradation rate k on modified nano TiO_2 surface conforms to the Langmuir–Hinshelwood kinetic model:

$$\ln(C/C_0) = -kt \quad (2)$$

Therefore, the photocatalytic reaction rate constant k (slope of the fitted straight line) can be calculated by the primary reaction rate (Equation (2)). The size of the photocatalytic reaction rate constant k reflects the photocatalytic activity, i.e., the larger the value of k, the higher the efficiency of the photocatalytic reaction.

Figure 9a shows the catalytic degradation of NO_x concentration in the exhaust gas with the multifunctional road marking paint in the tunnel. As can be seen from the figure, the concentration of NO_x generally shows a decreasing trend with the extension of time, and the NO_x concentration decreases faster in the first 20 min at the beginning, after which the decrease rate tends to stabilize. This is because the photocatalytic oxidation process of NO_x is $NO \rightarrow NO_2^- \rightarrow NO_2 \rightarrow NO_3^-$, the initial period of multi-functional coatings has physical adsorption of NO_x, NOx is adsorbed on the surface of the coating and the modified nano TiO_2 on the surface of the coating have better contact, with the degradation of the catalytic reaction and the degradation products of NO_3^- adhering to the surface of TiO_2, which prevents the photocatalytic reaction from proceeding further. In addition, the high concentration of NO_x at the initial stage of the reaction and the decrease in NO_x concentration as the reaction proceeds are also factors that affect the NO_x degradation rate. Figure 9b shows that the concentration of HC decreases with time because the photocatalytic oxidation process of HC is HC + $O_2 \rightarrow H_2O + CO_3^{2-}$, which is further decomposed to form CO_2 and water (a layer of water film is attached to the wall of the reaction box after the experiment), and it is an irreversible process, with the degradation rate being faster at the beginning of the 10 min, and then it tends to be stabilized within 10–60 min. Multifunctional road marking paint in the tunnel under incandescent light can degrade automobile exhaust fumes mainly because the N–V co-doping reduces the forbidden band energy gap of TiO_2, widens the range of nano TiO_2 absorption of visible light, and can absorb more light energy, generating more electron–hole pairs, and the excited state of the electrons are more prone to jump, accelerating the rate of electron interfacial transfer to improve the efficiency of visible photocatalysts so that the coatings have a photocatalytic function. In addition, the degradation rates k of the four harmful components of automobile exhaust fumes, NO_x, HC, CO, and CO_2, were 0.0044 min^{-1}, 0.0014 min^{-1}, 0.00044 min^{-1}, and 0.00051 min^{-1}, respectively.

As shown in Figure 9c,d, it can be seen that the degradation effect of multifunctional road marking paint on CO and CO_2 in the tunnel is not obvious, and the decrease in the two gases in the first 30 min should be the adsorption effect of the multifunctional paint, with the concentration of CO and CO_2 being almost unchanged after 30 min. This is because the degradation process of CO and CO_2 by nano TiO_2 is mutually reversible, and the specific mechanism is as follows:

$$2H^+ + 2e^- \longrightarrow H_2 \quad E^0_{redoc} = -0.41 \text{ eV}$$

$$CO_2 + 2H^+ + 2e^- \longrightarrow CO + H_2O \quad E^0_{redoc} = -0.53 \text{ eV}$$

$$CO_2 + 8H^+ + 8e^- \longrightarrow CH_4 + 2H_2O \quad E^0_{redoc} = -0.24 \text{ eV}$$

$$CO + 6H^+ + 6e^- \longrightarrow CH_4 + H_2O$$

The hydrogenation process and hydrogen precipitation process need to consume a lot of electrons and energy, and the priority of their degradation reaction is much lower than that of NO_x and HC. Incorporating TiO_2 nanoparticles into coatings will reduce the photocatalytic performance of TiO_2 nanoparticles under the influence of other components so that the multifunctional coatings have poorer degradation effects on CO and CO_2.

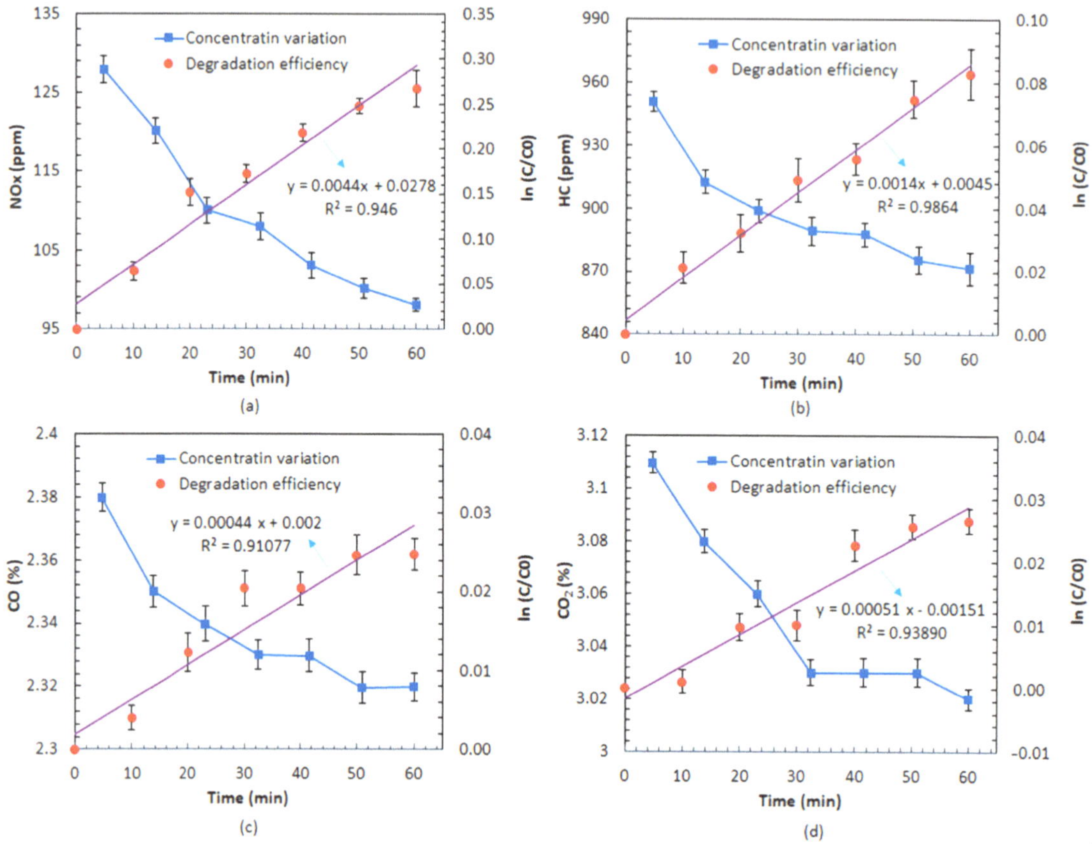

Figure 9. Degradation patterns of multifunctional road marking coatings in tunnels on four harmful components of vehicle exhaust: (**a**) NO_x; (**b**) HC; (**c**) CO; and (**d**) CO_2.

Finally, Table 4 shows the degradation efficiency of multifunctional road marking coating in tunnels under incandescent light for the four harmful components of automobile exhaust fumes, NO_x, HC, CO, and CO_2. The degradation efficiencies of NO_x, HC, CO, and CO_2 reached 23.4%, 8.3%, 2.5%, and 2.9%, respectively.

Table 4. Degradation efficiency of multifunctional road marking coatings for vehicle exhaust fumes in tunnels.

Exhaust Gas Composition	NO_x	HC	CO	CO_2
Photodegradation efficiency (%)	23.4%	8.3%	2.5%	2.9%

Self-Cleaning Performance Evaluation

To investigate the self-cleaning performance of multifunctional road marking coating in tunnels, this study adopts the self-cleaning performance test and water droplet contact angle test to obtain the relevant indicators for evaluation. Figure 10 shows the macro-hydrophobicity of multifunctional road marking coating in tunnels; water droplets on the surface of the coating in the testing process present obvious lotus leaf hydrophobic performance, and, after the test, it can be found that the tunnel multifunctional road marking coatings' surface color almost does not change, and no obvious stains are remaining, according to GB/T31815-2015 in the appendix standards and visual assessment of the test plate of the sewage stain grade: a grade without rainwater stains.

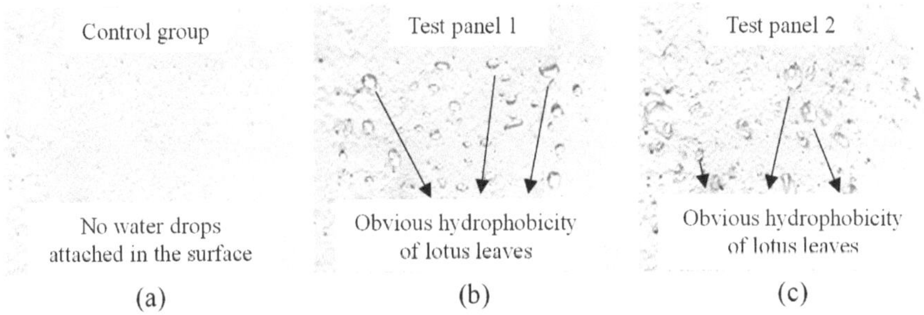

Figure 10. Macroscopic hydrophobicity of multifunctional road marking coating in tunnels: (**a**) comparison panel; (**b**) test panel 1; (**c**) test panel 2.

In addition, the SEM microstructure of the surface of the multifunctional road marking coating in the tunnel is demonstrated in Figure 11. The SEM image of the surface of the multifunctional coating at a scale length of 20 μm is shown in Figure 11a. It can be seen that the surface of the multifunctional coating has a high roughness. Protrusions of about 5–30 μm in size are uniformly distributed on the surface, while smaller milky bumps of varying spacings of about 1–4 μm in size are attached to the protrusions. This composite structure is embedded in the interior of the coating in varying degrees, which forms a rough morphology structure similar to that of a volcano. It can also be observed that the distribution of these micrometer-sized protrusions is surrounded by submicrometer-scaled protrusions. This is mainly due to the solid–liquid phase separation in the coating with the evaporation of water, which leads to the formation of the buildup of nanosized SiO_2 and nanosized TiO_2 particles on the coating surface, which in turn constructs the composite structure. Figure 11b shows the SEM image in Figure 11a when the scale length is 5 μm after continued magnification. The protrusions have a size of 5–30 μm, and around the protrusions, several hundred-nanometer sub-micron protrusions are uniformly distributed, and tens of nanometers of milky bumps are also attached on the individual sub-micron protrusions. The microstructure of this composite structure is similar to that of the lotus leaf surface, which is a "micrometer–nanometer micro-convex" microstructure. This composite structure is similar to the "micro–nanometer micro-convex" microstructure on the surface of the lotus leaf, which has the effect of bionic self-cleaning.

Finally, the contact angle of the multifunctional road marking coating in tunnels was determined as shown in Figure 12. The contact area between the water droplets and the surface of the multifunctional road marking coating in the tunnel is very small, and the water droplets will slide off the surface of the coating carrier when it is gently shaken. The static hydrophobic angle of the tested water droplets on the coating amounted to 134.2°. Therefore, this multi-functional road marking coating in the tunnel has good hydrophobic performance.

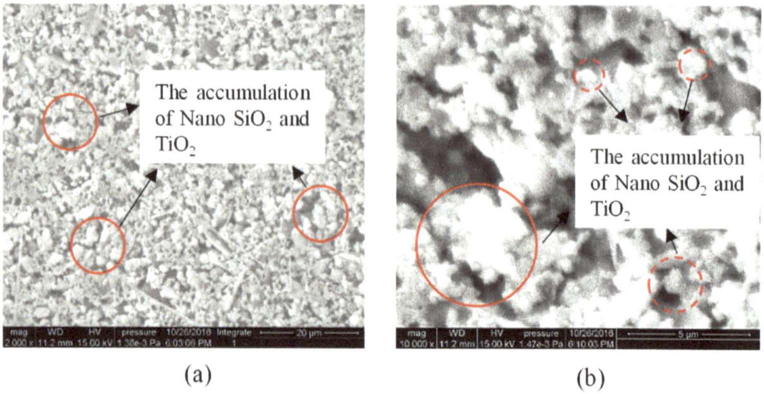

Figure 11. Microstructure of the surface of multifunctional road marking coatings in tunnels: (**a**) scale length of 20 μm; (**b**) scale length of 5 μm.

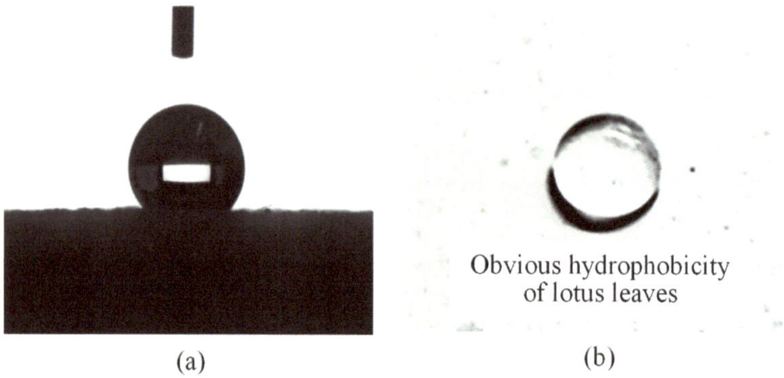

Figure 12. Hydrophobicity test of multifunctional road marking coatings in tunnels: (**a**) contact angle determination graph; (**b**) water droplets on the coating surface.

5. Conclusions

In this study, a two-component road marking material fabricated using active acrylic resin and hydrophobic organosilicon was selected as a base material. Insulation fillers (expanded vermiculite and glass microspheres) and flame-resistant materials (ammonium polyphosphate, pentaerythritol, melamine and aluminum hydroxide) were added into the base marking material. Then, nano SiO_2 and modified nano TiO_2 were added as modifiers to prepare a multifunctional road marking coating for tunnels. The physical and chemical properties of multifunctional road marking coating were evaluating based on laboratory tests including TG-DTG, DSC, FTIR, SEM, exhaust degradation, and contact angle tests, and the main conclusions are as follows:

1. The multifunctional road marking coating in the tunnel developed in this study effectively reduces the thermal conductivity of the carbon layer during the combustion process through the physical changes in the flame retardant system. The heat resistance is formed by nano SiO_2 with high fracture bonding energy and has good high-temperature stability by forming titanium pyrophosphate with a ceramic-like structure with nano TiO_2, which is conducive to the enhancement of high-temperature stability whilst not producing hazardous volatile gases.
2. N–V co-doping reduces the forbidden band energy gap of TiO_2, broadens the absorption range of visible light by nano TiO_2, and improves the catalytic efficiency of visible light. The degradation rates of multifunctional road marking coating in tunnels for

the four harmful components NO_x, HC, CO, and CO_2 in automobile exhaust fumes are 0.0044 min^{-1}, 0.0014 min^{-1}, 0.00044 min^{-1}, and 0.00051 min^{-1}, respectively. The degradation efficiencies of NO_x, HC, CO, and CO_2 reached 23.4%, 8.3%, 2.5%, and 2.9%, respectively, with better absorption of automobile exhaust.

3. The solid–liquid phase separation in the multifunctional road marking coating in the tunnel causes the nano SiO_2 and nano TiO_2 particles to form a pile up on the surface of the paint, which in turn constructs a microstructure similar to the "micro–nanometer micro-convex" on the surface of the lotus leaf. The contact angle of water droplets on the coating surface reaches 134.2°, which has better hydrophobic self-cleaning performance.

Future work will focus on the durability of the developed road marking materials, through the relevant standards to test, compared with the traditional marking materials, so as to apply the long-term service conditions of the road in tunnels.

Author Contributions: Study conception and design: L.Y.; data collection: X.Q.; analysis and interpretation of results: Y.C. and S.S.; draft manuscript preparation: H.L. All authors have read and agreed to the published version of the manuscript.

Funding: This work was undertaken with funding from the National Natural Science Foundation of China (No. 52108404), Natural Science Foundation of Jiangsu Province (BK20210251) and the fellowship of China Postdoctoral Science Foundation (2021M690613). The opinions, findings, and conclusions expressed in this publication are those of the authors and are not necessarily those of any organization.

Data Availability Statement: The data presented in this study are available on request from the corresponding author. The data are not publicly available due to their containing information that could compromise the privacy of research participants.

Conflicts of Interest: Author Liang Yang was employed by the company Suqian Highway Business Development Center. The remaining authors declare that the research was conducted in the absence of any commercial or financial relationships that could be construed as a potential conflict of interest.

References

1. Lu, Y.; Wang, B.; Jiang, J.; Li, H.; Chen, W.; Ye, J. Study on evaluation model of residual bearing capacity of shield tunnel lining structures after fire. *China J. Highw. Transp.* **2023**, 1–18. Available online: http://kns.cnki.net/kcms/detail/61.1313.U.20231110.0911.004.html (accessed on 1 February 2024).
2. Chen, H.; Liu, T.; You, X.; Yuan, D.; Ping, Y.; Zhang, Q. Experimental investigation on fire damage to staggered segmental lining of shield tunnel. *Tunn. Undergr. Space Technol.* **2023**, *141*, 105359. [CrossRef]
3. Li, Q.; Yao, J.; Liang, T. Experimental study on the effect of fireproof coating and cooling methods on the mechanical properties of concrete exposed to high temperature. *Constr. Build. Mater.* **2023**, *376*, 131045. [CrossRef]
4. Nakamura, T.; Nomura, Y.; Tanaka, A. Study on Spraying Method Using Fireproof Coating Spraying Robot. *AIJ J. Technol. Des.* **2023**, *29*, 639–644. [CrossRef]
5. Moosaei, H.R.; Zareei, A.R.; Salemi, N. Elevated Temperature Performance of Concrete Reinforced with Steel, Glass, and Polypropylene Fibers and Fire-proofed with Coating. *Int. J. Eng. Trans. A Basics* **2022**, *35*, 917–930. [CrossRef]
6. Li, Q.; Li, H.; Liang, T. Effect of non-intumescent fireproof coating on the durability of concrete exposed to elevated temperature. *Constr. Build. Mater.* **2023**, *404*, 133211. [CrossRef]
7. Wang, Z.F.; Cheng, W.C.; Wang, Y.Q. Investigation into geohazards during urbanization process of Xi'an, China. *Nat. Hazards* **2018**, *92*, 1937–1953. [CrossRef]
8. Qiao, R.; Shao, Z.; Wei, W.; Zhang, Y.Y. Theoretical Investigation into the Thermo-Mechanical Behaviours of Tunnel Lining during RABT Fire Development. *Arab. J. Sci. Eng.* **2019**, *44*, 4807–4818. [CrossRef]
9. Zhang, Y.; Weng, X.; Song, Z.; Sun, Y. Modeling of Loess Soaking Induced Impacts on a Metro Tunnel Using a Water Soaking System in Centrifuge. *Geofluids* **2019**, *2019*, 5487952. [CrossRef]
10. Qiu, J.; Yang, T.; Wang, X.; Wang, L.; Zhang, G. Review of the flame retardancy on highway tunnel asphalt pavement. *Constr. Build. Mater.* **2019**, *195*, 468–482. [CrossRef]
11. Zheng, Y.; Zhong, H.; Fang, Y.; Zhang, W.; Liu, K.; Fang, J. Rockburst Prediction Model Based on Entropy Weight Integrated with Grey Relational BP Neural Network. *Adv. Civil. Eng.* **2019**, *2019*, 3453614. [CrossRef]
12. Kim, J.H.J.; Lim, Y.M.; Won, J.P.; Park, H.G. Fire resistant behavior of newly developed bottom-ash-based cementitious coating applied concrete tunnel lining under RABT fire loading. *Constr. Build. Mater.* **2010**, *24*, 1984–1994. [CrossRef]

13. Alhawat, H.; Hamid, R.; Baharom, S.; Azmi, M.R.; Kaish, A.B.M.A. Thermal behaviour of unloaded concrete tunnel lining through an innovative large-scale tunnel fire experimental testing setup. *Constr. Build. Mater.* **2021**, *283*, 122718. [CrossRef]
14. Radzi, N.A.M.; Hamid, R.; Mutalib, A.A. A Review of Methods, Issues and Challenges of Small-scale Fire Testing of Tunnel Lining Concrete. *J. Appl. Sci.* **2016**, *16*, 293–301. [CrossRef]
15. Ibrahim, R.K.; Hamid, R.; Taha, M.R. Fire resistance of high-volume fly ash mortars with nanosilica addition. *Constr. Build. Mater.* **2012**, *36*, 779–786. [CrossRef]
16. Abhilash, P.P.; Nayak, D.K.; Sangoju, B.; Kumar, R.; Kumar, V. Effect of nano silica in concrete; a review. *Constr. Build. Mater.* **2021**, *278*, 122347. [CrossRef]
17. Pereyra, A.M.; Canosa, G.; Giudice, C.A. Nanostructured protective coating systems, fireproof and environmentally friendly, suitable for the protection of metallic substrates. *Ind. Eng. Chem. Res.* **2010**, *49*, 2740–2746. [CrossRef]
18. Zhang, H.J.; Wang, Y.C.; Li, F.; Zhao, J.P. In-situ polymerized zinc phytate chelated Si-C-P geopolymer hybrid coating constructed by incorporating chitosan oligosaccharide and DOPO for flame-retardant plywood. *Constr. Build. Mater.* **2023**, *397*, 132416. [CrossRef]
19. Ng, Y.H.; Suri, A.; Dasari, A.; Tan, K.H. Thermal decomposition and fire response of non-halogenated polymer-based thermal coatings for concrete structures. *Surf. Coat. Technol.* **2017**, *320*, 396–403. [CrossRef]
20. Abniki, M.; Shirkavand Hadavand, B.; Najafi, F.; Ghasedi, I. Synthesis of the effective flame retardant via modification of epoxy resin with phenylboronic acid. *J. Macromol. Sci. Part. A* **2022**, *59*, 411–420. [CrossRef]
21. Wu, J.; Luo, Y.; Qin, Z. Composite-modified nano TiO_2 for the degradation of automobile exhaust in tunnels. *Constr. Build. Mater.* **2023**, *408*, 133805. [CrossRef]
22. Oliveira, H.G.; Silva, E.D.; Longo, C. Electrochemical and photocatalytic properties of TiO_2/WO_3 photoelectrodes. In *Solar Hydrogen and Nanotechnology V*; SPIE: San Diego, CA, USA, 2010; Volume 7770, pp. 11–19. [CrossRef]
23. Luu, C.L.; Nguyen, Q.T.; Ho, S.T. Synthesis and characterization of Fe-doped TiO_2 photocatalyst by the sol-gel method. *Adv. Nat. Sci. Nanosci. Nanotechnol.* **2010**, *1*, 015008. [CrossRef]
24. Li, Y.; Sun, H.; Peng, T.; You, H.; Zeng, L.; Qin, Y. Preparation and Visible Photocatalytic Properties of N-Doped TiO_2/Muscovite Nanocomposites. *Clays Clay Miner.* **2021**, *69*, 254–262. [CrossRef]
25. Zhang, Z.; Liu, K.; Chong, D.; Niu, D.; Lin, P.; Liu, X.; Niu, Y.; Jing, R. Evaluation of photocatalytic micro-surfacing mixture: Road performance, vehicle exhaust gas degradation capacity, and environmental impacts. *Constr. Build. Mater.* **2022**, *345*, 128367. [CrossRef]
26. Liu, G.; Xia, H.; Niu, Y.; Zhao, X.; Zhang, G.; Song, L.; Chen, H. Photocatalytic performance of doped TiO_2/AC coating and its UV stability research. *Prog. Org. Coat.* **2020**, *148*, 105882. [CrossRef]
27. Xia, H.; Liu, G.; Zhang, R.; Song, L.; Chen, H. The photocatalytic degradation of vehicle exhausts by an Fe/N/Co-TiO_2 waterborne coating under visible light. *Materials* **2019**, *12*, 3378. [CrossRef]
28. Liu, W.; Wang, S.; Zhang, J.; Fan, J. Photocatalytic degradation of vehicle exhausts on asphalt pavement by TiO_2/rubber composite structure. *Constr. Build. Mater.* **2015**, *81*, 224–232. [CrossRef]
29. Lu, C.; Zheng, M.; Liu, J.; Zhu, R.; Su, Y. Characterization of self-cleaning pavement coatings with catalytic-hydrophobic synergistic effects. *Constr. Build. Mater.* **2023**, *397*, 132246. [CrossRef]
30. Jiang, B.; Zu, X.; Tang, F.; Wu, Z.; Lu, J.; Wei, Q.; Zhang, X. Surface modification on nanoscale titanium dioxide by radiation: Preparation and characterization. *J. Appl. Polym. Sci.* **2006**, *100*, 3510–3518. [CrossRef]
31. Qi, Y.; Chen, S.; Sun, H.; Zhang, J. A multifunctional cool material with contamination-resistant and anti-icing characters based on the application of fluorinated TiO_2. *Sol. Energy Mater. Sol. Cells* **2019**, *201*, 110091. [CrossRef]
32. Qi, Y.; Chen, S.; Zhang, J. Fluorine modification on titanium dioxide particles: Improving the anti-icing performance through a very hydrophobic surface. *Appl. Surf. Sci.* **2019**, *476*, 161–173. [CrossRef]
33. Qi, Y.; Xiang, B.; Zhang, J. Effect of titanium dioxide (TiO_2) with different crystal forms and surface modifications on cooling property and surface wettability of cool roofing materials. *Sol. Energy Mater. Sol. Cells* **2017**, *172*, 34–43. [CrossRef]
34. Wang, F.; Zhang, H.; Zhu, H.; Guo, Y. Ultra-hydrophobic modification of TiO_2 nanoparticles via thermal decomposition of polytetrafluoroethylene. *Powder Technol.* **2014**, *253*, 548–552. [CrossRef]
35. Xi, Y.; Qi, Y.; Mao, Z.; Yang, Z.; Zhang, J. Surface hydrophobic modification of TiO_2 and its application to preparing PMMA/TiO_2 composite cool material with improved hydrophobicity and anti-icing property. *Constr. Build. Mater.* **2021**, *266*, 120916. [CrossRef]
36. *GB/T31815-2015*; Self-Cleaning Coatings for Exterior Surfaces of Buildings. Code of China: Beijing, China, 2015.

buildings

MDPI

Article

Improvement of the Cracking Moment-Based Asphalt Mixture Splitting Test Method and Splitting Strength Research

Hongxin Guan [1,*], Wang Pan [1], Hairong Yang [1] and Yuxuan Yang [1,2]

1 School of Traffic & Transportation Engineering, Changsha University of Science & Technology, Changsha 410114, China; wangpan@stu.csust.edu.cn (W.P.); yhr@csust.edu.cn (H.Y.); buhuiba55@stu.csust.edu.cn (Y.Y.)

2 Wuhan Comprehensive Transportation Design and Research Institute, Wuhan 430012, China

* Correspondence: ghx_cs@csust.edu.cn

Abstract: The asphalt mixture splitting test is one of the most important methods for measuring asphalt's tensile properties. To characterize the limitations of the traditional splitting test and the influence of the specimen size and loading conditions on the accuracy of the test, the factors affecting the strength of the splitting test were analyzed to reveal the splitting failure state and establish a unified representation model between the splitting and direct tensile tests. Initially, the moment of specimen cracking was taken as a key indicator, combined with image processing technology, to establish a set of criteria to judge the splitting test. Subsequently, standardized splitting tests were conducted and compared to tests of different specimen sizes and loading methods. Based on the octahedral strength theory, the stress points before and after the improved test were compared to the existing failure criteria. Direct tensile and splitting tests were conducted at different rates, and a unified strength–rate function model was established, realizing the unified representation of direct tensile and splitting tests. The research results indicate that the standardized splitting test is prone to the phenomenon wherein the specimen end face cracks before the center, affecting the accuracy of the test and potentially leading to redundant material strength evaluations. Using a loading method with a "35 mm specimen thickness" and a "0.3 mm rubber gasket + 12.7 mm arc-shaped batten" can essentially achieve the test hypothesis of "cracking at the center first", resulting in less discrete outcomes and closer alignment to the three-dimensional stress failure state. The tensile and splitting strengths are both power function relationships with the rate as the independent variable, establishing a unified function model of the tensile and failure strengths. The research provides a more reliable testing method and calculation model for asphalt pavement structure design, and it also provides an effective basis for the improvement of splitting tests on materials such as concrete and rock.

Keywords: asphalt mixture; splitting test; cracking moment; loading methods; damage guidelines

Citation: Guan, H.; Pan, W.; Yang, H.; Yang, Y. Improvement of the Cracking Moment-Based Asphalt Mixture Splitting Test Method and Splitting Strength Research. *Buildings* **2024**, *14*, 457. https://doi.org/10.3390/buildings14020457

Academic Editor: Antonio Caggiano

Received: 29 December 2023
Revised: 29 January 2024
Accepted: 2 February 2024
Published: 7 February 2024

1. Introduction

In the loading zone, the stress state on the surface of the asphalt pavement is quite complex. Many scholars have improved the road performance of asphalt mixtures through the improvement of asphalt materials [1]. Based on the influence of different rubber sizes on the waste rubber binder, Ren studied the continuous expansion degradation behavior and chemical rheological properties of crumb-rubber-modified asphalt [2]; Li discussed the influence of different regenerants on the performance of an asphalt mixture through self-developed equipment and proposed the best maintenance conditions [3]. Their research provides a basis for the design of asphalt mixtures. Although many scholars have made great improvements to the performance of asphalt mixtures from the material point of view, their mechanical testing still relies on the traditional test methods. The highest tensile stress and strain occur at the outer boundary of the loading zone, where there is rapid stretching and compression changes in the tangential stress and strain, transitioning from compression to tension [4]. Within the loading area, the pavement mainly experiences compressive stress; outside the loading area, the pavement undergoes radial compression, as well as tangential stretching and compression. Therefore, it is crucial to study the strength characteristics of asphalt mixtures under tensile and compressive stress. Among the strength testing techniques, the splitting test reflects the stress conditions outside the loading zone. Consequently, it is imperative to conduct research on the splitting test of asphalt mixtures. However, due to the stress concentration effect at the loading end during the splitting test, the recorded strength values may not reflect the true tensile strength of the material, necessitating the improvement of the testing techniques.

A common technique for the estimation of the strength and low-temperature cracking resistance of asphalt pavements is the asphalt mix splitting test, often referred to as the indirect tensile test [5,6]. In the same year, Carneiro [7] and Akazawa [8] independently proposed the Brazilian disc test, which led to the development of the splitting test. The theoretical basis of the splitting test method is that, when a cylindrical specimen is subjected to radial compression, a relatively uniform transversal tensile zone appears at its center. A. Fahimifar [9] studied the impact of the boundary conditions and aspect ratios on the tensile splitting strength. By constructing finite element models under various boundary conditions and simulating fatigue tests and direct tensile tests, it was found that the strength of the splitting test was also affected by the location of the load application. Ma Yszko et al. [10] conducted splitting tests on cubic specimens at different load positions. According to the test results, the load applied along the height of the cube produced better results than that applied along the diagonal. When the location of the load application was changed, there was no significant difference in the splitting test results. Some researchers use soft wood spacers (cushions) between the specimen and the loading plate to reduce the impact of stress concentration at the loading end [11]. The 18th Road Conference Proceedings and the Japan Road Association suggest that, in practical engineering applications of the splitting test, rigid pressure plates should be immediately used for the splitting test, which is also the simplest loading technique [12]. Some scholars have found that, when a uniformly distributed load is applied to the specimen and the loading angle between the loading end and the specimen is controlled at around 20°, the specimen will start cracking from the middle [13,14]. Based on these conclusions, some researchers have developed an arc-shaped loading clamp with the same curvature as the specimen for the splitting test [15], and they have conducted a large number of laboratory tests, confirming that, when this loading method is used, some specimens' cracks start from the center point. However, when the arc-shaped loading clamp does not establish full contact with the circumference of the specimen, stress concentration at the loading end still exists. Therefore, with this improved technique, it is necessary to ensure the processing accuracy of the clamp and the specimen, as well as the alignment of the upper and lower loading clamps. The cost of this test is high, and it has not been widely promoted and accepted. Sarfarazi et al. [16] developed a new instrument that can directly apply tensile stress to the test piece. The disadvantage is that the overall structure of the equipment is complex and the stress at the failure point of the test piece is uneven, making it difficult to obtain accurate strength measurement results. Some researchers have found that the splitting test cannot satisfy the center point cracking hypothesis in theoretical derivation due to stress concentration at the actual loading end, and the strength value obtained from the splitting test stress concentration under cracking does not reflect the real tensile strength [17].

Although many researchers have discussed how to enhance the splitting test, most of these studies only look at one or two criteria and do not comprehensively evaluate multiple elements for a quantitative analysis and comparison. Furthermore, due to a lack of proper assessment means, objectively evaluating the test improvement outcomes is challenging.

This study addresses the question of splitting test accuracy by utilizing digital image processing technology, employing the moment of cracking as a judgement indicator and conducting an improvement study on the splitting test with respect to the specimen size and test loading method, among other aspects. The impact of the improvement is confirmed by comparing the results with the strength theories that are currently in use. Lastly, a suggested splitting test program is provided, which may effectively lower the test error and data dispersion, allow the more precise assessment of asphalt materials' tensile strength, and serve as a useful guide for the testing of other materials' tensile strength. The specific research plan is shown in Figure 1.

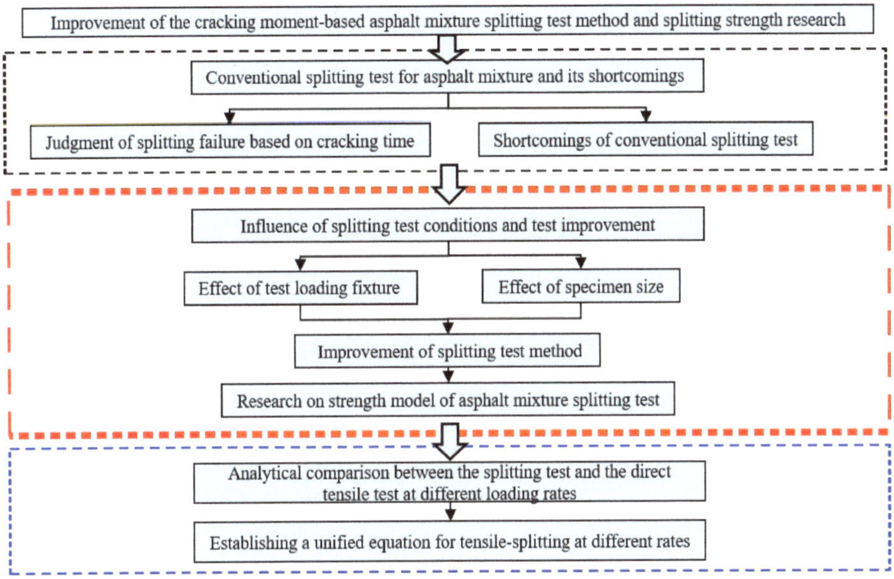

Figure 1. Research plan.

2. Materials and Testing Methods

2.1. Materials

The objective of this work was to determine the efficacy of an AC-13 fine asphalt mixture. The asphalt employed in this work was SBS (I-D)-modified asphalt; the coarse aggregate was basalt, and the fine aggregate was limestone. Refer to Tables 1–4 for the results of the asphalt performance index test, aggregate index test, mineral aggregate gradation, and Marshall test at the ideal asphalt aggregate ratio.

Table 1. Technical indexes and test results of SBS-modified asphalt.

Technical Indicators		Technical Requirement	Unit	Detection Result
Penetration (25 °C, 100 g, 5 s)		30~55	0.1 mm	53
Penetration index PI		≥0	—	0.21
Softening point (global method)		≥75	°C	86
Ductility (5 °C, 5 cm/min)		≥25	cm	31
Kinematic viscosity 135 °C		≤3	Pa·s	2.2
Flash point (COC)		≥230	°C	293
Solubility (trichloroethylene)		≥99	%	99.6
Elastic recovery (25 °C)		≥85	%	98
Rotating film oven test (163 °C, 85 min)	Ductility (5 °C)	≥15	cm	17
	Mass loss	≤±1.0	%	0.1
	Penetration ratio (25 °C)	≥65	%	67

Table 2. Technical indexes and test results of coarse aggregate.

Test Item	Unit	Technical Requirement	Test Result			
			2.36~4.75 mm	4.75~9.5 mm	9.5~13.2 mm	13.2~16 mm
Crushing value	%	≤26	10.1	10.7	11.3	11.8
Los Angeles abrasion value	%	≤28	11.2	12.8	13.9	14.2
Apparent relative density	—	≥2.6	2.701	2.703	2.709	2.720
Gross volume Relative density	—	—	2.649	2.660	2.686	2.698
Water absorption	%	≤2	1.35	1.23	0.98	1.11
Robustness	%	≤12	2.1	2.9	3.1	3.2
Needle and flake content	%	≤15	11.3	9.0	8.8	7.1
Polish value	PSV	≥42	45.9	47.2	47.6	46.5
Soft rock content	%	≤3	1.2	1.6	2.1	1.3

Table 3. AC-13 grading of the asphalt mixture.

Gradation Type	Mass Percentage (%) Passing through the Following Sieve Openings (Square Sieve, mm)									
	16	13.2	9.5	4.75	2.36	1.18	0.6	0.3	0.15	0.075
AC-13	100	95	74	48.5	34	23.5	15	11	8.5	6

Table 4. SBS Marshall test results of the modified asphalt AC-13 mixture under the optimum asphalt aggregate ratio.

Oil Stone Ratio/%	Gross Volume Relative Density/g·cm^{-3}	Void Fraction/%	Saturation/%	Mineral Aggregate Porosity/%	Stability/kN	Stream Value/0.1 mm
4.77	2.465	3.8	72.1	14.1	13.3	29.2

2.2. Test Equipment and Process

The test pieces in this research were compacted using an Italian CONTROLS 76-B0252 rotary compactor (Italian Matest Company, Arcore, Italy). The generated test pieces were φ100 mm × 90 mm AC-13 mixed test pieces. Using the surface drying technique, the gross bulk density, porosity, and other aspects of the cut test piece were determined in accordance with the Highway Engineering T0706-2011 Test Code for Asphalt and Asphalt Mixtures [18]. To reach the requisite specimen thickness, the cylindrical specimen created by rotary compaction was further cut with an infrared-controlled cutting machine with a cutting error of less than 0.1 mm. The specimen fabrication process is illustrated in Figure 2.

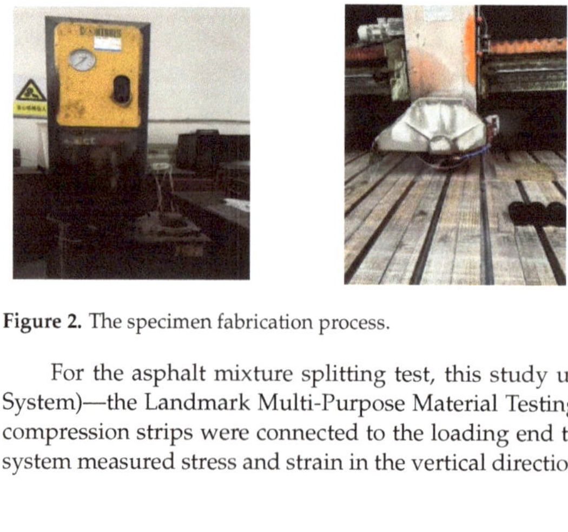

Figure 2. The specimen fabrication process.

For the asphalt mixture splitting test, this study utilized a MTS (Material Testing System)—the Landmark Multi-Purpose Material Testing System. The upper and lower compression strips were connected to the loading end to load the specimens. The MTS system measured stress and strain in the vertical direction, while a JM3812 wireless static strain gauge measured the strain on the specimen surface. According to the specifications, the test was loaded utilizing controlled displacement at a rate of 60 mm/min, and the test loading fixture was 12.7 mm arc-shaped batten. The specimen was loaded as shown in Figure 3.

Figure 3. Example of specimen splitting and loading.

2.3. Judgment of Splitting Failure

The split test is frequently used as an indirect tensile test to assess the tensile qualities of materials, including asphalt, concrete, and rock. During the loading process, the top of the specimen is pressurized, material deformation results in tensile force at the center of the circle, and the material is damaged in tension. Many academic studies have demonstrated that, during the splitting process, once the end of the specimen is loaded, the fracture is always created at the center of the circle first, and it soon penetrates through the entire specimen. At this point, the specimen's overall damage shape is transformed from perfect tensile damage to compressive damage. This test's inaccuracy will have a direct impact on the precision of the stress computation, which will result in mistakes when assessing the mechanical characteristics of the materials. Through the collection of the cracking times at the end and center of the specimen, as well as the calculation of the difference between the cracking times of the two, this study aimed to ascertain whether the specimen displayed the damage form of center-first cracking in the ideal scenario.

We pasted strain gauges along the horizontal direction at the center of the circle (channel 1), where the specimen was theoretically subjected to tensile force and the first to crack, and at the end of the specimen, which was prone to stress concentration cracking (channel 2), in order to more intuitively analyze the cracking state of the split specimen. We also pasted strain gauges at the same positions on the other side of the specimen (channel 3 and channel 4). The top portion of the strain gauges for channels 2 and 4 was approximately 9 mm away from the apex of the vertical axis, and the strain gauges were pasted on the horizontal axis at the round center points of the two end faces, as shown in Figure 4. In order to facilitate the strain collection procedure and determine the crack initiation point's position and timing, as well as the crack penetration sequence, the test additionally employed video equipment to film the whole specimen splitting test process.

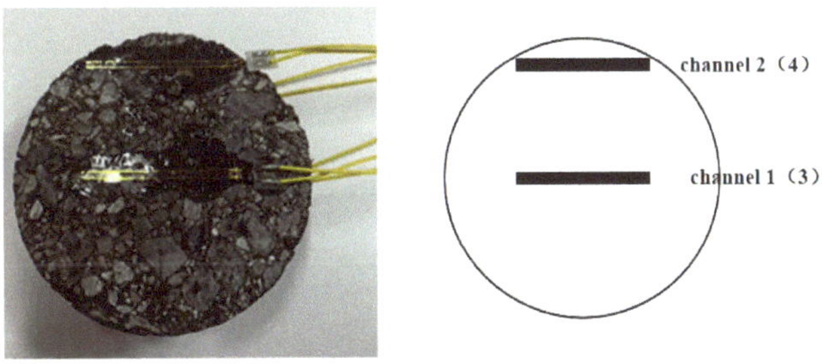

Figure 4. Distribution of the strain gauges on the specimen.

In ideal circumstances, the strain gauges at channels 1 and 3 would fail first, followed by the end strain gauges at channels 2 and 4, and finally, the middle of the circle would break when the specimen achieves its maximal strength. The MTS recorded the peak load at this point. In summary, this study led to the following conclusions about the specimen judgement method: ① the moment of failure of the strain gauges at the center of the two end faces is approximately equal; ② the moment of failure of the strain gauges at the center of the circle is approximately equal to the moment of the peak of the MTS force; ③ the strain gauges at the center of the same end face fail before the end strain gauges.

2.4. Results of the Conventional Splitting Test

In line with JTG E20-2011, Chinese Test Procedures for Asphalt and Asphalt Mixtures in Highway Engineering [18], the test was carried out in accordance with Chinese road engineering specifications. A test specimen with a diameter of 100 mm was chosen, using a 12.7 mm arc bead, and the test temperature was set at 15 °C ± 0.5 °C. The loading rate was 50 mm/min. It was advised that the split test piece be 63.5 mm ± 1.3 mm thick. The thickness of this piece was 63.5 mm to be as close to the plane stress condition as feasible. The findings of the tests, which were run in 15 groups in parallel, are displayed in Table 5.

Table 5. Summary table of the MTS data and failure time of the strain gauge of the standard test method.

Specimen Number	MTS Peak Load/kN	MTS Moment of Peak Force/s	Channel 1 Failure Time of Strain Gauge/s	Channel 2 Failure Time of Strain Gauge/s	Channel 3 Failure Time of Strain Gauge/s	Channel 4 Failure Time of Strain Gauge/s
I-1.1-3	20.268	2.051	3.063	3.125	3.594	4.125
I-1.1-8	24.928	1.924	3.344	3.344	4.125	3.313
I-1.1-13	23.754	2.520	4.125	3.938	3.625	3.125
I-1.1-35	23.569	2.178	3.439	3.758	3.495	3.652
I-1.1-40	24.167	2.264	3.650	3.794	3.738	3.813
I-1.1-42	22.222	3.018	4.031	3.625	4.844	4.313
I-1.1-43	21.501	2.305	3.719	3.438	4.094	4.125
I-1.1-49	23.445	2.334	6.250	7.469	3.063	3.125
I-1.1-55	23.217	2.676	6.188	0.875	6.031	4.281
I-1.1-57	23.609	1.885	2.974	3.530	2.986	3.717
I-1.1-65	23.223	2.061	3.553	4.178	3.703	3.678
I-1.1-68	23.485	1.963	—	3.656	3.199	3.594
I-1.1-91	23.647	2.898	4.293	4.356	4.168	4.449
I-1.1-93	21.830	2.051	5.031	4.969	3.719	3.813
I-1.1-98	22.989	2.879	4.594	4.794	4.644	4.648

When using the standard test conditions for the split test, it is not always the case that the strain gauges at channels 1 and 3 in the middle of the end face fracture first; in some cases, the fracture time of test specimens close to channels 2 and 4 in the upper loading position is earlier. This information was gathered from the failure times of strain gauges at different positions by the acquisition instrument. In 15 parallel tests, 6 of the split test specimens began to fracture first at the center of the circle and 9 at the location adjacent to the loading point, with considerable dispersion compared to the time of center cracking and the time of failure of the upper strain gauge.

The test process was observed using the VIC-3D digital speckle technique (as shown in Figures 5 and 6), and it was discovered that the test pieces mostly displayed two failure modes; these two failure modes also coincided with the two crack initiation patterns recorded by the digital speckle technique. According to the distance between the circle's center and the crack tip, Figure 5 demonstrates that, when the center cracks, it is less affected by the stress concentration at the crack tip because of the expansion of the upper end. The center of the circle in Figure 5 cracked "passively", whereas the center of the circle in Figure 6 cracked actively.

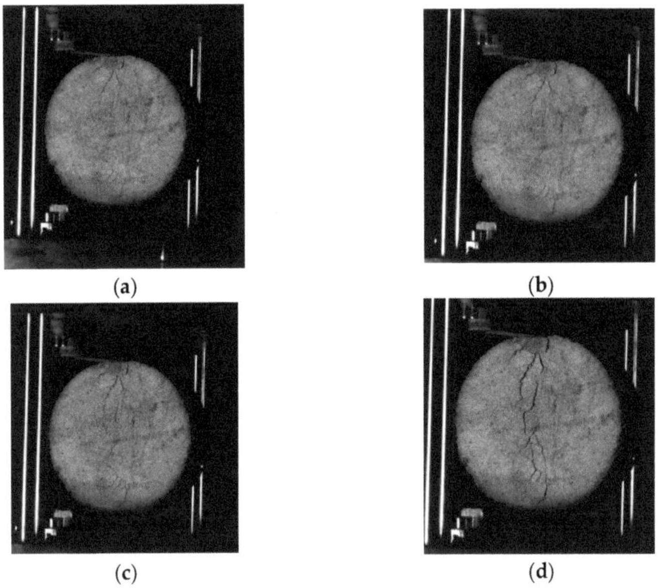

Figure 5. Splitting test of test piece 1. (a) Initial stage end stress concentration cracking. (b) The crack starts to expand. (c) The crack extends to the center area. (d) Formation of through cracks.

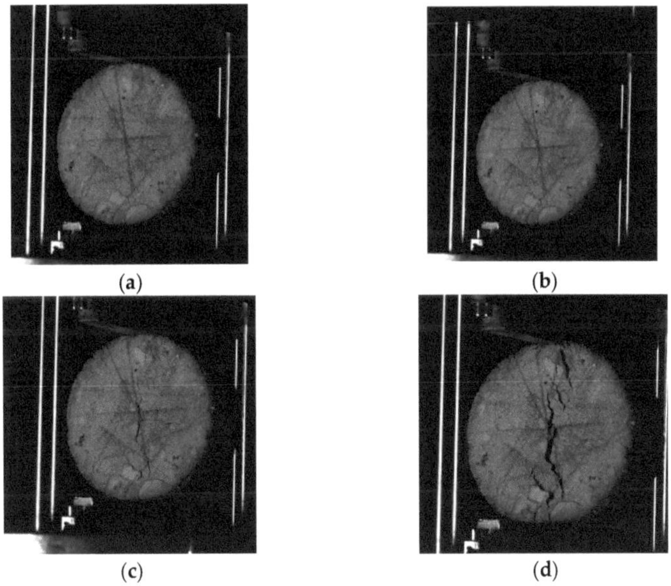

Figure 6. Splitting test of test piece 2. (a) Initial stage end stress concentration cracking. (b) Cracking at the center of the circle. (c) Crack propagation stage. (d) Formation of through cracks.

2.5. Shortcomings of the Conventional Splitting Test

According to the previous definition of specimen failure, the data in Table 5 are processed, and the test statistical results of central cracking, top cracking, and undifferentiated central or top cracking are listed in Table 6.

Table 6. Statistical table of the test data under different working conditions.

Number of Specimens	Representative Working Conditions	Statistical Results of Peak Load			
		Average Value/kn	Standard Deviation	Variance	Variation Coefficient
15	Undifferentiated central or top cracks	23.057	1.171	1.372	0.050
9	Top cracks first	22.739	1.407	1.979	0.062
6	Central cracks first	23.534	0.403	0.162	0.017

It can be seen from Table 6 that, if the specimen is cracked first from the top, the average strength tested will be reduced to the strength tested by the center crack due to the stress concentration at the top. More importantly, the failure time of the specimen is not fixed, which leads to the large discreteness of the measured strength value, and its standard deviation and coefficient of variation are increased by 248.15% and 264.70% compared to the specimen with the center crack first. Due to the problem of the test method, the test results show errors, which further affects the judgment of researchers on the material properties.

According to the previous definition of specimen failure, it can be seen from the data in Table 5 that, out of the 15 sets of tests that were administered using the conventional test circumstances, only 6 sets (with a pass rate of only 40%) met the assumption of "first crack at the center of the circle". The test yielded a maximum peak load of 24.928 KN and a minimum load of 20.268 KN, with a 22.97% variation in both values. This indicates that, on the one hand, it is challenging to achieve the optimal damage state within the existing specifications, and, on the other hand, there is a significant amount of fluctuation in the test data, which will result in a significant calculation error.

Table 5 shows that the peak time of the MTS load was later than the crack initiation time of the strain gauge in the center area of each test specimen, even those specimens that cracked first in the center of the circle. The cause of this phenomenon was the stress concentration at the point of contact between the loading strip's edge and the test specimen, which caused the local asphalt mixture to fail before it did so in the loading strip's center. To some extent, the magnitude of this erroneous impact can be reflected in the time difference. A larger discrepancy means that the peak load value that the MTS records will be higher, since the middle of the circle is shattered later than the end.

This study then calculated the difference between the strain gauge failure time and the MTS peak load time and the MTS peak load for 6 groups' central cracks first specimens, as well as their corresponding MTS peak loads, which were plotted as shown in Figure 7.

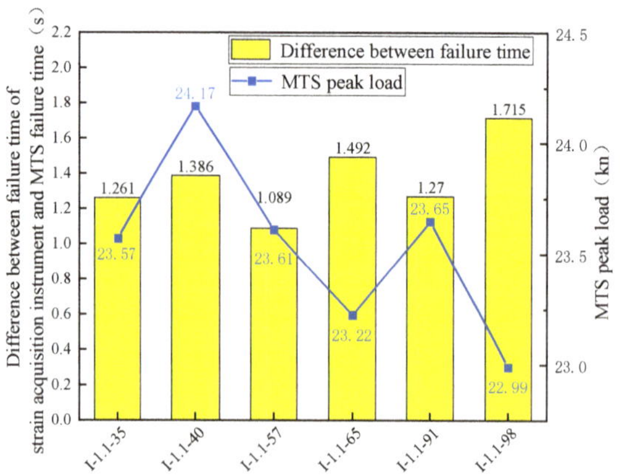

Figure 7. Time difference and MTS peak load in the specimens of "central cracks first".

Figure 7 shows that the average value of the difference between the peak time of the MTS load and the crack initiation time of the effective data center as measured by the strain acquisition instrument was 0.97 s. It is clear that there are issues with the traditional split test method, because previous studies have uniformly taken the average value of the split test data. For the specimen in which the end cracks before the center of the circle, the data are more discrete, and the value obtained will be larger than the true stress value of the specimen. Due to the test circumstances, most researchers pay little attention to this area, which results in a wide range of asphalt mixture splitting test findings that can only be averaged during analysis. As a consequence, the evaluation of asphalt mixtures' functioning is impacted.

3. Influence of the Splitting Test Conditions and Test Improvement

3.1. Effect of the Specimen Size

According to the Chinese highway specifications, when the nominal maximum particle size of the aggregate is less than or equal to 26.5 mm, the thickness of the test specimen should be taken as 63.5 mm. Some scholars have discovered that the splitting test exhibits significant size effects. Fahimifar A employed numerical simulations to conduct splitting and direct tensile tests under various sizes. The findings indicated that, the smaller the ratio of specimen thickness to diameter, the closer the test results approach those of direct tensile tests [9]. In order to explore the influence of specimens of different thickness on the time of cracking at the center of the circle, 35-mm- and 50-mm-thick cylindrical asphalt mixture specimens were selected for the splitting tests. The test conditions remained the same as in Section 2.3, and six valid data points were obtained. The differences between their strain failure times and the MTS failure time were summarized and processed, as shown in Figure 8 below.

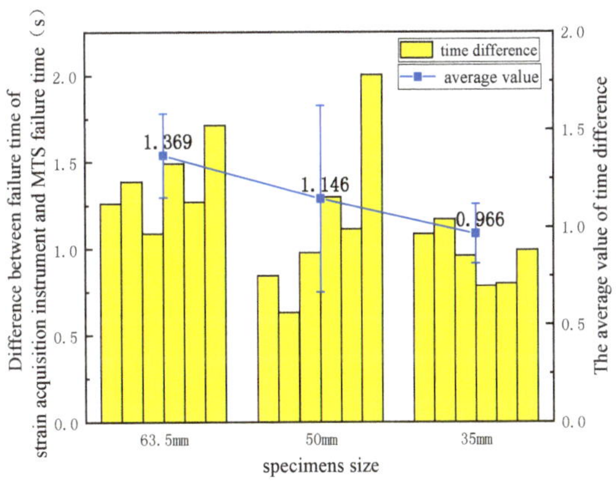

Figure 8. Time differences for differently sized specimens.

As shown in Figure 8 and Table 7, the average value of the "difference between the moment of strain gauge failure and the moment of peak MTS load" was approximately 0.966 s for a specimen thickness of 35 mm, which was 18.14% less than the specimen thickness of 50 mm and 41.03% less than the specimen thickness of 63.5 mm. The test results show that the difference between the cracking moment at the center of the specimen with a thickness of 35 mm and the peak moment of the MTS load was smaller, and the damage form of the specimen was also closer to the ideal state assumption in which the "center of the circle is the first to crack". Furthermore, the processing of the test data yielded a coefficient of variation of 0.0178 for the peak MTS load for a specimen thickness of 35 mm, which was 123% lower than the value of 0.0398 for the 50 mm specimen. This suggests that the data dispersion of specimens with a 35 mm size was smaller and more reliable.

Table 7. Statistical table of the test data under different specimen sizes.

Specimens Size (mm)	Statistical Results of Time Difference			
	Average Value	Standard Deviation	Variance	Variation Coefficient
63.5	1.369	0.216	0.047	0.158
50	1.146	0.478	0.229	0.417
35	0.966	0.152	0.023	0.157

The analysis revealed that, the thinner the specimen, the closer it is to the plane stress condition indicated by the theory of the splitting test, resulting in the most satisfying result of a 35 mm splitting specimen, based on our study. When compared to the 50-mm- and 63.5-mm-thick specimens, the time delay from the breakage at the center of the circle to cracking at the edge of the compression strip was shorter in the 35-mm-thick specimen, which is more favorable for further strength analysis and also more conducive to material savings. In conclusion, the suggested size for the splitting test in this study was 35 mm.

3.2. *Effect of the Test Loading Fixture*

The strength of the split test is significantly impacted by the test fixture as well. Previous array parallel tests on 12.7 mm arc bars have all revealed that the terminal stress concentration eventually cracks. P. Jonsén [17] conducted a rock splitting test by varying the arc of the arc splint. The outcomes demonstrated that altering the clamp's size could successfully reduce the stress concentration at the end. Therefore, 15 sets of parallel tests were conducted for each of the 19 mm arc bar splitting and rigid pressing plate splitting specimens in this section at the loading rate of 60 mm/min, chosen via earlier testing, maintaining the test specimen thickness of 35 mm and the test temperature of 15 °C. Figures 9–12 illustrate the cracking moments and typical damage diagrams of the 19 mm curved compression bar splitting and rigid compression plate splitting specimens.

Figure 9. Starting state of the 19 mm arc bead splitting end face.

Figure 10. State of the splitting end face of the rigid pressing plate at the time of starting.

The rigid plates and the split specimens were in linear contact at the start of the test, and the type of loading was linear loading. Because of the low load value in this state, even if there was stress concentration at the loading end, it was insufficient to induce a macro break. The load gradually increased as the test progressed, and the split specimen was converted from a linear load to a surface load. The uneven deformation of the asphalt mixture itself caused a high degree of stress concentration at the loading end, complex damage occurred at the end face, and the 19 mm curved compression strip specimen was similar to the rigid compression plate specimen. It was found that merely altering the loading fixture did not completely reduce the impact of stress concentration at the loading end.

Figure 11. The 19 mm arc bead splitting failure state.

Figure 12. Splitting failure state of the rigid pressing plate.

In line with the valid data points determined in Section 2.2, Figure 13 presents the differences between the MTS peak load and the strain acquisition instrument failure time for the valid data points in both loading modes.

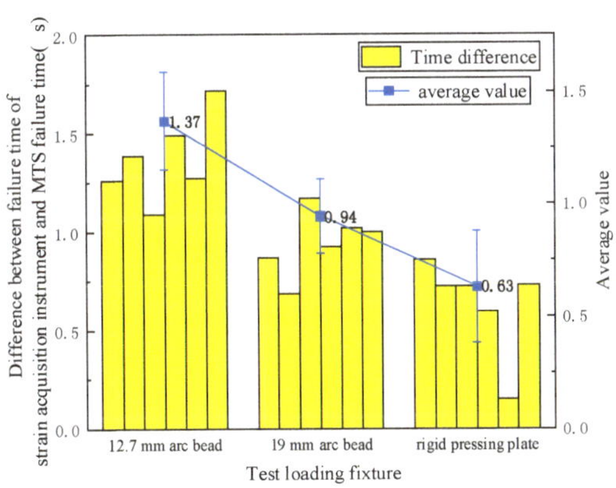

Figure 13. Time difference in various test loading fixtures.

Fifteen parallel splitting experiments were performed using each of the two enhanced loading fixtures, with only six achieving center-first cracking, for a 15% success rate. The success rate was not significantly increased over the specification loading technique, and the test still carried a degree of unpredictability. When the difference between the failure time of the strain gauge and the peak load time of the MTS was considered, the results were decreased by 2.5% for the 19 mm curved bar condition and 34.9% for the rigid plate condition when compared to the specification's 12.7 mm curved bar. This suggests that the rigid plate, among the three loading fixtures used in this work, may more readily achieve the ideal result of center-first cracking.

3.3. Improvement of the Splitting Test Method

In the previous stage, conventional methods were used to analyze the effects of changing the test conditions and specimen dimensions on the splitting test, and it was discovered that the assumption that the center of the circle would be the first to start splitting in the test process, as deduced by theory, was impossible to achieve. Some researchers in the field of rock and granular launchers have performed the Brazilian disc test by inserting rubber gaskets into the loading end and specimen, which effectively solves the problem of stress concentration in the contact area of the loading end and specimen and ensures that the cleavage test begins from the center of the cleavage [8]. The rubber gasket is intended to improve the splitting outcome at the loading end and asphalt mixture; after loading the rubber gasket, deformation occurs, and the asphalt mixture forms a certain type of curvature. However, it is not possible to apply rubber gaskets of different thicknesses at the loading end of a specific specimen in order to conduct an in-depth analysis. To conduct the asphalt mixture splitting test in this stage, this study used the loading method of 12.7 mm curved compression strips and rigid compression plates, which worked well in the previous experiments. Table 8 lists the individual test improvement techniques, whereas Figure 14 displays the test scheme. The remaining experimental conditions remained the same.

Table 8. Improvement of the splitting test with a rubber gasket and the number of parallel tests.

Test Number	Improvement Strategy	Number of Parallel Tests	Test Success Rate
I	3 mm rubber gasket + rigid pressing plate	6	33.3%
II	1 mm rubber gasket + rigid pressing plate	7	28.6%
III	10 mm rubber gasket + 12.7 mm arc-shaped batten	15	40%
IV	1 mm rubber gasket + 12.7 mm arc-shaped batten	7	28.6%
V	0.5 mm rubber gasket + 12.7 mm arc-shaped batten	12	16.7%
VI	0.3 mm rubber gasket + 12.7 mm arc-shaped batten	12	41.7%

Figure 14. Principle of rubber gasket splitting test. (**a**) Initial loading stage of the 12.7 mm arc-shaped batten. (**b**) Initial loading stage of the rigid pressing plate.

According to the Chinese road engineering standard, the stress at the center point of the split test circle may be determined using the elastic mechanics theory from Equation (1) [18].

$$\begin{cases} \sigma_x = -\frac{2P}{\pi A} = -0.637\frac{P}{A} \\ \sigma_y = \frac{6P}{\pi A} \\ \tau_{xy} = 0 \end{cases} \quad (1)$$

where σ_x and σ_y signify the tension in the x and y directions at the center of the specimen's circle. The peak load applied to the specimen (KN) is denoted by P, and the area of the specimen exposed to the force is denoted by A (m^2).

Because the compressive strength of the asphalt mixture is substantially larger than the tensile strength, the specimen's center point first exceeds the tensile stress limit to achieve destruction, and the destructive stress equals the tensile stress, as indicated in Equation (2):

$$R_t = \sigma_t = \sigma_x = 0.637\frac{P}{A} \quad (2)$$

where σ_T signifies the splitting strength of the specimen.

As can be seen in Figures 15 and 16, this study analyzed the test data and obtained results for the splitting strength, cracking duration, and cracking moment difference of valid specimens under six distinct loading scenarios, in accordance with the judgement method of specimen failure presented in Section 2.3.

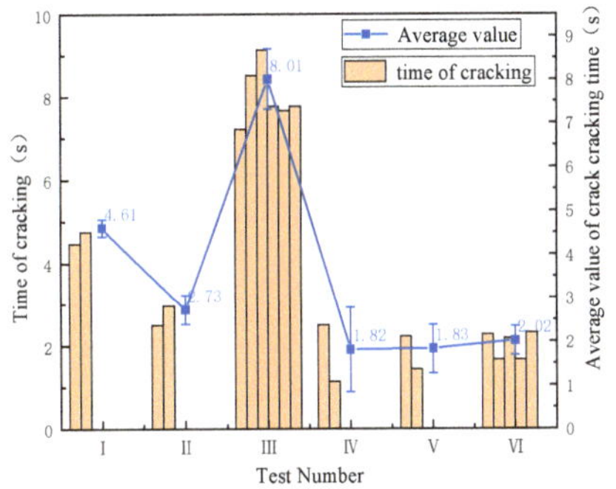

Figure 15. Scatter plot of the data at the time of cracking.

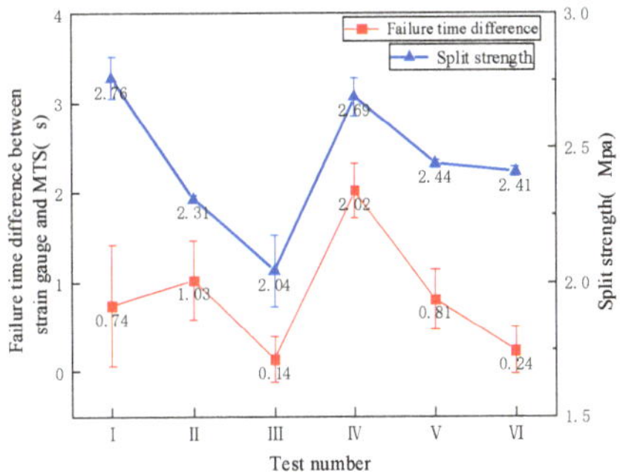

Figure 16. Failure time difference between the strain gauge and MTS under different test conditions.

The data in Figure 15 reveal that, when the rubber pad thickness increases from 1 mm to 10 mm, there is a significant increase in the failure time, and both the success rates of the test show improvement. Conversely, when the rubber pad thickness increases from 0.3 mm to 1 mm, there is no substantial change in the failure time, but both the success rate of the test and the dispersion of the data decrease. This suggests that both overly thick and overly thin rubber pads can reduce the experimental error. This is a result of the rubber gasket's buffering effect on the split test force. The stronger the buffering effect, the thicker the rubber gasket. However, if the rubber pad is excessively thick, there will be a loss in the force transmission process, which will result in the measured strength value being lower. As a result, the longer the split failure time and the longer the load transmission time on the rubber gasket, the greater the data dispersion. Compared to rigid loading, the loading method of 0.3 mm rubber mat + curved compression strip can increase the success rate of the test and reduce the dispersion of the data.

The statistical difference between the MTS failure time and the strain gauge failure time at the effective data point is depicted in Figure 16. Although the average time obtained by splitting the 12.7 mm arc bead is shortened to 0.140 s, which significantly lessens the impact of stress concentration, the starting time of the strain gauge is still later than the peak time of the MTS load. This finding suggests that the introduction of rubber gaskets can indeed resolve the issue at the loading end. The starting point of the splitting test can be transferred from the edge of the loading strip to the center area after the thickness of the rubber gasket is reduced, which is consistent with one of the theoretical assumptions of the splitting test. In comparison to the other loading circumstances, the splitting strength values demonstrated a considerable decrease under the loading of a 10 mm rubber gasket + 12.7 mm arc-shaped batten. This shows that, while the rubber mat helps the material to break first in the center of the circle, it also absorbs some of the energy, lowering the splitting strength. As a result, it is important to regulate the rubber mat's thickness as much as possible within a range that does not impact its splitting strength.

This study processed the test findings using the principle of valid data screening, which is explained in Section 2.3, to create a stress–strain curve for each valid specimen, as shown in Figure 17, in order to better assess the stress patterns of the specimens.

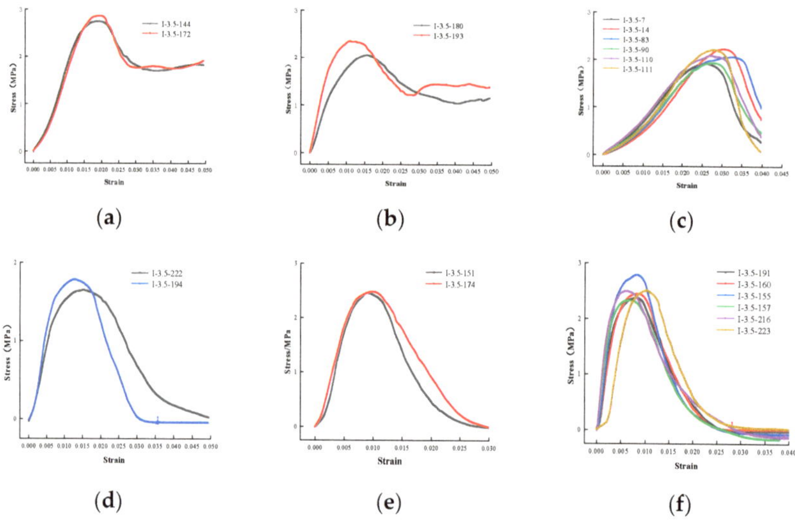

Figure 17. Stress–strain at the center of a circle with different improvement strategies using rubber gaskets. (**a**) A 3 mm rubber gasket + rigid pressing plate; (**b**) 1 mm rubber gasket + rigid pressing plate; (**c**) 10 mm rubber gasket + 12.7 mm arc-shaped batten; (**d**) 1 mm rubber gasket + 12.7 mm arc-shaped batten; (**e**) 0.5 mm rubber gasket + 12.7 mm arc-shaped batten; (**f**) 0.3 mm rubber gasket + 12.7 mm arc-shaped batten.

The test findings suggest the "0.3 mm rubber gasket + 12.7 mm arc-shaped batten" combination as a superior splitting test technique for the following reasons:

(1) It is clear from Figure 17 that the enhanced splitting result of the "0.3 mm rubber gasket + 12.7 mm arc-shaped batten" strategy better characterizes the elastic deformation stage in the stress–strain curve than previous techniques, and the test results are generally more trustworthy.
(2) In the cracking test, the "0.3 mm rubber gasket + 12.7 mm arc-shaped batten" can successfully lessen the failure time difference and accomplish the ideal result in which the "center of the circle cracks first". It also has the most logical strength data, a high success rate, and little data dispersion from the perspective of data analysis.

(3) By comparing the splitting test results for rubber gaskets of various thicknesses, it is discovered that "0.3 mm rubber gasket + 12.7 mm arc-shaped batten" is the most economical and cost-effective.

3.4. Research on the Strength Model of the Asphalt Mixture Splitting Test

This study employed sophisticated digital image techniques to monitor the surface strain during the test in order to gain a better understanding of the failure mode of the split specimen. First, the specimen had to be pretreated with a size grid and dispersant. A high-speed camera was utilized to record the specimen's deformation during the test; then, the surface strain was processed, and the results were plotted on a graph. Among them, the surface strain of the specimen obtained in the loading mode of "0.3 mm rubber gasket + 12.7 mm arc-shaped batten" is shown in Figure 18.

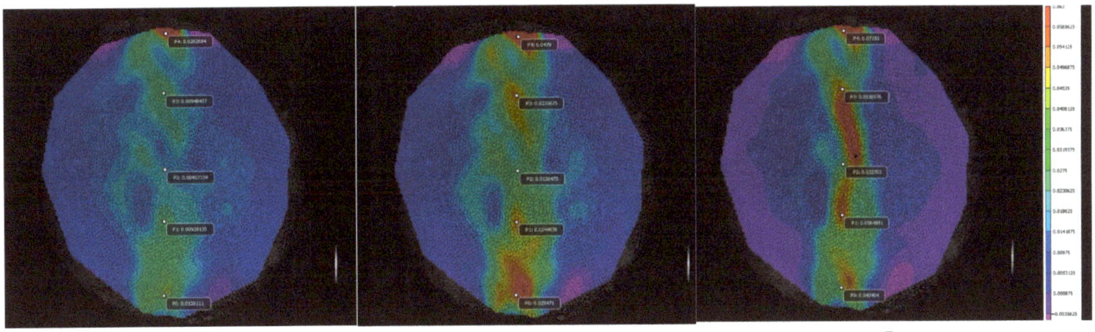

3 s 4 s 5 s

Figure 18. Displacement cloud maps at different time points.

Figure 18 illustrates that the specimen's end pressure caused substantial deformation during the initial loading stage. At the core of the specimen's circle, progressive tensile strain started to form during the middle loading phase. Once the specimen reached a later stage of loading, the center of the circle reached the destructive strength of the material before the end, causing cracking (the point of failure as seen on the strain cloud). The specimen's interior was subjected to both the pressure descending from the tip and the tensile force caused by the material's deformation.

According to the current Chinese code, damage occurs when the tensile stress in the center of the circle approaches the material's maximum tensile limit. The associated strength theory is a one-dimensional strength criterion [19]. However, it is clearly incorrect to simply consider the tensile limit at the split specimen's center as the destructive load, since the split specimen's core is, in fact, in a two-dimensional stress state of vertical pressure and transverse tension. The strength model for the splitting test should take into account the impacts of tensile and compressive stresses, since the damage point at the end face is under both of these types of stress.

The octahedral shear stress strength theory takes into account the three-way principal stresses, σ_1, σ_2, and σ_3, in the specimen and is often used to describe the damage of materials under complex stress states, such as asphalt mixtures and concrete. The relationship between the relative normal stress and the relative shear stress of the octahedron of asphalt mixtures can be determined as follows by analyzing the test data points in order to fit the damage meridian equation in the form of a functional equation for the damage criterion of asphalt mixtures [20]:

$$\frac{\tau_{otc}}{f_c} = A\frac{\tau_{otc}}{f_c} + B \tag{3}$$

where coefficients A, B, and f_c are the material parameters and uniaxial compressive strength of asphalt concrete (Mpa). $(\frac{\tau_{otc}}{f_c})$ and $(\frac{\tau_{otc}}{f_c})$ are the values of normal stress and shear stress on the compression meridian. σ_{oct} and τ_{otc} are the octahedral normal stress and shear stress and are calculated by the formulas below.

$$\sigma_{oct} = \frac{1}{3}(\sigma_1 + \sigma_2 + \sigma_3) \tag{4}$$

$$\tau_{otc} = \frac{1}{\sqrt{3}}\sqrt{(\sigma_1 - \sigma_2)^2 + (\sigma_1 - \sigma_3)^2 + (\sigma_2 - \sigma_3)^2} \tag{5}$$

Many scholars have studied the failure mechanism of asphalt mixtures through triaxial tests. Huang et al. produced an AC-13C asphalt mixture and conducted triaxial tensile and compressive tests under different loading modes to obtain the octahedral strength theory failure criterion for this type of asphalt mixture, as follows [21]:

$$\frac{\tau_{otc}}{f_c} = -0.572\frac{\tau_{otc}}{f_c} + 0.295 \tag{6}$$

Four representative splitting test data of 63.5 mm, 50 mm, 35 mm, and 35 mm + 0.3 mm rubber mats under the loading of 12.7 mm curved compression strips were selected and substituted into Formulas (3) and (4) in order to verify the improvement effect of the splitting test in this study. f_c was obtained in accordance with the requirements of the uniaxial compression test of asphalt mixtures, and the average was 11.4 Mpa. After processing the data points, this study considered them alongside Huang's suggested damage criterion [21] for the AC-13C asphalt mixture on the hydrostatic pressure surface. Figure 19 presents the findings.

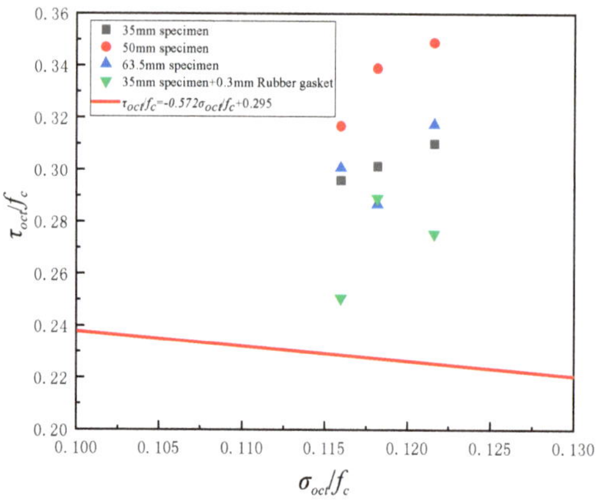

Figure 19. Damage curves and stress points.

As shown in Figure 19, after optimizing the loading technique and split test specimen size, the data obtained were closer to the damage curve obtained from the triaxial test under the strength criterion of a tensile and compressive composite, thereby reducing the error and increasing the test accuracy when compared to the pre-improvement case.

Although, compared to other test methods, the improved data are closer to the ideal failure state, due to the facts that the splitting tests conducted in this study are all two-dimensional tension compression tests and Huang's formula is obtained from the tension compression composite test under a three-dimensional stress state, the test data points and failure model still cannot fit well in the stress space, which will also be the focus of future research.

4. Analytical Comparison between the Splitting Test and the Direct Tensile Test at Different Loading Rates

Since the split test center is under mixed compressive and tensile stress, its strength result cannot be purely interpreted as a direct tensile strength value. Nevertheless, the split test is an indirect tensile test often used to assess the tensile qualities of materials. The straightforward splitting test may take the place of the problematic direct tensile test, greatly aiding in the tensile property testing of materials, provided that the splitting test results can be connected to the direct tensile results.

Using the same conditions as in the previous experiment, rectangular specimens of the AC-13C asphalt mixture were cut, measuring 50 mm × 50 mm × 250 mm. Strain gauges were attached to the specimens' four outside surfaces in order to record the specimens' breakage moments, and the ends of the specimens were attached to the MTS loading plate using epoxy glue. In order to more accurately analyze the effect of different loading rates on the damage of the specimen, tension force was applied in the form of stress control, and the MTS tension rates were 2 kn/s, 4 kn/s, 6 kn/s, and 8 kn/s, respectively. The splitting test used the recommended loading method described earlier, also in the form of a stress control, with loading rates of 2 kn/s, 4 kn/s, 6 kn/s, and 8 kn/s. The test temperature was 15 °C, as indicated in the specifications. The loading process of the direct tensile test is illustrated in Figure 20, and the specimen after tensile failure is shown in Figure 21.

Figure 20. Tensile test loading process.

Figure 21. Specimen after loading for the tensile test.

According to the research by Lv [22], the tensile strength of asphalt mixtures exhibits a power function relationship with the rate, as illustrated in Equation (7):

$$R = \alpha \times V^{\beta} \tag{7}$$

where R is the strength (Mpa), V is the loading rate (Mpa/s), and α and β are the material parameters.

Substituting the direct tensile test results into Equation (7), the fitting gives the relationship between the strength and loading rate, as shown in Equation (8):

$$R_t = \alpha_t \times V^\beta = 3.575 \times V^{0.339} \left(R^2 = 0.968\right) \tag{8}$$

where R_t is the tensile strength (Mpa), and α_t denotes the tensile material parameters (3.575).

The splitting test results were fitted using Equation (7) to determine the strength–loading rate relationship, which is represented in Equation (9). Using the result from Equation (8), β is considered to be 0.339.

$$R_s = \alpha_s \times V^\beta = 2.724 \times V^{0.339} \left(R^2 = 0.989\right) \tag{9}$$

where R_s is the split strength (Mpa), and α_s denotes the split material parameters (2.724).

Based on Equations (8) and (9), the relationship between the tensile strength (R^T) and splitting strength (R^s) can be derived, as shown in Equation (10):

$$R_t = 1.312R_s(V) \tag{10}$$

The model hypothesis test results of Equations (8) and (9) are shown in Table 9, and the model has good accuracy.

Table 9. Model t-value test.

Name	Value	Standard Error	T-Value	Probability > |t|
α_t	3.575	0.039	91.631	0
α_s	2.724	0.020	132.638	0
β	0.339	0.016	20.754	1.844×10^{-12}

The results of Formulas (8) and (9), as well as all the splitting and tensile tests, are plotted in the strength–rate relationship diagram shown in Figure 22.

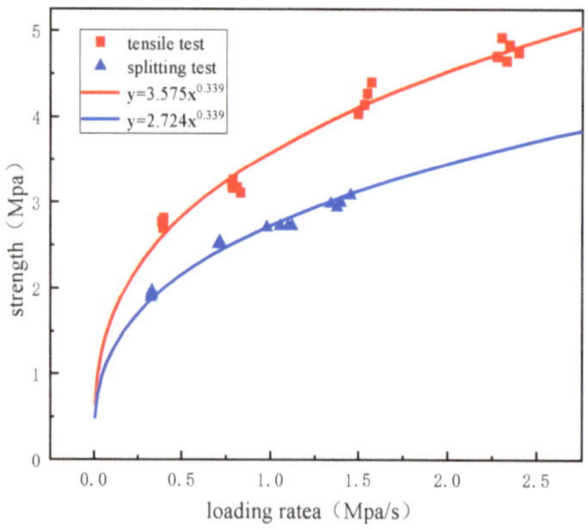

Figure 22. Strength–rate relationship diagram.

Figure 22 shows that the split strength and tensile strength increase initially with an increasing loading rate; the growth rate is first great and then modest, and it then gradually stabilizes, demonstrating a power function trend. The tensile strength is always larger than the splitting strength, as seen in Figure 22, and the two parameters are linearly connected. This also verifies the research presented in Section 3.4. The split test theory regards the transverse ultimate tensile stress at the center of the circle as the indirect tensile breaking strength, but, in reality, the split test is a combination of transverse tensile and longitudinal compression, and the asphalt mixture is more easily damaged than in the direct tensile case. As a result, the indirect tensile strength attained in accordance with the specification will be lower than the direct tensile strength. To minimize errors, the strength–rate unification model derived from the split and tensile test data can be used to quickly determine the direct tensile strength of asphalt mixtures using the split test, greatly simplifying the test procedure and facilitating engineering applications.

5. Conclusions and Recommendations

Based on the analysis of the specimen cracking time, this study used digital image technology to analyze and improve the splitting test of asphalt mixtures. The results can provide a reference for the indirect tensile testing of asphalt mixtures, concrete, and rock.

1. According to the "China highway code", the splitting test of asphalt mixtures was carried out. The end of the sample often breaks before the center of the circle, resulting in stress concentration, and it deviates from the perfect stress state specified in the specifications, resulting in significant errors in the calculation of the splitting strength.
2. When using the test strategies of a "35 mm specimen thickness" and "0.3 mm rubber gasket + 12.7 mm arc-shaped batten", the outcome of the "first crack at the center of the circle" is essentially realized.
3. The octahedral strength theory proves that the improved stress point is closer to the existing failure criterion, indicating that the improved splitting test is closer to the ideal failure stress state. In addition, it also successfully reduces the discreteness of the data and improves the success rate of the testing.
4. Based on the test results, it is found that there is a power function relationship between the strength and the loading rate, and the unified equations of the tensile and splitting strength rate are constructed. This not only improves the operability but also provides a reference for engineering applications, so that researchers can use the splitting test to quickly estimate the tensile strength of asphalt mixtures.

In future work, the effects of different test loading rates and loading modes on direct and indirect tensile tests will be analyzed. At the same time, the corresponding relationship between the splitting strength and direct tensile strength will be discussed using different strength theories. Finally, the unified direct tension and splitting and fatigue equations will be further established.

Author Contributions: Methodology, H.G. and W.P.; Investigation, H.Y.; Data curation, H.G. and Y.Y.; Writing—original draft, W.P.; Writing—review & editing, H.G. and W.P. All authors have read and agreed to the published version of the manuscript.

Funding: This research and the APC were funded by National Natural Science Foundation of China, grant number: 52078062.

Data Availability Statement: The data presented in this study are available on request from the corresponding author.

Conflicts of Interest: The authors declare no conflicts of interest.

References

1. He, Z.; Xie, T.; Yu, H.; Ge, J.; Dai, W.J.C.; Materials, B. Evaluation of quantum dot composite graphene/Titanium oxide enhanced UV aging resistance modified asphalt. *Constr. Build. Mater.* **2023**, *408*, 133732. [CrossRef]
2. Ren, S.; Liu, X.; Lin, P.; Wang, H.; Fan, W.; Erkens, S.J.C.; Materials, B. The continuous swelling-degradation behaviors and chemo-rheological properties of waste crumb rubber modified bitumen considering the effect of rubber size. *Constr. Build. Mater.* **2021**, *307*, 124966. [CrossRef]
3. Li, M.; Ren, S.; Liu, X.; Wu, Z.; Zhang, H.; Fan Xu, J.J.M. A Comprehensive Study on the Rejuvenation Efficiency of Compound Rejuvenators for the Characterization of the Bituminous Binder, Mortar, and Mixture. *Materials* **2022**, *15*, 5458. [CrossRef] [PubMed]
4. Li, S.; Sun, Y.; Xu, L.; Yu, S.; Liang, X.; Ye, J.J.B. Asphalt Layer Cracking Behavior and Thickness Control of Continuously Reinforced Concrete and Asphalt Concrete Composite Pavement. *Buildings* **2022**, *12*, 1138. [CrossRef]

5. Takaikaew, T.; Hoy, M.; Horpibulsuk, S.; Arulrajah, A.; Mohammadinia, A.; Horpibulsuk, J.J.H. Performance improvement of asphalt concretes using fiber reinforcement. *Heliyon* **2021**, *7*, e07015. [CrossRef]
6. Ahmed, K.U.; Geremew, A.; Jemal, A.J.H. The comparative study on the performance of bamboo fiber and sugarcane bagasse fiber as modifiers in asphalt concrete production. *Heliyon* **2022**, *8*, e09842. [CrossRef]
7. Carneiro, F.C. A new method to determine the tensile strength of concrete. In Proceedings of the 5th Meeting of the Brazilian Association for Technical Rules, Lisbon, Portugal, 16 September 1943.
8. Akazawa, T. New test method for evaluating internal stress due to compression of concrete (the splitting tension test) (part 1). *J. Jpn. Soc. Civ. Eng.* **1943**, *29*, 777–787.
9. Fahimifar, A.; Malekpour, M. Experimental and numerical analysis of indirect and direct tensile strength using fracture mechanics concepts. *Bull. Eng. Geol. Environ.* **2012**, *71*, 269–283. [CrossRef]
10. Małyszko, L.; Kowalska, E.; Bilko, P. Splitting tensile behavior of autoclaved aerated concrete: Comparison of different specimens' results. *Constr. Build. Mater.* **2017**, *157*, 1190–1198. [CrossRef]
11. *ASTM D3967-16*; Standard Test Method for Splitting Tensile Strength of Intact Rock Core Specimens. ASTM International: West Conshohocken, PA, USA, 2008.
12. Garcia-Fernandez, C.; Gonzalez-Nicieza, C.; Alvarez-Fernandez, M.; Gutierrez-Moizant, R. Analytical and experimental study of failure onset during a Brazilian test. *Int. J. Rock Mech. Min. Sci.* **2018**, *103*, 254–265. [CrossRef]
13. Gutiérrez-Moizant, R.; Ramírez-Berasategui, M.; Santos-Cuadros, S.; García-Fernández, C.J.R.M.; Engineering, R. Computational verification of the optimum boundary condition of the Brazilian tensile test. *Rock Mech. Rock Eng.* **2018**, *51*, 3505–3519. [CrossRef]
14. Erarslan, N.; Williams, D.J. Experimental, numerical and analytical studies on tensile strength of rocks. *Int. J. Rock Mech. Min. Sci.* **2012**, *49*, 21–30. [CrossRef]
15. Gutiérrez-Moizant, R.; Ramírez-Berasategui, M.; Sánchez-Sanz, S.; Santos-Cuadros, S. Experimental verification of the boundary conditions in the success of the Brazilian test with loading arcs. An uncertainty approach using concrete disks. *Int. J. Rock Mech. Min. Sci.* **2020**, *132*, 104380. [CrossRef]
16. Sarfarazi, V.; Haeri, H.; Ebneabbasi, P.; Shemirani, A.B.; Hedayat, A. Determination of tensile strength of concrete using a novel apparatus. *Constr. Build. Mater.* **2018**, *166*, 817–832. [CrossRef]
17. Jonsén, P.; Häggblad, H.; Sommer, K. Tensile strength and fracture energy of pressed metal powder by diametral compression test. *Powder Technol.* **2007**, *176*, 148–155. [CrossRef]
18. Zhang, C.; Yu, H.; Zhu, X.; Yao, D.; Peng, X.; Fan, X. Unified Characterization of Rubber Asphalt Mixture Strength under Different Stress Loading Paths. *J. Mater. Civ. Eng.* **2024**, *36*, 04023498. [CrossRef]
19. Qian, G.; Luo, X.; Yu, H.; Shi, C.; Zhang, C.; Li, P. Evaluation of Material Composition on the Shear Performance of Asphalt Mixtures under Different Test Conditions. *Buildings* **2023**, *13*, 936. [CrossRef]
20. Wang, P.; Elchalakani, M.; Zhou, Y.; Yan, S.-T.; Li, S.-B. A study on the influences of a hygrothermal environment on the compressive strength and failure criteria of asphalt mixtures based on true triaxial tests. *Heliyon* **2022**, *8*, e10060. [CrossRef] [PubMed]
21. Huang, T.; Tang, Y.; Li, M.; Chen, J.; Xie, J.; Lv, S.; Qian, G.; Liu, H. Study on multiple failure criteria of asphalt mixtures in complex stress states. *J. Mater. Civ. Eng.* **2021**, *33*, 04021272. [CrossRef]
22. He, Y.; Lv, S.; Wang, Z.; Ma, H.; Lei, W.; Pu, C.; Meng, H.; Xie, N.; Peng, X. Research on the Modulus Decay Model under a Three-Dimensional Stress State of Asphalt Mixture during Fatigue Damage. *Buildings* **2023**, *13*, 2570. [CrossRef]

buildings

MDPI

Article

Study of Low-Content Epoxy Asphalt Mixture Applied to the Road

Xiaodong Li [1], Chuanxi Luo [2,3], Shaohuai Wang [2], Xiang Long [1], Yan Wang [1], Jian Li [3,*] and Mu He [1]

1 Guangzhou North Second Ring Road Traffic Technology Co., Ltd., Guangzhou 510540, China; li.xiaodong@yuexiutransport.com (X.L.); long.xiang@yuexiutransport.com (X.L.); wang.yan4@yuexiutransport.com (Y.W.); he.mu@yuexiutransport.com (M.H.)
2 School of Civil Engineering and Transportation, South China University of Technology, Guangzhou 510641, China; 18825181976@163.com (C.L.); 15391104229@163.com (S.W.)
3 Xiaoning Institute of Roadway Engineering, Guangzhou 510000, China
* Correspondence: m15738523861@163.com; Tel.: +86-157-3852-3861

Abstract: To realize the application of epoxy asphalt concrete on roads and solve the problem of the high cost of epoxy asphalt concrete and the causes of bulging in the construction process, a low-content epoxy asphalt mixture was experimentally studied. Rheological and microscopic tests were carried out to study its performance. At the same time, 17 kinds of low-content asphalt mixture Marshall test pieces and rut plate test pieces with different oil stone ratios were made. Their road performances were tested, and a watertight breathable epoxy asphalt mixture was studied to solve the bulging problem. The research shows that, for ordinary roads, a content of epoxy resin of 10–15% can meet both the high-temperature and the low-temperature requirements. For sections with special rutting resistance requirements, a controlled epoxy resin content between 15 and 30% is recommended. When the content of epoxy resin is greater than 30%, epoxy asphalt initially forms a crosslinked spatial network. An epoxy asphalt with a complex structure from asphaltene to epoxy, with ultra-high performance, can be used for small steel bridge pavements. A BBR test showed that, with an increase in epoxy resin content, the low-temperature performance of asphalt gradually weakens. When the content was 20%, epoxy asphalt's low-temperature performance was weaker than that of SBS-modified asphalt. Under extremely low-temperature conditions, the performance of epoxy asphalt could not meet the specifications. When the voids of low equivalent epoxy asphalt mixture are controlled at 4.1–4.5%, it is watertight and breathable; this can solve the problem of bulging and greatly reduce the cost of projects.

Keywords: pavement engineering; low content; epoxy asphalt; road performance

Citation: Li, X.; Luo, C.; Wang, S.; Long, X.; Wang, Y.; Li, J.; He, M. Study of Low-Content Epoxy Asphalt Mixture Applied to the Road. *Buildings* **2024**, *14*, 443. https://doi.org/10.3390/buildings14020443

Academic Editor: Pengfei Liu

Received: 5 December 2023
Revised: 8 January 2024
Accepted: 19 January 2024
Published: 6 February 2024

1. Introduction

The service life of asphalt pavements is 15 years, and some long-life pavements have achieved a service life of 30 years; however, high-temperature rutting issues in specialized sections, such as climbing sections and level intersections, are still frequent due to the unreasonable functional design of asphalt pavement materials [1–4]. The proportion of high-temperature shear instability rutting in asphalt pavement rutting is over 90% [5]. At home and abroad, the high-temperature rutting problem caused by unreasonable pavement function design is mainly compensated for by improving the material performance standards. These methods include using high-performance-modified asphalt and function-enhanced additives to improve the material's performance; for example, one can use PG82 asphalt in long steep slope sections and in pouring semi-flexible pavement structures at level intersections [6–8]. The construction viscosity of PG82-modified asphalt is too high, construction quality control is difficult, and the high construction temperature is also problematic due to the high level of energy consumption. Experience of applying PG82-modified asphalt shows that the rutting problem of long steep slopes has not been completely solved [9,10]. Poured, semi-flexible pavement structures are not wear-resistant and cannot play the function of a wear layer in pavement structures [11–13]. The research shows that epoxy asphalt pavements exhibit excellent performances, high-temperature stability, fatigue durability, and aging resistance [14–16]. The thermosetting characteristics of epoxy asphalt can, in theory, fundamentally solve the rutting problem experienced by asphalt pavements. It will become an advantageous pavement material for dealing with long steep slopes, service areas, bridge decks, and other working conditions.

Epoxy asphalt is prepared by adding epoxy resin, curing agent, and other auxiliaries to the mixture. The epoxy group of epoxy resin is crosslinked with the reaction site of the amine curing agent through a ring-opening procedure; here, an irreversible curing epoxy crosslinked system is formed after the hardening reaction. Epoxy resin is a thermoplastic polymer containing two or more epoxy groups [17], which can be cured using various curing agents to form a thermosetting, three-dimensional, crosslinked network structure. Such polymers have strong adhesion, high-temperature and oil corrosion resistance, and are often used as adhesives and waterproof materials in the industry. Its molecular structure contains rigid groups such as benzene rings and isopropyl; this gives this kind of epoxy resin excellent mechanical properties and heat resistance. Polar groups such as the ester group and the hydroxyl group give it excellent bonding properties for steel and glass, and it can be used as an adhesive [18]. Due to these properties of epoxy resin, asphalt's strength and high-temperature performance can be greatly improved. A curing agent is another factor that enhances the excellent properties of epoxy asphalt. The epoxy groups in the structure of epoxy resin crosslink with the reaction site of the curing agent molecules through the open ring; here, a cyclo-oxygen crosslinked system with sufficient mechanical strength is formed during the curing reaction. The crosslinked system has high tensile strength, thermal stability, and other properties [19,20]. Therefore, it can effectively improve asphalt pavement performance, such as its high-temperature stability and durability. The Shell Oil Company of the United States has begun to develop an epoxy asphalt mixture. To address the issue of aircraft fuel and jet incursions on airport runways, the Shell Oil Company developed a thermosetting epoxy asphalt in the 1950s. Epoxy resins and curing agents were used to form a three-dimensional mesh structure by crosslinking during the curing reaction. Asphalt, as a dispersed phase, was filled in the three-dimensional mesh structure formed by crosslinking the epoxy resin, forming a two-phase, three-dimensional mesh structure [21–24]. Epoxy asphalt was first applied to the steel deck of the San Mateo Hayward Bridge in 1967 [25,26]. With the successful application of epoxy asphalt, epoxy asphalt has been widely used in orthotropic steel bridge deck pavements with complex stresses, such as the San Diego Coronado Bridge in California [27–30]. In the past three decades, epoxy asphalt mixtures have been used in steel bridge deck paving, mainly in the United States, Canada, the Netherlands, and Australia, with the United States being the most widely used. In the 1970s, Japanese pavement scholars systematically studied epoxy asphalt mixture's configuration, modulus, stress relaxation, and destruction performance [31,32]. Japan's "Asphalt Pavement Outline", 1988, presented detailed provisions for epoxy pavement technologies, from design to construction [33,34]. In the same period, epoxy asphalt, such as KD-BEP, produced in Japan, was widely used and successfully applied in pavement projects, such as the Shanghai Yangtze River Bridge, the Humen Bridge, and the Nansha Bridge in China [35]. Because steel bridge pavements need a good collaborative deformation ability, ordinary asphalt pavement materials cannot be applied. In contrast, epoxy asphalt has an ultra-high bonding performance and can be tightly bonded with steel plates. Therefore, it is widely used in steel bridge deck paving. However, due to the addition of epoxy materials, the expense of epoxy asphalt mixtures is 10–20 times higher than that of general asphalt mixtures. When using expensive epoxy asphalt mixtures on general roads, the problem of imbalance between input and output is very prominent.

Due to the special properties of epoxy asphalt mixtures, researchers have attempted to research general road applications [36,37]. The proportion of epoxy resin will directly affect epoxy asphalt's structure and mechanical performance [38–40]. Li et al. prepared epoxy asphalts with different epoxy proportions and performed microanalysis, which showed that when the proportion of epoxy was more than 30%, the epoxy resin formed a skeleton. Epoxy asphalt's softening point and tensile strength, compared to ordinary asphalt, have significantly increased, mainly affected by the epoxy resin crosslinking the formation of the network structure [41]. Huang Hongming observed the cured hot-mix epoxy asphalt with a fluorescence microscope. The research showed that the phase of epoxy asphalt changed from asphalt to an epoxy resin network structure when the proportion of epoxy was between 30% and 40%. An SEM test showed that the cured epoxy asphalt formed a network structure when the content was 40%. The cured particles of epoxy resin were distributed evenly when the content was 50%, and the epoxy asphalt formed a stable and dense network structure [42]. Li Xiaoxu observed the epoxy asphalt with epoxy content between 31 and 40% by SEM and found that 37% was the critical value for epoxy asphalt to form a spatial network structure [43]. Xu Peijun et al. studied the relationship between the curing process and the mechanical performance of epoxy through three aspects: different curing agent content, temperature, and time. The research found that epoxy's fracture elongation and toughness increased with the raising of curing agent proportion. In addition, its tensile strength increased and its elongation at break decreased with increasing curing time. The above studies analyzed the spatial network structure of epoxy asphalt. They investigated limiting the dosage of epoxy, but there is no practical solution for the mixture performance and construction convenience after reducing the dosage of epoxy [44]. Epoxy asphalt is a thermosetting material that changes from a fluid to a solid state with increasing temperature and time. It is still difficult to grasp the time and temperature window of epoxy asphalt pavement construction in engineering applications, and the control of construction quality and rolling technology lacks solid data support [45–47]. In addition, the gradation design of epoxy asphalt mixture adopts a continuous dense gradation form in steel bridge deck pavement, which belongs to a suspended dense structure to prolong the fatigue performance and waterproof performance. Due to being too dense, epoxy asphalt pavements easily form bulges, which causes great construction difficulty to directly quote the research on steel bridge deck pavement systems [48–51]. The bulging problem of epoxy asphalt concrete has not been substantially solved.

The demand for epoxy asphalt mixtures is very high due to their excellent rutting resistance, fatigue durability, and aging resistance. However, epoxy asphalt mixtures are more costly and unusually difficult to construct due to the problem of rumble, which prevents them from being widely used. In this study, by carrying out tests on rheological properties and microscopic appearance of epoxy asphalt with different epoxy resin content, the limiting content of epoxy resin applicable to different places was summarized, and a low-content epoxy asphalt system was established. The low-content epoxy asphalt mixture with dense water and air permeability is proposed to solve the application problem of epoxy asphalt mixture on ordinary roads. The results of this study will contribute to the widespread use of epoxy asphalt on ordinary roads.

2. Experimental Plan

2.1. Preparation

The epoxy resin and curing agent were preheated at 60 °C for about 1 h and then mixed according to a certain proportion and stirred manually for about 3 min. The mixed epoxy resin was mixed with 165 °C basis asphalt according to a certain proportion, and then shearing and dispersing were carried out by a high-speed shearing instrument at 5000 r/min for 5 min to mix epoxy asphalt at different epoxy proportions. See Tables 1 and 2 for performance indicators of epoxy resin and basis asphalt.

Table 1. Technical indicator of epoxy resin.

Item	Technical Indicator	Test Method
Tensile strength (23 °C, MPa)	3.83	GB/T 16777-2008 [52]
Elongation at break (23 °C, %)	252	GB/T 16777-2008
Viscosity (120 °C) increased to 5000 mPa·s (min)	45	JTG E20-2011 [53]

Table 2. Technical indicator of basis asphalt.

Item	Indicators	Test Method
Penetration (25 °C, 100 g, 0.1 mm)	64.5	JTG E20-2011
Softening point TR and B (°C)	47.7	JTG E20-2011
Ductility (10 °C, 5 cm/min, cm)	31.5	JTG E20-2011
Flashpoint (°C)	356	JTG E20-2011
Solubility (%)	99.96	JTG E20-2011
Density (15 °C, g/cm^3)	1.04	JTG E20-2011
Dynamic viscosity at 60 °C (Pa viscosity)	191	JTG E20-2011
	Residue after RTFOT	
Quality change (%)	−0.3	JTG E20-2011
Residual ductility (10 °C, cm)	8.5	JTG E20-2011

2.2. Fluorescence Microscope Test

A fluorescence microscope was used to irradiate the tested sample with a specific light source to make it emit certain fluorescence. Because different materials have different reflection characteristics on fluorescence, the shape and specific position of samples can be observed under a microscope. Under the irradiation of blue light (wavelength 488 nm), epoxy resin is strongly excited, and the morphology of epoxy in epoxy asphalt can be directly observed without adding a fluorescent indicator. A small amount of epoxy asphalt was dipped and placed on a glass slide, which was then covered with a coverslip, and the epoxy asphalt was flattened with appropriate force. The sample was put in a constant-temperature oven at 60 °C for 4 days and at 25 °C for 1 day. After curing, fluorescence microscopy was used to observe the epoxy asphalt samples and to analyze the structural morphology of the epoxy resin and the base asphalt in different epoxy asphalts. The proportions of epoxy resin were 0%, 10%, 20%, 30%, 40% and 50%, respectively.

2.3. Rheological Test

The dynamic shear rheometer is a general instrument for evaluating the rheological properties of materials in the polymer materials industry, and it is also an important piece of test equipment for asphalt of middle and high-temperature properties in the SHRP program of the United States. The rheological performance of SBS-modified asphalt (SBS content 4.5%) and epoxy asphalt was tested using an AR2000 rheometer (TA Corporation, New York, NY, USA). The proportions of epoxy resin were 5%, 10%, 15%, 20%, 25%, 30% and 40%, respectively. The test temperatures were 58 °C, 64 °C, 70 °C, 76 °C, 82 °C, 88 °C, 94 °C, 100 °C, 106 °C and 112 °C, respectively.

2.4. Watertight Breathable Mixture Test

Seventeen kinds of epoxy asphalt mixture Marshall specimens with different asphalt–aggregate ratios were prepared, and Marshall tests were carried out, with the mixture gradation as shown in Table 3; the ratio of asphalt mass to aggregate mass was 5.5–7.0%, and the interval was 0.1%. The ratio of asphalt mass to aggregate mass was increased by 5.0% as a supplementary group to expand the test boundary. The mold of the rut plate specimen was modified to a sealed tray with the specification of 300 mm × 300 mm × 55 mm. The bottom plate was evenly coated with epoxy resin binder at a 0.5 kg/m^2 dosage. After curing with epoxy resin binder for 1 day, 3 mL of water was dripped in the center of the tray to simulate the basic environment of bulging. The 17 kinds of epoxy asphalt concrete with different asphalt–aggregate ratios were prepared to form rutting test specimens that were placed in an oven at 100 °C for 12 h. The height between the top of the bulge's center and the surface of the specimen was measured, and the water seepage test was carried out for the rutting plate after the temperature dropped. Low-content epoxy–asphalt-mixture-related performance tests, including rutting test, low-temperature bending test, immersion Marshall test, etc. were carried out.

Table 3. Mixture gradation.

Sieve Hole (mm)	13.2	9.5	4.75	2.36	1.18	0.6	0.3	0.15	0.075
Pass rate (%)	100	97.6	75.3	60	43.5	34.6	25.4	17.5	10.8

3. Methods

The test chart of this research study is shown in Figure 1. The tests were divided into three main parts: rheological performance tests, microscopic characteristics tests, and road performance tests of asphalt mixture. Rheological performance tests were conducted according to SHRP specifications. A dynamic shear rheometer conducted temperature sweep (TS) and multiple stress creep recovery (MSCR) tests. The TS test was performed on the asphalt sample, and the temperature was 42–80 °C. The MSCR test was performed on the short-term aged asphalt sample. The bending beam rheometer (BBR) test proceeded at −18 °C and −24 °C. A scanning electron microscope (SEM) test was carried out to reflect the morphology characteristics of materials. The performance test of asphalt mixture is carried out following the standard "Test Regulations for Asphalt and Asphalt Mixture of Highway Engineering" (JTG E20-2011) [52], which mainly includes the dynamic stability test, trabecular splitting test and freeze–thaw splitting test.

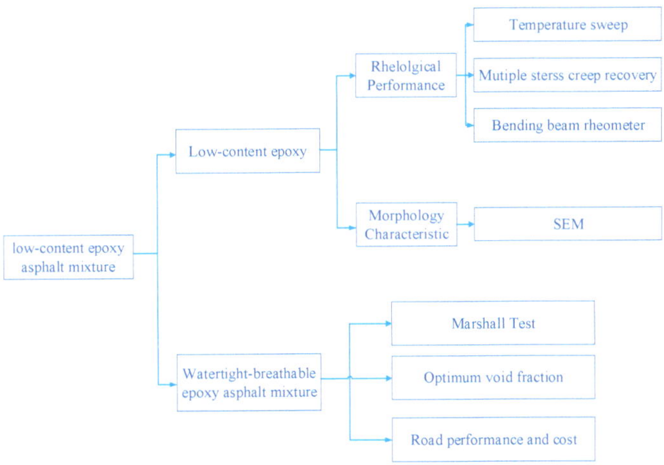

Figure 1. Tests chart of low-content epoxy asphalt and its mixtures.

4. Performance of Different Epoxy Asphalt

4.1. Structure of Different Epoxy Asphalt

The fluorescence microscope images are summarised in Figure 1. Yellow represents epoxy resin, and black represents basic asphalt. In Figure 2, with the gradual increase in the epoxy resin content, the yellow particles in the picture gradually increased, and the size of the particles also enlarged. When the proportion of epoxy reaches 10%, the epoxy is just doped into the asphalt base in a tiny medium, which is only equivalent to the modifier role. When the proportion of epoxy resin reaches 20%, the yellow particles increase significantly. When the proportion of epoxy resin is increased to 30%, the resin aggregates with each other and forms larger epoxy resin particles. The whole mixture is still mainly coated with asphalt epoxy resin. When the proportion of epoxy resin reaches 40%, the epoxy asphalt changes from bitumen-based to epoxy-resin-based. The epoxy resin has formed a skeleton, and the asphalt is distributed in the epoxy skeleton, but the particle size of the asphalt is not uniform. When the proportion of epoxy resin reaches 50%, the asphalt particles are evenly distributed in epoxy. The properties of epoxy asphalt change qualitatively as the epoxy proportion increases. When the proportion of epoxy is less than 30%, asphalt is the major phase, and epoxy is the auxiliary phase in epoxy asphalt. When the proportion of epoxy is more than 40%, epoxy is the main phase. Therefore, epoxy resin proportions between 30% and 40% are the two values of phase inversion in epoxy asphalt.

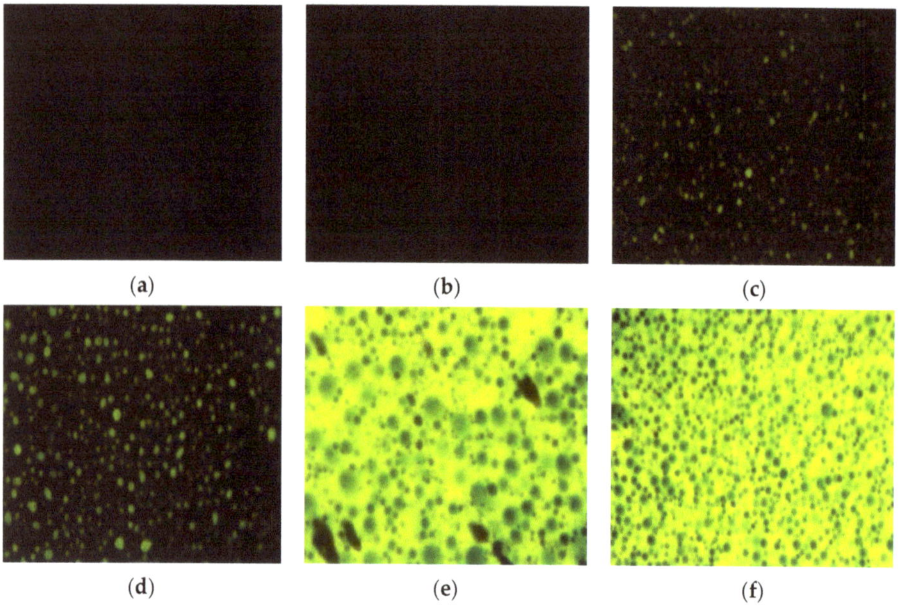

Figure 2. Fluorescence microscope photo: (**a**) 0%; (**b**) 10%; (**c**) 20%; (**d**) 30%; (**e**) 40%; (**f**) 50%.

In a steel bridge deck, the optimum dosage ratio is obtained when epoxy resin appears in a continuous phase and forms a spatial network structure. Under the optimum dosage state, the continuous phase formed by epoxy resin can effectively exert the characteristics of epoxy resin and change thermoplastic materials into thermosetting materials. As the asphalt is uniformly distributed in the epoxy resin structure, epoxy asphalt has excellent toughness, while the crosslinking network formed by the epoxy resin endows epoxy asphalt with high strength and thermosetting. For a long-span steel bridge deck pavement, to ensure good follow-up, the epoxy content in epoxy asphalt is usually 50%. There will not be huge flexural deformation on ordinary roads, and the epoxy asphalt mixture does not need to have excellent following and thermosetting. The proportion of epoxy should be determined according to the required performance for common roads without excessively pursuing the performance of epoxy asphalt.

4.2. *Viscoelastic Properties of Different Epoxy Asphalt*

4.2.1. TS Test

The rheological test results are summarized in Figures 3 and 4, in which Figure 3 shows phase angle parameters and Figure 4 shows complex shear modulus parameters. From Figure 3, the phase angle decreases with increasing epoxy proportion at the same temperature. The phase angle of SBS asphalt is similar to epoxy asphalt with 10–15% epoxy resin. When the temperature rises above 70 °C, the phase angle of epoxy asphalt with 15% epoxy resin is lower than that of SBS asphalt, which indicates that the elasticity of epoxy asphalt with 15% epoxy resin is better than that of SBS asphalt. When the epoxy resin proportion is lower than 25%, the epoxy asphalt's phase angle increases with the increase in the test temperature. The phase angle of epoxidized asphalt did not increase with increasing test temperature when the proportion of epoxy was greater than 30%. When the proportion of epoxy asphalt reaches 40%, the phase angle of epoxy asphalt decreases and then increases with increasing temperature.

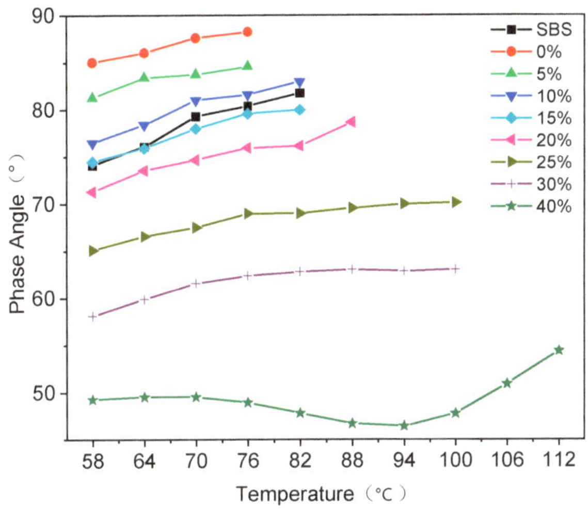

Figure 3. Phase angle parameters.

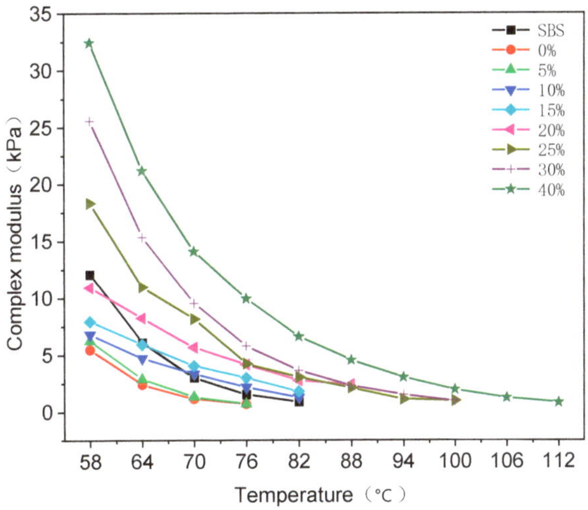

Figure 4. Complex shear modulus parameters.

Figure 5 shows that the composite shear modulus increases with the increase in epoxy proportion at the same temperature. The complex modulus of epoxy asphalt with 5% epoxy resin is similar to basis bitumen. It can be seen that the addition of 5% epoxy has relatively little effect on the composite modulus of basis asphalt. The composite modulus of SBS asphalt is greater than that of epoxy asphalt with 20% epoxy resin at 58 °C. The complex modulus of SBS asphalt decreased significantly as the test temperature increased. When the test temperature reaches 76 °C, the complex modulus of epoxy asphalt with a 10% epoxy resin proportion is higher than that of SBS asphalt.

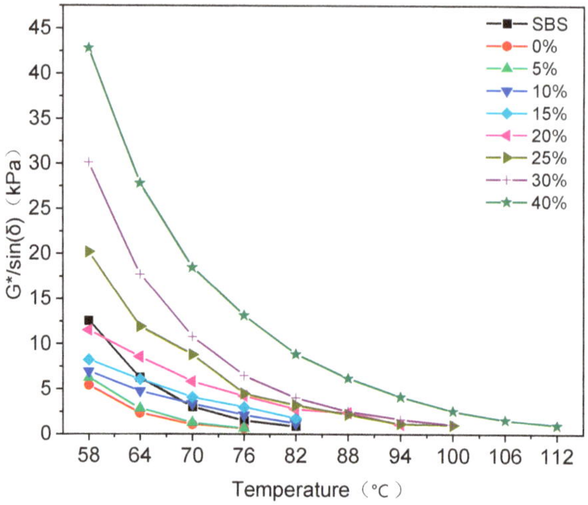

Figure 5. $G^*/\sin\delta$ parameter.

In summary, epoxy asphalt cannot fully reflect its anti-deformation ability in the middle-temperature stage. When the temperature rises due to its thermosetting, the anti-deformation ability of epoxy asphalt will be fully reflected. To further investigate the anti-deformation ability of epoxy asphalt, the rutting factor is calculated by using complex modulus and phase angle parameters at different test temperatures, and the results are summarised in Figure 5.

From Figure 5, the change law of the rutting factor is basically consistent with the change law of the complex modulus. The $G^*/\sin(\alpha)$ before aging is required to be greater than or equal to 1 kPa in PG grade of asphalt, and the temperature at which epoxy asphalt with different proportions of epoxy meets PG grade requirements is shown in Figure 6. From Figure 6, the PG high-temperature grade of epoxy asphalt gradually increases with the rising epoxy proportion. According to PG grading, the high-temperature grade of basic asphalt is 70 °C, while epoxy asphalt with 5% epoxy can only reach 70 °C. The performance improvement is not obvious. Epoxy asphalts with 10–15% epoxy have a high PG temperature grade of 82 °C, which is higher than SBS asphalts. The high-temperature grade of epoxy asphalt with 25–30% epoxy resin reaches 100 °C, and the high-temperature grade of epoxy asphalt with 40% epoxy resin reaches 112 °C. It shows that adding 5% epoxy resin improves the high-temperature performance of asphalt, while more than 10% epoxy resin can greatly improve the high-temperature properties of asphalt. In the PG grade system, 6 °C is a grade span, so certain epoxy asphalts have the same high-temperature grade, but it does not mean that their anti-rutting is the same. In order to further study the influence of epoxy asphalt with different proportions of epoxy on rutting resistance, rutting factors of epoxy asphalt under different proportions were compared to study the difference in rutting factor decline rate. The specific calculation formula is shown in Equation (1), and the results are shown in Table 4 and Figure 6.

$$L_{i-j} = G_i^*/\sin(\alpha_i) - G_j^*/\sin(\alpha_j) \quad (1)$$

where L_{i-j} is the decrease range of rut factor of epoxy asphalt when the proportion of epoxy decreases from i to j, kPa; i and j are epoxy resin content, where i is 5%, 10%, 15%, 20%, 25%, 30%, 40%, and j is 0%, 5%, 10%, 15%, 20%, 25%, 30%; $G_i^*/\sin(\alpha_i)$ is the rut factor value of epoxy asphalt when the epoxy proportion is i, and kPa; $G_j^*/\sin(\alpha_j)$ is the rut factor value of epoxy asphalt when the epoxy proportion is j, kPa.

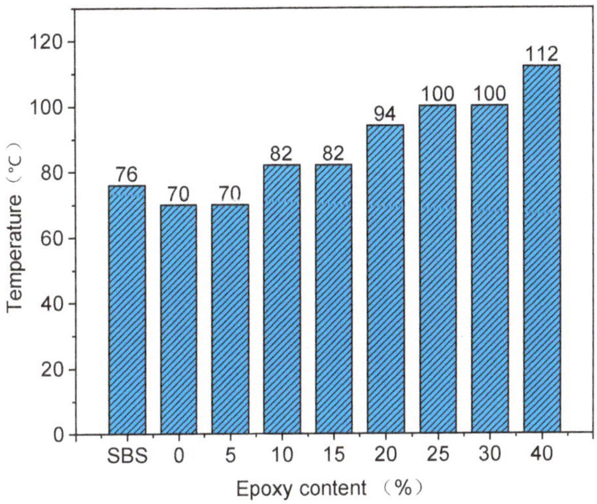

Figure 6. High-temperature grade.

Table 4. Rutting factor decline rate table.

Temperature (°C)	Amount of Addition						
	$L_{5\%–0\%}$	$L_{10\%–5\%}$	$L_{15\%–10\%}$	$L_{20\%–15\%}$	$L_{25\%–20\%}$	$L_{30\%–25\%}$	$L_{40\%–30\%}$
58	−0.82	−0.67	−1.26	−3.30	−8.69	−9.90	−12.68
64	−0.50	−1.90	−1.30	−2.48	−3.37	−5.80	−10.09
70	−0.21	−2.04	−0.76	−1.72	−2.97	−2.00	−7.69
76	0.00	−1.49	−0.85	−1.24	−0.25	−2.00	−6.66
82			−0.56	−1.00	−0.44	−0.80	−4.84
88					0.18	−0.33	−3.66
94						−0.52	−2.48
100						−0.06	−1.53
Average	−0.38	−1.53	−0.95	−1.95	−2.59	−2.68	−6.20

From Table 4 and Figure 7, when the test temperature is 58 °C, the rutting factor of epoxy asphalt with a proportion less than 15% does not decrease rapidly with the increase in epoxy content. When the proportion of epoxy is 15–25%, the rutting factor decreases rapidly; when the content exceeds 25%, the decline rate slows down. When the test temperature is higher than 58 °C, the decline rate of the rutting factor of epoxy asphalt with less than 25% content oscillates. When the content of the epoxy asphalt is more than 25%, the decline rate of the rutting factor begins to increase with the improvement of epoxy resin content. In order to further reflect the change rule of anti-rutting of epoxy asphalt with different proportions of epoxy, the average rutting factors at all temperatures were calculated. From Figure 6, with the improvement of epoxy proportion, the decline rate of the rutting factor of epoxy asphalt oscillates less than 15%. When the proportion of epoxy is 15–30%, the decline rate of the rutting factor increases slowly. When the proportion of epoxy exceeds 30%, the decline rate of the rutting factor increases sharply. In summary, with the improvement of epoxy proportion, the rate of rutting factor decline is increasing. That is, the rate of anti-rutting performance improvement is increasing.

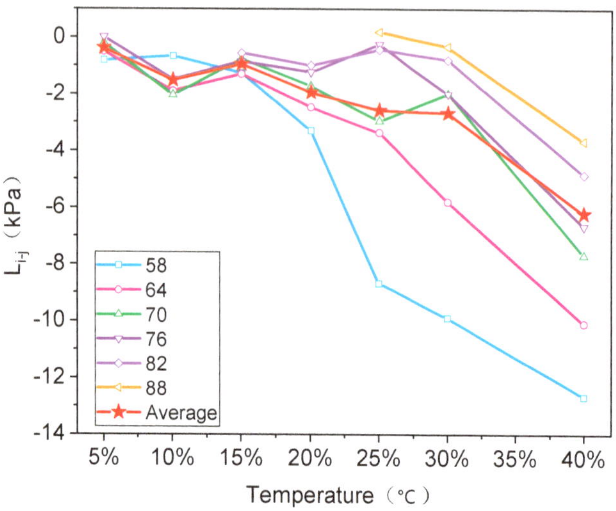

Figure 7. Rutting factor decline rate.

The rutting factor is analyzed logarithmically, and the relationship between $\lg(G^*/\sin(\alpha))$ it and the test temperature is shown in Figure 8. Figure 8 $\lg(G^*/\sin(\alpha))$ has a good linear relationship with temperature, which the linear model fits. The results are shown in Table 5, where K is the slope, and C is the intercept. From Table 5, the correlation coefficient R^2 of linear fitting is higher than 0.99, which has a good fitting effect and can be used as a standard curve to predict other rutting factors. The slope represents the temperature sensitivity of the logarithmic rutting factor of epoxy asphalt. The greater the slope, the greater the decrease in the logarithmic rutting factor between different temperatures, i.e., the higher the temperature sensitivity. From Table 5, the slope changes significantly when the epoxy content is 5−10%. When the epoxy proportion is greater than 10%, the slope is relatively flat, indicating that the temperature sensitivity of the rutting factor is relatively small. That is, the temperature sensitivity of the anti-rutting of the epoxy asphalt is relatively small.

Table 5. Linear fitting parameters of $\lg(G^*/\sin\delta)$.

Parameter	SBS	0%	5%	10%	15%	20%	25%	30%	40%
C	8.96	8.33	8.94	6.04	5.75	5.59	7.2	7.92	7.64
K	−0.11	−0.12	−0.12	−0.07	−0.06	−0.05	−0.07	−0.08	−0.07
R^2	0.997	0.992	0.996	0.993	0.994	0.993	0.994	0.997	0.998

According to the above analysis, the rutting factor of epoxy asphalt is increasing with the increase in epoxy proportion. From Figure 7, the rutting factor decline rate increases significantly when the epoxy proportion is greater than 15%. The results show that the rutting resistance of epoxy asphalt is greatly improved when the epoxy proportion is greater than 15%. When the proportion of epoxy is greater than 30%, the rate of improving anti-rutting properties of epoxy asphalt increases further.

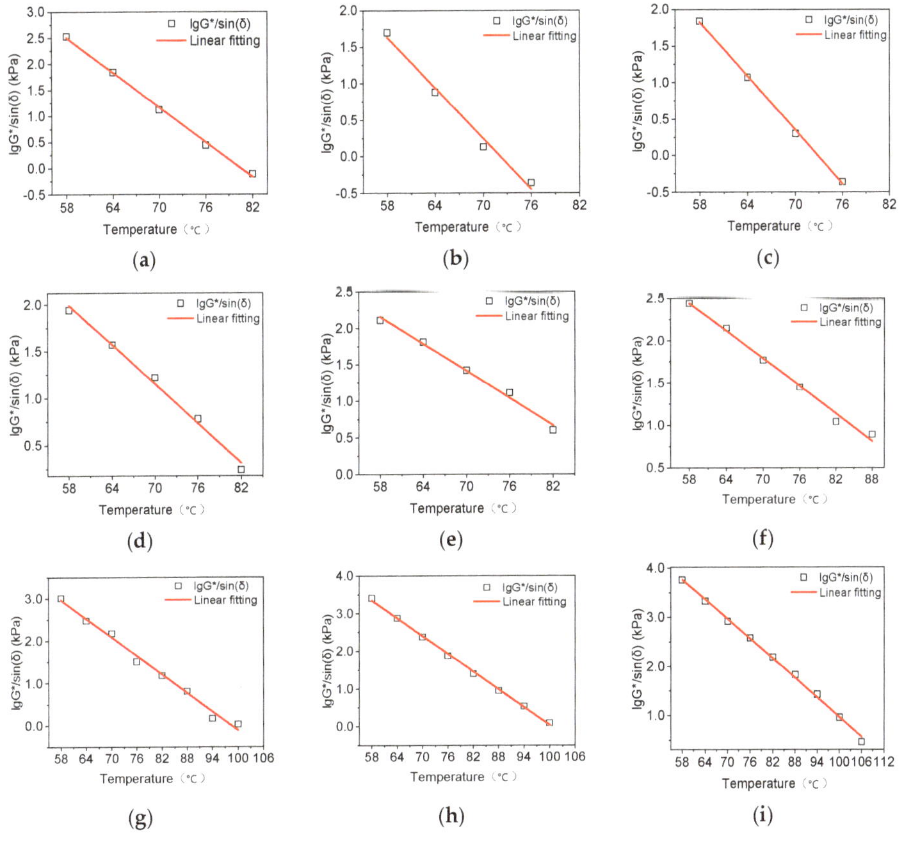

Figure 8. The influence of $\lg(G^*/\sin(\alpha))$ under different conditions. (**a**) SBS; (**b**) 0; (**c**) 5%; (**d**) 10%; (**e**) 15%; (**f**) 20%; (**g**) 25%; (**h**) 30%; (**i**) 40%.

4.2.2. MSCR Test

The creep and recovery curves of 10 kinds of bitumen (matrix bitumen, SBS-modified bitumen, and eight different epoxy bitumen mixtures) under 0.1 kPa and 3.2 kPa stress were obtained through the MSCR test. The DSR rheometer automatically recorded the initial strain, creep strain, and recovery strain of each loading cycle under different stress and each one after applying stress for 1 s. The creep strain correction value within the loading cycle and the total strain correction value within each loading cycle after 10 s of stress are applied, and the unrecoverable creep compliance Jnr and the average strain recovery rate R are obtained through calculation. The test temperature is 58 °C, and the results are shown in Figure 9.

It can be seen from the above figure that under different stress levels, the high-temperature performance of epoxy asphalt gradually becomes stronger with the increase in epoxy resin content. It can be seen that the addition of epoxy resin can change the high-temperature performance of epoxy asphalt. At 3.2 kPa stress, the R-value is greater than the M value, indicating that epoxy asphalt's recovery ability and rutting resistance are decreased at high stress levels. Under the stress condition of 0.1 kPa, the high-temperature performance of epoxy asphalt with a content of 10% is basically the same as that of SBS-modified asphalt. When the content of the epoxy resin is increased from 10% to 15%, the strain recovery rate increases by 31.2%, indicating that the high-temperature stability of epoxy asphalt has been greatly improved. At the stress level of 3.2 kPa, when the content of epoxy resin reaches 15%, the high-temperature performance can be equal to that of SBS-modified asphalt. Therefore, the content of epoxy resin in high temperature and high load environments should be greater than 15%.

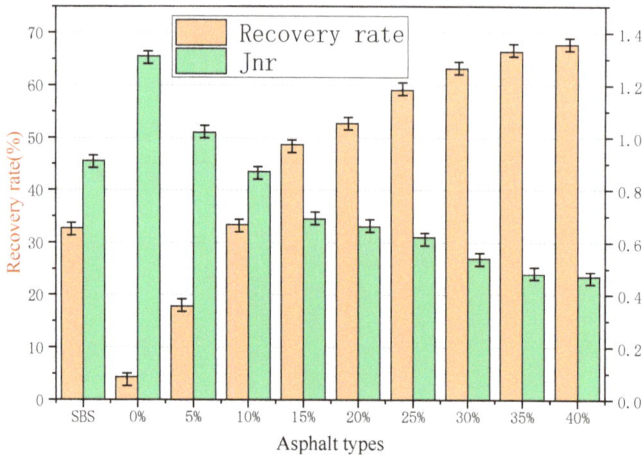

(a)

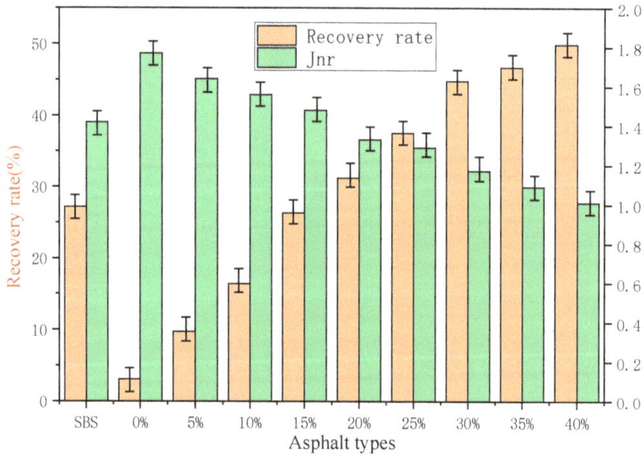

(b)

Figure 9. MSCR test result: (**a**) 0.1 kPa, (**b**) 3.2 kPa.

The results of temperature scanning and MSCR were analyzed. When the content was increased to 30%, the improvement rate of high-temperature performance of epoxy asphalt was further increased. According to the results of fluorescence microscope analysis, the content was 30%. Epoxy asphalt has undergone qualitative changes. Epoxy asphalt initially formed a crosslinked spatial network structure. Epoxy asphalt began to change from asphaltene to epoxy, with ultra-high performance, and can be used in small steel bridge pavement. For sections with special anti-rutting performance requirements, when the content is greater than 15%, the anti-rutting performance improvement rate is significantly increased, so the epoxy resin content can be controlled at 15–30%. For ordinary roads, the performance of epoxy asphalt should at least reach the performance of SBS-modified asphalt, so the content of epoxy resin can be controlled at 10−15%.

4.2.3. BBR Test

The low-temperature bending beam rheology (BBR) test can characterize the performance of asphalt under low-temperature conditions. According to the requirements of low-temperature grade of road asphalt in most areas of China, 10 kinds of asphalt (matrix asphalt, SBS modified asphalt, eight different kinds of epoxy asphalt) were tested at −12 °C and −18 °C, respectively, and the creep modulus (S) and creep rate (M) indexes were extracted It is an evaluation index. Generally speaking, the value of s is less than 300 MPa, and the value of M is greater than 0.3. The results are summarized in Figure 10.

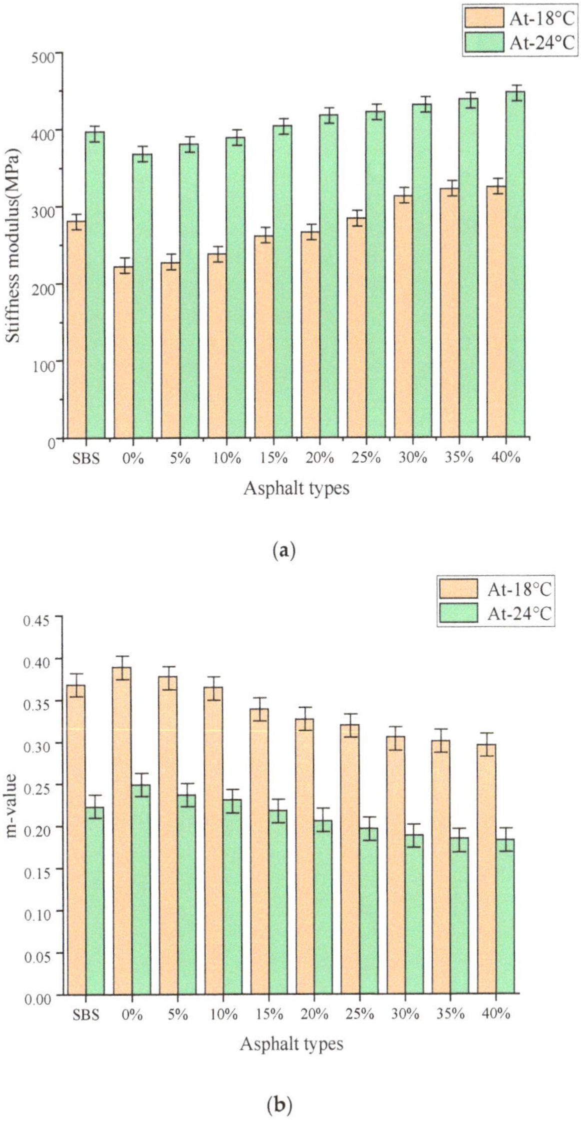

Figure 10. BBR test result: (**a**) Stiffness modulus, (**b**) m−value.

It can be seen from Figure 10 that with the increase in epoxy resin content, S gradually increases and M gradually decreases, indicating that the greater the epoxy resin content, the more susceptible asphalt is to low-temperature stress. The analysis is because the addition of epoxy resin reacts with the curing agent when the curing product is brittle and hard at low temperatures, and the deformation ability is poor, leading to epoxy asphalt's low-temperature performance. Under the condition of −18 °C, when the epoxy resin content increases from 10% to 15%, the value of S increases by 8.8%. When the epoxy resin content does not exceed 25%, both s value and M value meet the specification requirements. When the epoxy resin content does not exceed 20%, the low-temperature performance of epoxy asphalt is better than that of SBS modified asphalt. When the temperature drops to −24 °C, the s value of epoxy asphalt with different dosages does not meet the specification requirements. In summary, when the temperature is low, the content of epoxy resin can be appropriately reduced to improve its low-temperature performance, and in an extremely low-temperature environment, the low-temperature performance of epoxy asphalt cannot meet the requirements.

4.2.4. Applicable Conditions for Different Epoxy Resin Content

Table 6 summarizes different application scenarios corresponding to different epoxy resin contents based on the above research results. For ordinary roads, when the content of epoxy resin is 10−15%, it can meet both high-temperature and low-temperature requirements. For sections with special rutting performance requirements, the content of epoxy resin is recommended to be controlled between 15 and 30%. When the content of epoxy resin is greater than 30%, the epoxy asphalt initially forms a crosslinked spatial network structure, and the epoxy asphalt starts to change from asphaltene to epoxy, which has ultra-high performance and can be applied to small steel bridge deck pavements. The low-temperature performance of epoxy asphalt gradually decreases with the increase in temperature. Therefore, the epoxy resin content of the road section should not exceed 20% in the low-temperature environment, and the epoxy asphalt cannot meet the specification requirements under extreme low-temperature conditions.

Table 6. Different dosages corresponding to different use places.

Epoxy Resin Content	Applicable Place
10~15%	Ordinary road
15~30%	Sections with special rutting resistance requirements
>30%	Small steel deck pavement
<20%	Road section in low-temperature environment

5. Watertight Breathable Epoxy Asphalt Mixture

5.1. Marshall Test

To further study the performance of low-content epoxy asphalt concrete, 17 kinds of Marshall specimens with different asphalt–aggregate ratios were prepared using epoxy asphalt with 20% epoxy resin. The Marshall test was carried out, which is shown in Figure 11. From Figure 11, with the rising asphalt–aggregate ratio, the bulk density and stability first increase and then tend to be stable. When the asphalt–aggregate ratio reaches 6.5%, the bulk density and stability reach the peak. The porosity decreases and the flow value increases with the increase in asphalt–aggregate ratio. Because of the need to prevent rain from corroding steel decks, epoxy asphalt concrete with less than 3% air voids is usually selected for steel deck pavement. Because epoxy asphalt concrete is too dense, the vaporized water in the construction process cannot be discharged, resulting in bulging. Anticorrosion is unnecessary in conventional asphalt pavement, so it does not need extreme compactness. However, the property of epoxy asphalt pavement is closely related to its compactness. Excessive porosity will cause water seepage in epoxy asphalt pavement, and then early water damage will occur. Therefore, achieving watertight ventilation has become a key point in the design of epoxy asphalt concrete.

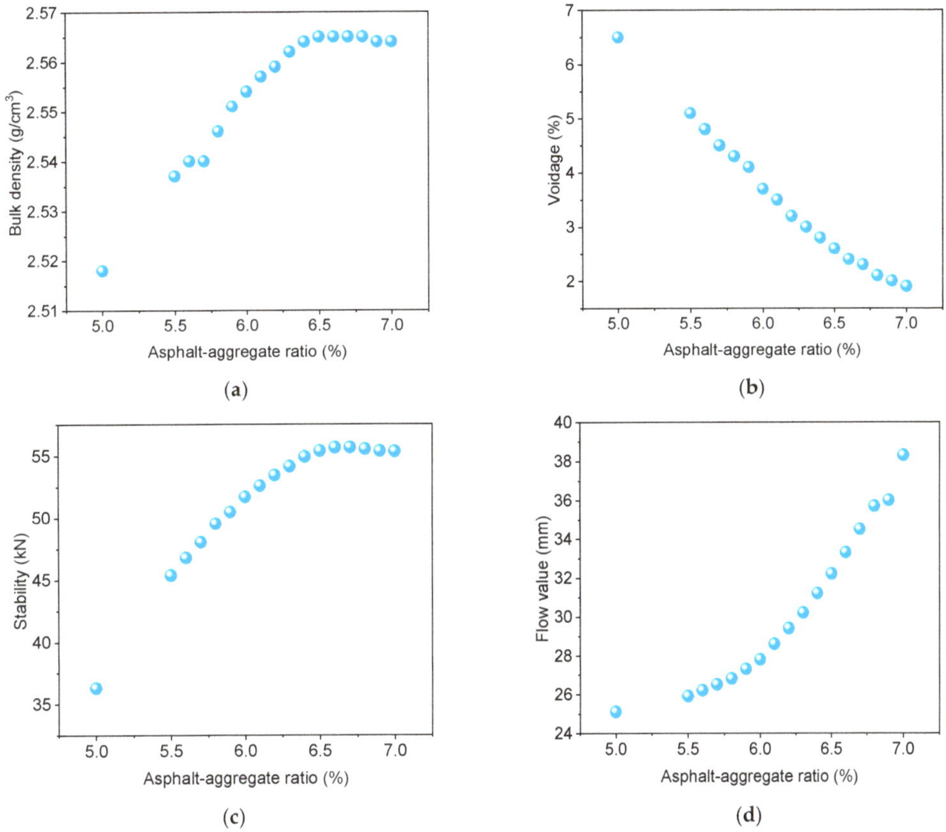

Figure 11. Marshall test results: (**a**) bulk density; (**b**) porosity; (**c**) stability; (**d**) flow values.

5.2. Optimum Void Fraction

The 17 kinds of epoxy asphalt concrete with different asphalt–aggregate ratios were prepared to form rutting test specimens. The specimens were placed in an oven at 100 °C for 12 h. The specimen was taken out, and the height between the center bulge apex and the top of the specimen was measured. After the temperature of the epoxy asphalt mixture specimen drops, the water seepage test is carried out. The results are summarized in Figures 12 and 13.

From Figure 12, the epoxy asphalt mixture with air voids in the range of 4.1–6.5% has no bulging, which indicates that the epoxy asphalt concrete in this range can discharge water vapor. That is, there will be no bulging. With the decrease in air voids, the height of the bulge of the epoxy asphalt concrete gradually increases. During the molding process of epoxy asphalt concrete specimens, due to the high temperature, the liquid in the bearing layer forms a gasification state. Then, the epoxy asphalt concrete is directly jacked up. The deformation process changes the gas-tight state of epoxy asphalt concrete. Gas is discharged from the interlayer along the internal channel of the concrete. With the decrease in air voids, the compactness of the epoxy asphalt mixture increases, and the gas that can be discharged is less. That is, more and more gaseous or liquid water remains between layers. When the epoxy asphalt mixture specimen molding work is completed, the interlayer water vapor jacks up the pavement again under the oven heating. Then, it forms the bulge, and the bulge height increases with the increase in interlayer water vapor content, i.e., a decrease in air voids leads to an increase in bulge height. When the air voids are less than

2.4%, the height of the bulge does not increase with the decrease in air voids, which shows that the epoxy asphalt mixture is completely dense (even if a bulge occurs in the molding process, the epoxy mixture is also dense). Interlayer gas cannot escape from the epoxy asphalt concrete during the molding process, so the height of the bulge is determined by the pressure formed by interlayer gas. As can be seen from Figure 13, an epoxy asphalt mixture with air voids of 1.9–4.5% is completely water-dense without water seepage. Therefore, the air voids of watertight breathable epoxy asphalt concrete are 4.1–4.5%, corresponding to an asphalt–aggregate ratio of 5.1–5.9%.

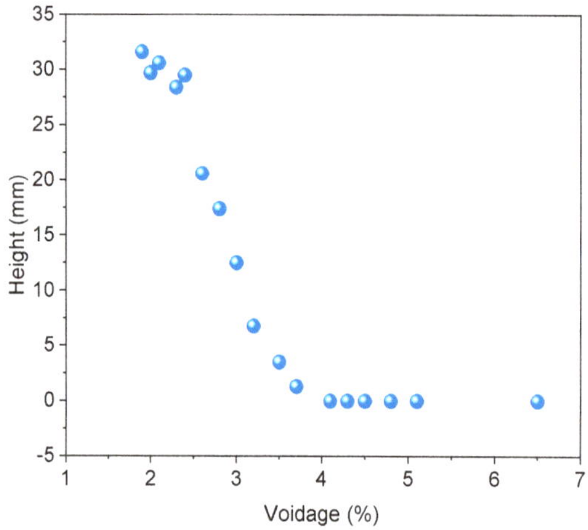

Figure 12. Height of bulge.

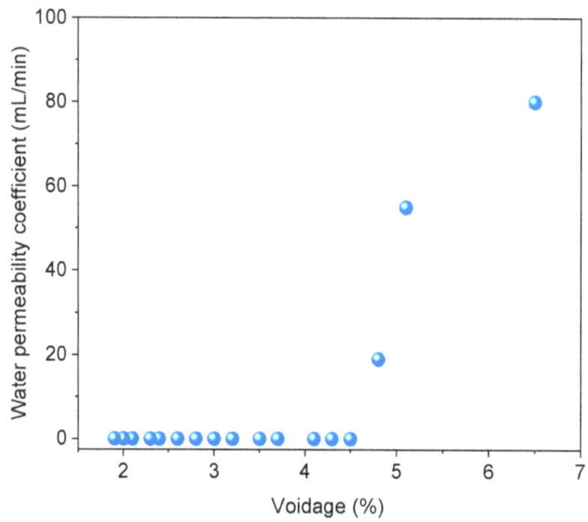

Figure 13. Permeability coefficient.

5.3. Road Performance and Cost

Taking the porosity of 4.1% watertight breathable epoxy asphalt mixture as an example, its road performance is shown in Table 6. From Table 7, the deformation of the solidified specimen is very small at high temperatures, and the rutting test results far exceed the technical requirements of the modified asphalt mixture 3000 times/mm, which has good anti-rutting stability. The low-temperature crack resistance also meets technical requirements that the flexural–tensile strain of modified asphalt concrete is greater than 0.0025. In the aspect of water stability, the dense and breathable asphalt mixture also has good performance, in which the ratio of immersion residual stability to residual splitting strength reaches 96.74% and 87.8%, respectively, which meets the technical requirements of modified asphalt mixture greater than 85%, and has better water damage resistance.

Table 7. Test results of mixture performance.

Indicators	Dynamic Stability, 70 °C (Times/mm)	Ultimate Flexural-Tensile Strain	Residual Stability (%)	Splitting Strength (%)
Results	12,444	0.0031	96.74	87.8

Epoxy resin content is 20%.

The cost of the low-content epoxy asphalt mixture is summarized in Tables 8 and 9. From Table 8, the unit price of low-content epoxy asphalt that needs to resist rutting is 12,600 CNY/t, and low-content epoxy asphalt for ordinary sections is 8100 CNY/t, which is 49% and 67% lower than the epoxy asphalt used for the conventional steel bridge deck, respectively. Due to the relatively small epoxy resin content in low-content epoxy asphalt, the cost of low-content epoxy asphalt has decreased significantly. From Table 8, the unit price of the low-content epoxy asphalt mixture that needs to resist rutting is 95 CNY/m^2, and that of the low-content epoxy asphalt mixture on ordinary road sections is 61.5 CNY/m^2, which is 46% and 65% lower than epoxy asphalt used for the conventional steel bridge deck, respectively. The costs of low-content epoxy asphalt pavement (ordinary section) and SBS asphalt pavement are similar. Although the cost of the low-content epoxy asphalt pavement (Special section) is 58% higher than the cost of the SBS asphalt pavement, the low-content epoxy asphalt pavement (Special section) can effectively resist rutting and reduce the amount of maintenance. Although the initial cost of the SBS asphalt pavement is low, it requires frequent maintenance in special sections, and the comprehensive cost is much higher than the application of a low-content epoxy asphalt pavement, and it can reduce interference to traffic. In addition, because the low-content epoxy asphalt mixture has eliminated the bulge, the construction cost is greatly reduced compared with the conventional steel bridge deck pavement. Therefore, the cost of the low-content epoxy asphalt mixture in common sections is the same as that of the conventional SBS asphalt pavement, which solves the problem of the high cost of promoting epoxy asphalt pavement in communal sections.

Table 8. Unit price of binder combination.

Application	Conventional Steel Deck	Special Section (Rutting Resistance)	Ordinary Section
Asphalt unit price (CNY/t)	4500	4500	4500
Epoxy unit price (CNY/t)	45,000	45,000	45,000
Asphalt proportion	0.5	0.8	0.8
Epoxy proportion	0.5	0.2	0.1
Epoxy asphalt unit price (CNY/t)	24,750	12,600	8100
The reduction ratio of Binder Material cost (compared with conventional steel deck)	--	49%	67%

Table 9. The cost of different roads.

Application	SBS Asphalt	Conventional Steel Deck	Special Section	Ordinary Section
Thickness (m)	0.04	0.04	0.04	0.04
Mixture weight (Kg/m^2)	102	102	102	102
Aggregate weight (Kg)	95.7	95.7	95.7	95.7

Table 9. *Cont.*

Application	SBS Asphalt	Conventional Steel Deck	Special Section	Ordinary Section
Cement weight (Kg)	6.32	6.32	6.02	5.2
Aggregate unit price (CNY/t)	200	200	200	200
Binder unit price (CNY/t)	6500	24,750	12,600	8100
Material cost (CNY/m^2)	60.2	175.7	95	61.5
Cost reduction ratio (compared with conventional steel deck)	--	--	46%	65%
Cost increase ratio (compared with SBS asphalt pavement)	--	--	58%	2%

SBS asphalt pavement: SBS asphalt content of 6.2%; conventional steel bridge deck: epoxy asphalt content is 6.2%; special sections: the content of epoxy asphalt is 5.9%; ordinary section: the content of epoxy asphalt is 5.1%.

6. Conclusions

In order to make the low equivalent epoxy asphalt more widely used in road engineering, the rheological performance and morphological characteristics of epoxy asphalt at different dosages were tested. A Marshall test was performed to study its road performance, and its cost was analyzed. In summary, we can reach the following conclusion.

(1) Due to the excellent heat resistance of epoxy resin, the rate of improvement in anti-rutting performance is increasing with the increase in epoxy resin content. When the proportion of epoxy resin is increased to 30%, the resin aggregates and forms larger epoxy resin particles, and high-temperature performance is further enhanced.
(2) Compared with the arbitrary selection of epoxy asphalt content in epoxy asphalt pavement, this paper puts the epoxy asphalt content in different application scenarios based on the starting point of road performance and cost. For road sections with special anti-rutting requirements, epoxy content should be limited to 15–30%. For ordinary roads, the performance of epoxy asphalt should at least reach the performance of SBS asphalt, and the proportion of epoxy should be limited to 10–15%. The low-temperature performance of epoxy asphalt gradually decreases with the increase in temperature. Therefore, in a low-temperature environment, the content of epoxy resin in the lower section should not exceed 20%. Under extremely low-temperature conditions, epoxy asphalt does not meet the requirements of the code.
(3) The traditional epoxy asphalt mixture cannot solve the problem of bulging when applied to ordinary roads. This paper solves the problem of bulging by studying the watertight and breathable properties of epoxy asphalt mixture with different voids. When the void of low-content epoxy asphalt mixture is 4.1–4.5%, it has the function of water tightness and air permeability and will not form a bulge. Moreover, the low-content epoxy asphalt mixture meets the requirements of various indicators of a modified asphalt mixture, and its deformation resistance far exceeds the technical requirements of a modified asphalt mixture.
(4) The low-content epoxy asphalt mixture proposed in this paper, which is suitable for rutting-resistant pavement and ordinary road sections, reduces the cost of the mixture by 46% and 65% compared with steel bridge deck pavement.

In addition, the watertight breathable mixture system solves the problem of swelling, the construction difficulty of epoxy asphalt concrete is greatly reduced, and the comprehensive cost will be further reduced.

A low-equivalent epoxy asphalt mixture has not been widely used on road surfaces. The tests conducted in this study are all conducted based on the Marshall test, and its durability is unknown. Therefore, the next stage of research should focus on the accelerated loading test of low-content epoxy asphalt mixture to further verify the applicability of low-content epoxy asphalt mixture while carrying out the application research of solid engineering and finally determining the application technology of a low-content epoxy asphalt mixture.

Author Contributions: Conceptualization, X.L. (Xiaodong Li) and X.L. (Xiang Long); Methodology, X.L. (Xiaodong Li); Software, J.L.; Validation, X.L. (Xiaodong Li), X.L. (Xiang Long), and J.L.; Formal analysis, M.H.; Investigation, J.L.; Resources, Y.W.; data curation, J.L.; writing—original draft preparation, X.L. (Xiaodong Li), C.L. and J.L.; writing—review and editing, X.L. (Xiaodong Li), M.H. and J.L.; visualization, X.L. (Xiaodong Li); supervision, Y.W. and S.W.; project administration, X.L. (Xiang Long), C.L. and S.W.; funding acquisition, X.L. (Xiang Long). All authors have read and agreed to the published version of the manuscript.

Funding: This research was funded by the Natural Science Fund of Guangdong Province, grant 427 number 2019A1515011965, and the Natural Science Foundation of China, grant number 51808228.

Data Availability Statement: The data presented in this study are available on request from the corresponding author.

Conflicts of Interest: Authors Xiaodong Li, Xiang Long, Yan Wang, and Mu He were employed by the company Guangzhou North Second Ring Road Traffic Technology Co., Ltd. The remaining authors declare that the research was conducted in the absence of any commercial or financial relationships that could be construed as a potential conflict of interest.

References

1. Ding, S.; Xing, Y.; Lang, H.; Wen, T.; Lu, J.J. Line-Structured Light Rut Detection of Asphalt Pavement with Pavement Markings Interference under Strong Light. *J. Transp. Eng. Part B Pavements* **2022**, *148*, 04022007. [CrossRef]
2. Rushing, J.F.; Little, D.N. Static Creep and Repeated Load as Rutting Performance Tests for Airport HMA Mix Design. *J. Mater. Civ. Eng.* **2014**, *26*, 04014055. [CrossRef]
3. Zhang, C.; Yu, H.; Zhu, X.; Yao, D.; Peng, X.; Fan, X. Unified Characterization of Rubber Asphalt Mixture Strength under Different Stress Loading Paths. *J. Mater. Civ. Eng.* **2024**, *36*, 04023498. [CrossRef]
4. Yuan, J.; Lv, S.; Peng, X.; Liu, K. Comprehensive Properties Evaluation of Modified Bio-Asphalt Mixture Based on Comparison Matrix. *J. Mater. Civ. Eng.* **2024**, *36*, 04023616. [CrossRef]
5. Ling, J.; Ren, L.; Tian, Y.; Gao, J.; Man, L. Analysis of airfield composite pavement rutting using full-scale accelerated pavement testing and finite element method. *Constr. Build. Mater.* **2021**, *303*, 124528. [CrossRef]
6. Yang, J.; Li, W.-N.; Chen, Z.-W.; Yu, B. Finite element analysis of asphalt pavement on long-steep longitudinal slope. *J. Traffic Transp. Eng.* **2010**, *10*, 20–24.
7. Li, L.; Huang, X.; Han, D.; Dong, M.; Zhu, D. Investigation of rutting behavior of asphalt pavement in long and steep section of mountainous highway with overloading. *Constr. Build. Mater.* **2015**, *93*, 635–643. [CrossRef]
8. Wang, H.; Zhao, Z.Q.; Zeng, J. Rutting Treatment Technology of Asphalt Pavement of Abrupt Slope Section in Heavy Load and High Temperature Zone. In Proceedings of the International Conference on Advanced Engineering Materials and Technology (AEMT2011), Sanya, China, 29–31 July 2011.
9. Chen, Y.-L.; Li, X.-L.; Zhou, Z.-G. Rutting of Asphalt Middle Course for Long and Steep Grade Expressway. In Proceedings of the International Workshop on Energy and Environment in the Development of Sustainable Asphalt Pavements, Changan University, Xi'an, China, 6–8 June 2010.
10. Salem, H.A.; Uzelac, D.; Matic, B. Temperature Zoning of Libya Desert for Asphalt Mix Design. In Proceedings of the 3rd International Conference on Civil, Architectural and Hydraulic Engineering (ICCAHE), Hangzhou, China, 30–31 July 2014.
11. Yang, Y.; Huang, S.L.; Ding, Q.J.; Peng, X.Y. The Property Research on Interfacial Modificated Semi-flexible Pavement Material. In Proceedings of the International Conference on Green Building, Materials and Civil Engineering (GBMCE 2011), Shangri La, China, 22–23 August 2011.
12. Zhang, R.K. Research and application on high-performance pouring semi-flexible pavement material. Master's Thesis, Wuhan University of Technology, Wuhan, China, 2009.
13. Hu, C.; Zhou, Z.; Luo, Y. Study on damage evolution and mechanism of semi-flexible pavement under acid rain erosion. *Case Stud. Constr. Mater.* **2023**, *19*, e02286. [CrossRef]
14. Chen, X.-H.; Chen, Y.; Huang, W. Flexure characters of epoxy asphalt surfacing on steel decks. *J. Highw. Transp. Res. Dev.* **2007**, *24*, 5–8.
15. Nie, W.; Wang, D.; Sun, Y.; Xu, W.; Xiao, X. Integrated Design of Structure and Material of Epoxy Asphalt Mixture Used in Steel Bridge Deck Pavement. *Buildings* **2021**, *12*, 9. [CrossRef]
16. Qin, K.; Yang, F.; Fu, J.; Han, T.; Liu, F. Study on Technology of Bulge and Crack Repair in Epoxy Asphalt Pavement of Steel Bridge Deck. *IOP Conf. Series Earth Environ. Sci.* **2021**, *804*, 042059. [CrossRef]
17. Wang, Y.D. Study on Synthesis and Curing of Epoxy Terminated PPESK. Master's Thesis, Dalian University of Technology, Dalian, China, 2008.

18. Zhang, L. Preparation and Properties of Asphalt Modified Epoxy Resin Composites. Master's Thesis, Xi'an University of Architecture and Technology, Xi'an, China, 2017.
19. Wunderle, B.; Dermitzaki, E.; Hölck, O.; Bauer, J.; Walter, H. Molecular dynamics approach to structure-property correlation in epoxy resins for thermo-mechanical lifetime modeling. *Microelectron. Reliab.* **2010**, *50*, 900–909. [CrossRef]
20. Girard-Reydet, E.; Riccardi, C.C.; Sautereau, H. Epoxy-aomatic diamine kinetics. 1. Modeling and influence of the diamine structure. *Macromolecules* **1995**, *28*, 7599–7607. [CrossRef]
21. Gao, L. Molecular Simulation of Curing Behavior of Epoxy Resin/Amine System and Design of Novel Carbon Hybrid Reinforcement. Master's Thesis, Beijing University of Chemical Technology, Beijing, China, 2017.
22. Wang, Z.; Zeng, L. Construction Control of TAF Epoxy Asphalt Mixture. *J. Highw. Transp. Res. Dev.* **2013**, *30*, 12–16.
23. Liu, Y.; Qian, Z.; Wang, Y.; Xue, Y. Development and Laboratory Evaluation of a Cold Mix High-Early-Strength Epoxy Asphalt Concrete for Steel Bridge Deck Pavements. *Materials* **2021**, *14*, 4555. [CrossRef]
24. Zhu, C. Japan TAF Epoxy Asphalt Concrete Design and Steel Bridge Deck Pavement Construction Technology. *Appl. Mech. Mater.* **2013**, *330*, 905–910. [CrossRef]
25. Luo, S.; Liu, Z.; Yang, X.; Lu, Q.; Yin, J. Construction Technology of Warm and Hot Mix Epoxy Asphalt Paving for Long-Span Steel Bridge. *J. Constr. Eng. Manag.* **2019**, *145*, 04019074. [CrossRef]
26. Chen, Z.; Xu, W.; Zhao, J.; An, L.; Wang, F.; Du, Z.; Chen, Q. Experimental Study of the Factors Influencing the Performance of the Bonding Interface between Epoxy Asphalt Concrete Pavement and a Steel Bridge Deck. *Buildings* **2022**, *12*, 477. [CrossRef]
27. Fan, X.; Luo, R. Deformation recovery characteristics of asphalt pavement material on steel bridge decks under nondestructive conditions. *Constr. Build. Mater.* **2021**, *313*, 125495. [CrossRef]
28. Luo, S.; Qian, Z. Preparation and Performance Evaluation of Novel High Durability Epoxy Asphalt Concrete for Bridge Deck Pavements. *Asian J. Chem.* **2014**, *26*, 5595–5598. [CrossRef]
29. Xu, P.; Cong, P.; Li, D.; Zhu, X. Toughness modification of hyperbranched polyester on epoxy asphalt. *Constr. Build. Mater.* **2016**, *122*, 473–477. [CrossRef]
30. Qian, Y.-C.; Chen, S.-F.; Cong, P.-L. Curing characteristics of epoxy asphalt studied by differential scanning calorimetry. *J. Wuhan Univ. Technol.* **2011**, *33*, 24–27.
31. Yang, S.-Q.; Hao, P.-W.; Wang, H.-N. The Accelerated Loading Test Research of Japan Epoxy Asphalt Paving Materials. In Proceedings of the International Workshop on Energy and Environment in the Development of Sustainable Asphalt Pavements, Changan Univ, Xi'an, China, 6–8 June 2010.
32. Hong, Y.Q. Studies of Epoxy Asphalt's Performances and Optimization of Design. Master's Thesis, South China University of Technology, Guangzhou, China, 2010.
33. Apostolidis, P.; Liu, X.; van de Ven, M.; Erkens, S.; Scarpas, T. Kinetic Viscoelasticity of Crosslinking Epoxy Asphalt. *Transp. Res. Rec. J. Transp. Res. Board* **2019**, *2673*, 551–560. [CrossRef]
34. Shao, W.-X.; Zhou, B.-J. Study on performance of new type cold mix epoxy asphalt and mixture for steel deck pavement. *IOP Conf. Series Earth Environ. Sci.* **2021**, *787*, 012004. [CrossRef]
35. Liu, Y.Q. Technical Research of Asphalt Concrete Pavement on Steel-Girder Bridge Deck. Master's Thesis, South China University of Technology, Guangzhou, China, 2010.
36. Zhang, R.; He, Y.; Ao, Z. An asphalt emulsion modified by compound of epoxy resin and styrene-butadiene rubber emulsion. In Proceedings of the Joint Conference 9th WSEAS Int Conf on Mathematical and Computational Methods in Sci and Engineering/6th WSEAS Int Conf on Data Networks, Commununications, Computers, Port Spain, Trinidad and Tobago, 5–7 November 2007.
37. Chang, Y.; Chen, Z. Study on the Road Performance of Epoxy Emulsified Asphalt Mixture. In Proceedings of the 1st International Conference on Transportation Infrastructure and Materials (ICTIM), Xi'an, China, 16–18 July 2016.
38. Cong, P.; Luo, W.; Xu, P.; Zhang, Y. Chemical and physical properties of hot mixing epoxy asphalt binders. *Constr. Build. Mater.* **2018**, *198*, 1–9. [CrossRef]
39. Cong, P.-L.; Yu, J.-Y.; Wu, S.-P. Investigation on Influence Factors of Epoxy Asphalt and Its Mixture Properties. *J. Wuhan Univ. Technol.* **2009**, *31*, 7–10.
40. Yin, H.; Jin, H.; Wang, C.; Sun, Y.; Yuan, Z.; Xie, H.; Wang, Z.; Cheng, R. Thermal, damping, and mechanical properties of thermosetting epoxy-modified asphalts. *J. Therm. Anal. Calorim.* **2013**, *115*, 1073–1080. [CrossRef]
41. Li, Y.; Cao, D.; Zhang, Y.; Jia, X. Performance of a dry-method-epoxy modifier and a modified epoxy-asphalt mixture. *Constr. Build. Mater.* **2020**, *266*, 120229. [CrossRef]
42. Huang, H. Study on Evaluation and Application of Hot-Mixed Epoxy Asphalt Steel Bridge Deck Pavement Materials. Master's Thesis, South China University of Technology, Guangzhou, China, 2013.
43. Li, X. Experimental Study on Evaluation and Application of Epoxy Asphalt Steel Bridge Deck Pavement Materials. Master's Thesis, South China University of Technology, Guangzhou, China, 2015.
44. Peijun, X.; Zhen, Z.; Yin, H. Phase structure control and toughening modification of epoxy asphalt based on linear epoxy resin. *Highway* **2022**, *67*, 295–301.
45. Chen, Y.; Hossiney, N.; Yang, X.; Wang, H.; You, Z. Application of Epoxy-Asphalt Composite in Asphalt Paving Industry: A Review with Emphasis on Physicochemical Properties and Pavement Performances. *Adv. Mater. Sci. Eng.* **2021**, *2021*, 1–35. [CrossRef]

46. Lv, Q.; Huang, M.; Huang, W.D. A Conception of Application Epoxy Asphalt Mixture on Airport Pavement. In Proceedings of the 2nd Global Conference on Civil, Structural and Environmental Engineering (GCCSEE 2013), Shenzhen, China, 28–29 September 2013.
47. Wang, J.; Shen, J.; Qian, Z. Construction controlling research of domestic epoxy asphalt mixture. *J. Southeast Univ.* **2009**, *9*, 1226–1230.
48. Li, Y.C.; Zhang, P.; Lv, R.; Xu, Y.Z.; Hou, G.J. Performance evaluation of Epoxy asphalt and study of Epoxy asphalt used on deck surfacing of concrete bridge. In Proceedings of the International Conference on Intelligent Materials, Applied Mechanics and Design Science (IMAMD 2011), Beijing, China, 24–25 December 2011.
49. Zhang, Y.J.; Cao, D.W.; Zhang, H.Y.; Wu, X.W. The study of curing characteristics of epoxy asphalt. In Proceedings of the 3rd International Conference on Chemical Engineering and Advanced Materials (CEAM 2013), Guangzhou, China, 6–7 July 2013.
50. Zhang, X.; Ren, Y.; Chi, F. Study of viscous-elastic properties of epoxy asphalt mixtures by using dynamic frequency sweep. *J. Huazhong Univ. Sci. Technol.* **2009**, *37*, 102–105.
51. Peng, X.; Xie, N.; Xia, C.; Zhou, X.; Zhao, P.; Ma, S.; Zhang, C.; Lv, S. Laboratory evaluation of different bio-oil recycled aged asphalts: Conventional performances and microscopic characteristics. *J. Clean. Prod.* **2023**, *428*, 139442. [CrossRef]
52. *GB/T 16777-2008*; Test method for building waterproof coatings. National Technical Committee for Standardization of Light and Decorative Building Materials (SAC/TC 195): Beijing, China, 2009.
53. *JTG E20-2011*; Test Code for Asphalt and Asphalt Mixtures for Highway engineering. Ministry of Transport: Beijing, China, 2011.

buildings

MDPI

Article

Experimental Study on Dynamic Modulus of High Content Rubber Asphalt Mixture

Guozhi Zheng [1], Naitian Zhang [2,3,*] and Songtao Lv [1]

[1] National Engineering Laboratory of Highway Maintenance Technology, Changsha University of Science & Technology, Changsha 410114, China; zgz@stu.csust.edu.cn (G.Z.); lst@csust.edu.cn (S.L.)

[2] Xinjiang Key Laboratory for Green Construction and Smart Traffic Control of Transportation Infrastructure, Xinjiang University, Urumqi 830017, China

[3] Key Laboratory of Special Environment Road Engineering of Hunan Province, Changsha University of Science & Technology, Changsha 410114, China

* Correspondence: znt@xju.edu.cn

Abstract: Currently, the research on the mechanical properties of rubber-modified asphalt mixtures primarily focuses on small-scale investigations, with insufficient exploration into the performance of rubber particles and their relationship with the mechanism and properties of modified asphalt mixtures. Limited studies have been conducted on large-scale rubber modification in asphalt mixtures. Due to frequent use and subsequent high damage to existing asphalt pavements, incorporating rubber-modified asphalt mixtures can partially alleviate premature deterioration. Dynamic modulus tests were conducted using MTS equipment under unconfined conditions to investigate the viscoelastic behavior of rubber-modified asphalt mixtures with high rubber content and elucidate the influence of rubber particle content on the elastic deformation and recovery capability. The dynamic mechanical properties of the mixtures were determined at different loading rates, temperatures, and types of rubber-modified asphalt mixtures. Based on the test data, variations in the dynamic modulus, phase angle, storage modulus, loss modulus, loss factor, and rut factor of the rubber-modified asphalt mixtures under different loading frequencies, temperatures, and types were analyzed. The results demonstrate the pronounced viscoelastic behavior of rubber-modified asphalt mixtures. The mixtures exhibit enhanced elasticity at low temperatures and high frequencies, while their viscosity becomes more prominent at high temperatures and low frequencies. Under constant test temperatures, an increase in load loading frequency leads to a higher dynamic modulus; conversely, a decrease in dynamic modulus is observed with increasing test temperatures. The dynamic modulus of ARHM-25 at a frequency of 10 Hz is found to be 12.99 times higher at 15 °C compared to that at 60 °C, while at 30 °C, the dynamic modulus at 25 Hz is observed to be 2.72 times greater than that at 0.1 Hz. Furthermore, the rutting resistance factors of the asphalt mixtures increase with loading frequency but decrease with temperature. The rutting factor for ARHM-13 at a frequency of 10 Hz is found to be 22.98 times higher at 15 °C compared to that at 60 °C, while at a temperature of 30 °C, the rutting factor for this material is observed to be 3.09 times greater at a frequency of 25 Hz than at 0.1 Hz. These findings suggest that rutting is most likely when vehicles drive at low speeds in hot weather conditions.

Keywords: road engineering; viscoelasticity; dynamic modulus; crumb rubber; asphalt mixture

Citation: Zheng, G.; Zhang, N.; Lv, S. Experimental Study on Dynamic Modulus of High Content Rubber Asphalt Mixture. *Buildings* **2024**, *14*, 434. https://doi.org/10.3390/buildings14020434

Academic Editor: Dan Bompa

Received: 10 January 2024
Revised: 25 January 2024
Accepted: 31 January 2024
Published: 5 February 2024

1. Introduction

Currently, the layered elastic system theory is utilized in calculating the structure in China's Code for Design of Asphalt Pavement Structure, with a double circular uniform load applied. However, in practical scenarios, asphalt pavement structures are influenced by various factors, such as vehicle loads, climate conditions, and environmental impacts. Therefore, a significant disparity exists between the actual working state and the behavior predicted by the layered system under a double circular uniform load. As a result, replacing the static modulus with the dynamic modulus has become an essential requirement for pavement materials based on empirical design.

As a viscoelastic material, the asphalt mixture exhibits its viscoelastic properties through the characterization of its dynamic modulus. Consequently, various temperature-dependent viscoelastic tests were conducted on the asphalt mixture. The findings indicate that [1] at low temperatures and under small deformations, the asphalt mixture demonstrates linear elastic behavior; at high temperatures and large deformations, it behaves as a viscoplastic elastomer, while in the transitional range of average temperatures, it displays true viscoelasticity. According to the design specifications of the American AASHTO2002, the dynamic modulus can effectively serve as an indicator of asphalt mixture performance [2]. A study conducted under the NCHRP9-19 project in the United States shows the correlation between dynamic modulus and rutting. Additionally, the anti-rutting factor exhibits a close relationship with the phase angle. At the same time, the phase angle represents the extent of strain lag and strain in asphalt [3–5]. During the Ninth Five-Year Plan period, the Ministry of Communications in China conducted numerous experimental studies on the factors influencing the dynamic modulus of asphalt mixtures. The research findings indicate that the test temperature, load frequency, and loading method are primary determinants affecting the dynamic modulus. Additionally, it was observed that there is a minimal impact on the dynamic modulus from load gap time. Consequently, based on these observations, domestic and foreign research institutions widely employ dynamic modulus testing to assess the high and low-temperature properties and viscoelastic characteristics of asphalt mixtures [5–7].

Reputable scholars have conducted extensive research on the dynamic modulus of matrix asphalt mixtures, encompassing testing methods for dynamic modulus and the analysis of influential factors. The changes in porosity, water permeability, and mechanical properties of three types of asphalt mixtures (ARHM-13, ARHM-20, and ARHM-25) were investigated by Yan Luo et al. [8] before and after undergoing freezing and thawing processes. Furthermore, the relationship between the porosity of rubber-modified asphalt mixtures with different grades and their corresponding water seepage rates was analyzed. To accurately predict the fatigue life (Nf) of rubber asphalt mixtures, Chenze Fang et al. [9] analyzed the correlation between Nf and stiffness modulus decay (SMD), as well as the relative rate of change of stiffness modulus decay (SMDR). They established a functional relationship between Nf and the number of fatigue loads, as well as a damage model based on the cumulative stiffness modulus decay relative rate of change (CSMDR). The fatigue equation was derived from the correlational analysis between SMDR and the damage evolution rate. Xinya Liu et al. [10] conducted uniaxial static load compression tests to investigate the creep characteristics of the mixture under various temperatures and confining pressure conditions. They used the viscoelastic Burgers model to examine the mechanical properties of a warm mixed-rubber-modified asphalt mixture. Zining Chen et al. [11] investigated the impact of various factors, including the recycled agent dosage, curing temperature, and curing time, on the dynamic modulus and bending stiffness modulus of hot mix asphalt incorporating reclaimed asphalt pavement (RAP). Hui Wang et al. [12] divided the samples of different asphalt content layers into the upper and lower parts and then compacted them together. SPT was used to conduct dynamic modulus tests to simulate the state of asphalt migration between layers of asphalt pavement and further study the influence of asphalt migration on the dynamic modulus of asphalt mixture. Ruan Li et al. [13] investigated the impact of square and semi-sine loads on the dynamic and resilience modulus of asphalt mixtures. The dynamic modulus tests were conducted using semi-sine and bell-shaped loads on AC-13, AC-16, AC-20, AC-25, and SUP-25 asphalt mixtures. Ding Han et al. [14] investigated the impact of aggregate morphology on the dynamic modulus of asphalt mixture using a digital image processing method. They developed a predictive model for the dynamic modulus of asphalt mixture under different temperatures and loading frequencies based on aggregate morphology and asphalt mortar material parameters. Xujie Li et al. [15] investigated the impact of long-term indoor oven aging on the stiffness of the asphalt mixture by conducting indoor dynamic modulus tests. Various conditions of long-term oven aging were applied to different materials, including varying aging temperatures and durations for loose mixtures and compacted samples. The principal curve of the dynamic modulus and data on the dynamic modulus ratio (DMR) indicate that aging the loose mixture at 95 °C is an effective method for characterizing the on-site long-term aging process with good efficiency and performance. Harnaeni SR et al. [16] assessed the viscoelastic properties, including phase angle and dynamic modulus, of asphalt concrete wear process (AC-WC) and hot rolled sheet wear process (HRS-WC) mixtures by establishing a principal curve based on dynamic modulus tests. Xiaochun Qin et al. [17] conducted a comparative study on two testing methods, namely indirect tensile and uniaxial compression, to measure the dynamic modulus of asphalt mixture with five different binder types and a nominal maximum aggregate particle size using a point measurement system. The dynamic modulus and phase angle principal curves at a reference temperature of 20 °C were obtained according to the superposition principle of time and temperature. The results revealed significant differences in the dynamic modulus, phase angle, and displacement coefficient between the two test modes. The fitting formula demonstrated a strong correlation between the dynamic moduli measured under uniaxial compression and indirect tensile conditions. Huining Xu et al. [18] investigated the application of Rayleigh wave technology for nondestructive measurement of the dynamic modulus of asphalt mixture by proposing a three-dimensional finite element numerical algorithm to simulate Rayleigh wave propagation in a multi-layer semi-space system. The feasibility of Rayleigh wave technology in capturing mechanical property variations within shallow structures of asphalt pavement was verified through numerical analysis. Furthermore, an experimental application of Rayleigh wave technology was conducted using five types of asphalt mixtures.

A substantial volume of discarded tires is generated with the rapid advancement of the socio-economic landscape and continuous enhancement of automobile capabilities. If not appropriately and effectively managed, waste tires, being chemical substances, can lead to significant environmental pollution. Aligned with the deepening green concept and emphasis on environmental preservation and energy conservation, rubber has been incorporated as an additive in asphalt to enhance its road performance. After decades of extensive research in rubber-modified asphalt technology conducted by worldwide researchers, it has been established that rubber-modified asphalt and its mixture serve as environmentally friendly pavement materials exhibiting commendable high/low-temperature properties, fatigue resistance, and noise reduction. Qingyu Zhang et al. [19] employed two modifiers, namely HDPE and TOR, to fabricate rubber powder/SBS-modified asphalt mixtures with rubber powder contents of 20% and 25%. A comparative analysis of technical performance was conducted between the rubber powder/SBS-modified asphalt mixtures and the SBS-modified asphalt mixture. Under repeated load, the triaxial dynamic creep test investigated the permanent deformation behavior of the three types of rubber powder/SBS-modified asphalt mixtures. Subsequently, a prediction model for permanent deformation under repeated loading was established based on the modified Burgers model. Zihao Zhou et al. [20] employed a UTM material testing machine to investigate the dynamic bending modulus and phase angle of a mixture comprising 70# matrix asphalt and rubber asphalt using two different test methods, namely, four-point bending and ladder beam tests. By applying the time-temperature equivalent principle, they calculated the shift factor of the principal curve. They generated dynamic modulus principal curves and phase angle principal curves to analyze the viscoelastic properties of the rubber asphalt mixtures. Furthermore, they compared and analyzed the effects of these two dynamic bending and tensile modulus test methods on both the dynamic bending/tensile moduli and the phase angles of the asphalt mixtures. To investigate the impact of rubber powder particle size and dosage on the viscoelastic properties of asphalt mixture at high temperatures, Zhichen Wang et al. [21] initially conducted a series of feed creep tests on dry rubber asphalt mixtures under 13 different working conditions with varying particle sizes and dosages. Subsequently, an improved four-element and five-parameter model was developed by incorporating static and dynamic load equivalence and applying the Boltzmann linear superposition principle. This study proposes a dynamic creep viscoelastic model for asphalt mixture that can be solved in a single iteration, while also analyzing the model parameters. The anti-fatigue cracking properties of various base materials and recycled tire rubber-modified asphalt mixtures were compared and evaluated by S. Kocak et al. [22] using pp-feed analysis and AASHTOWare Pavement ME (version 2.3) software. These materials included the innovative devulcanized rubber (DVR), rubber powder blend (CRTB), rubber wet process (CRWET), as well as the raw/base material PG58-28 and the softer PG58-34 binder. Fan Bai et al. [23] conducted an extensive series of uniaxial compression tests on both the ideal asphalt mixture and the fine and coarse rubber-modified (CRM) asphalt mixture. A viscoelastic viscoplastic random constitutive model incorporating the lognormal distribution of random parameters was proposed to depict the mechanical behavior accurately. Guoqing Wang et al. [24] conducted aging studies on three types of rubber-modified asphalt with rubber contents of 30%, 40%, and 50%, as well as SBS-modified asphalt and 70# matrix asphalt, using the RTFOT and PAV methods. The results revealed that while RTFOT failed to accurately differentiate the fatigue properties of modified asphalt, PAV successfully distinguished the fatigue properties of different types. The ranking order for fatigue properties was AR40 > AR30 > AR50 > SBS > 70#. Hua Tan et al. [25] employed a dynamic shear rheometer to conduct hysteresis tests on rubber-modified asphalt with various additives, thereby assessing the compound modification effect of the modifier on rubber asphalt and its impact on viscoelastic properties based on indicators such as residual deformation elastic energy retention, dissipated energy, elastic ratio, and composite elastic modulus. Adding the PE modifier enhances the resistance to permanent deformation in rubber asphalt but exhibits poor fatigue resistance. On the other hand, rock asphalt can enhance the stiffness, viscous flow behavior, and fatigue resistance of rubber asphalt. In order to investigate the impact of rubber asphalt mortar on the dynamic modulus of the mixture, Wang Xinqiang et al. [26,27] employed a rotary shear compaction test. The skeleton clearance rate was determined by gradually filling coarse aggregate until reaching 0.85, 1.00, and 1.10 times the minimum skeleton clearance volume, respectively. Subsequently, 30% rubber powder-modified asphalt mortar was added to form the mixture. Dynamic modulus specimens were prepared using the rotary shear compaction test method, and their dynamic moduli were measured through an asphalt mixture performance test. Zifeng Zhao et al. [28] investigated the impact of rubber powder and RAP material on the viscoelastic properties of asphalt mixture, revealing that at loading conditions of 1 Hz and 35 °C, the creep resistance of asphalt mixture exhibited an increase exceeding 10%. In comparison, the complex modulus and storage modulus experienced a growth surpassing 50%. The viscoelastic properties of asphalt mixture modified with rubber particles were investigated by Miao Yu et al. [29,30]. They examined the influence of rubber particle content on the asphalt mixture's elastic deformation and recovery ability. Dynamic modulus tests were conducted using an unconfined asphalt mixture performance testing machine. The dynamic properties of the mixture were measured at different loading frequencies, temperatures, and varying rubber particle contents. Guoqing Wang et al. [31] conducted an experimental study on the road performance of modified asphalt with a high dosage of rubber powder. The dynamic shear rheometer was employed to conduct frequency scanning and multi-stress creep recovery tests on matrix and rubber powder-modified asphalt with varying content. Through a comprehensive analysis of dynamic mechanical response under different loading and working conditions, the viscoelastic properties and high-temperature creep mechanical behavior of rubber powder-modified asphalt were systematically characterized. The effects of cup lump rubber and polyphosphoric acid on the mechanical properties of the asphalt mixture were investigated by Anwaar Hazoor Ansari et al. [32]. The results demonstrated that the combined modification of CLA and PPA significantly enhanced the dynamic resilience modulus of the asphalt mixture. The road performance of the SMA mixture with graphene composite rubber-modified asphalt was analyzed by Rui Guo et al. [33], revealing that the SMA mixture exhibited superior bending strength and dynamic stability compared to the AH-70# and SBS asphalt mixtures. JC Wang et al. [34] investigated the shear modulus and damping ratio of two modified asphalt samples with rubber contents of 20% and 10% under various temperature and pressure conditions. The shear modulus and damping ratio strongly correlate with temperature, while stiffness shows a slight decrease with increasing confining pressure. Temperature and confining pressure play significant roles in determining these properties. The stable rubber-plastic-modified asphalt mixture designed by R Ren et al. [35] was characterized by establishing a dynamic modulus master curve to evaluate its mechanical properties. Compared to the SBS-modified asphalt mixture, the stable rubber-plastic-modified asphalt mixture exhibited a 15% increase in low-temperature tensile strength and a more than 40% increase in high-temperature dynamic modulus. SA Kumar et al. investigated the dynamic mechanical properties of SBS and rubber powder composite-modified asphalt mixture [36]. A multivariate analysis of variance revealed that temperature, frequency, and type of asphalt binder significantly influenced the measured dynamic modulus. Furthermore, under high temperatures and prolonged loading conditions, the road performance of the composite modified asphalt mixture exhibited superior characteristics compared to the unmodified asphalt mixtures. KL Roja et al. [37] employed the Sasobit additive in a rubber crumb-modified binder to produce warm mixed asphalt. It was observed that the dynamic modulus and rutting resistance of the modified asphalt mixture exhibited an increasing trend with higher Sasobit content. At the same time, the fatigue life demonstrated a decreasing trend. The viscoelastic micromechanical model proposed by ZC Wang et al. [38] predicts the dynamic modulus of granular rubber-modified asphalt concrete (GCRMAC) based on existing two-phase and three-phase micromechanical models. Building upon this, an equation for the GCRMAC dynamic modulus is established, taking into account the viscoelastic effect of asphalt concrete and the content of rubber chips. Laboratory experiments confirm the effectiveness of this model in predicting the dynamic modulus of GCRMAC. In a study conducted by AA Abdulridha et al. [39], the influence of rubber scrap size and content on the elastic modulus, dynamic creep, and durability of rubber-modified asphalt mixture was investigated, revealing that reducing the rubber particle size had the most significant impact on enhancing the performance of asphalt mixture. Yunchao Tang et al. [40] utilized recycled rubber particles (RPs) and recycled aggregate (RA) to fabricate rubber-modified recycled aggregate concrete (RRAC). The RRAC characteristics, including compressive strength and peak strain, were investigated after subjecting the specimens to various temperatures while varying the amounts of RA and RPs. Tao Zhou et al. [41] utilized waste bio-oil and scrap rubber from waste tires to modify bitumen, producing bio-modified rubber bitumen (BMR). They investigated the time-dependent evolution of BMR properties, focusing on its rheological, chemical, and microscopic characteristics. Memduh Karalar et al. [42] investigated the impact of waste tire rubber (WTR) on the flexural behavior of reinforced concrete beams (RCB) through experimental tests, ANSYS, and SAP2000 simulations, as well as three-dimensional finite element model (FEM) visualization. Alexey Beskopylny et al. [43] examined the influence of preparation factors on the properties of concrete with partial replacement of coarse aggregate (CA) by rubber tree (RT) seed shells. They identified the optimal composition for achieving enhanced strength characteristics.

To enhance the recycling efficiency of waste rubber, it is imperative to augment the rubber content in rubber-modified asphalt and its mixture. The conventional utilization of rubber powder in rubber asphalt typically involves a 20% composition, with the United States adopting a range between 17% and 20%, while South Africa employs an approximate ratio of 25%. Furthermore, efforts have been made to prepare rubber-modified asphalt mixtures containing over 30% rubber powder. Although extensive research has been conducted on the mechanical properties of low-content rubber-modified asphalt and its mixtures, there remains substantial room for exploration regarding high-content rubber-modified asphalt and its mixtures. This article investigates the viscoelastic properties of high-content rubberized asphalt mixtures. The dynamic mechanical properties of high content rubber modified asphalt were comprehensively considered under different temperature and loading frequency conditions.

2. Materials and Methods

2.1. Test Methods

The Superpave rotary compactor was employed in this experiment to compact the specimen. This compactor allows for precise control of forming pressure and porosity, utilizing the lateral tilt angle of the mold cylinder and chassis to generate shear force within the mixture. Consequently, it can effectively simulate the stress state experienced during actual asphalt mixture spreading by roller forming. Compaction times are selected based on target porosity and compaction height data obtained during molding.

Laboratory testing methods for dynamic modulus include uniaxial compression, direct tensile, triaxial compression, indirect tensile, two-point bending, three-point bending, four-point bending tests, and more. The uniaxial compression test, indirect tensile test, and four-point bending test are commonly used for testing asphalt mixtures. Because pavements primarily experience compression under driving loads and China's Code for Design of Asphalt Pavement Structure uses the uniaxial compressive modulus as a design parameter, the study adopts the uniaxial compression test method to investigate the dynamic modulus of asphalt mixtures. Currently, there are three primary types of equipment utilized for conducting uniaxial compressive dynamic modulus tests on indoor asphalt mixtures: the MTS810 series servo-hydraulic material test system manufactured by MTS Systems Corporation in Eden Prairie, MN, USA, and Industrial Process Controls Ltd. in Glanmire, Queensland, Australia, the essential performance test instrument (SPT) for asphalt mixtures and the UTM series servo-hydraulic or pneumatic material test system produced by the company. In this study, we employed the MTS810 series servo-hydraulic material test system to perform dynamic modulus testing under uniaxial compression at temperatures of 15 °C, 30 °C, 45 °C, and 60 °C. The frequencies used for the deformation test on the dynamic modulus of asphalt mixture are 0.1 Hz, 0.5 Hz, 1 Hz, 5 Hz, 10 Hz, and 25 Hz. The extensometer is connected to the deformation input channel of MTS to ensure synchronization between the load and deformation measurements during testing. To accurately describe the relationship between load and deformation within a loading cycle, it is necessary to adjust the data acquisition frequency according to the loading frequency. The data acquisition frequency is typically ten times higher than the loading frequency. The general procedure for conducting a dynamic modulus test under uniaxial compression is as follows:

1. Before the test, place the specimen in the environmental chamber at the designated testing temperature and maintain a consistent temperature for 4–5 h until reaching the required test temperature within the specified timeframe.
2. Retrieve the specimen from the incubator and position the extensometer equidistantly along its circumference. The placement procedure is as follows: situate the extensometer at the midpoint of one side of the specimen, perpendicular to its cross-section.
3. Position a polytetrafluoroethylene film above and below the specimen to minimize friction resistance between the specimen and the loading plate, thereby reducing potential end effects. Subsequently, place it at the center of the loading plate, as depicted in Figure 1.
4. The specimen is subjected to a sine-wave axial compressive stress test load, and the repeated loading cycles specified in Table 1 are conducted from high frequency to low frequency at the designated temperature. At any two testing frequencies with an interval of 2 min, the load and deformation curves of the last five waveforms are collected while recording and calculating both the applied load during testing and the axial strain of the specimen.
5. The next temperature test on the specimen should be conducted, with the temperature selection ranging from low to high.

Figure 1. Dynamic modulus testing under uniaxial compression conditions.

Table 1. Loading times under different load frequencies.

Frequency (Hz)	Loading Times
25	200
10	200
5	100
1	20
0.5	15
0.1	15

2.2. Methods for Preparing Materials

The asphalt used in this study is a high-content rubber with rubber contents of 30%, 40%, and 50%, corresponding to the upper, middle, and lower surfaces of Jingde high-speed, respectively. The penetration, ductility, and softening point indices are presented in Table 2, while the PG classification performance is displayed in Table 3. All indices for rubber asphalt and basalt aggregate meet the requirements specified by the "Technical Specification for Highway Asphalt Pavement Construction" (JTG F40-2004).

Table 2. Test results for conventional properties of rubber asphalt.

Test Items		Asphalt Test Results for 30% Rubber Content	Asphalt Test Results for 40% Rubber Content	Asphalt Test Results for 50% Rubber Content
Penetration degree 25 °C, 100 g, 5 s (0.1 mm)		62	66	58
Ductility 5 cm/min, 5 °C (cm)		16	18	12.8
Softening point TR and B (°C)		72.5	72	78.7
Rotary viscosity 180 °C (Pa·s)		2.8	3.5	2.96
Elastic recovery 25 °C (%)		84	81	71
Storage stability segregation, 48 h softening point difference (°C)		2.0	0.5	6.4
Residue after TFOT	Quality loss (%)	−0.29	−0.3	0.3
	Residual penetration ratio 25 °C (%)	77	76	86.2
	Residual ductility 5 °C (cm)	11	12	10.4

Table 3. PG specification evaluation of rubber asphalt.

Test Items	Asphalt Test Results for 30% Rubber Content	Asphalt Test Results for 40% Rubber Content	Asphalt Test Results for 50% Rubber Content
PG specification evaluation	88–34	88–34	82–28
Rutting factor (88 °C)	1683.8	1421.6	1638
Rutting factor after aging (88 °C)	3429.5	2767.2	1672
Fatigue limit temperature (°C)	13	10	16
Creep stiffness S (−24 °C)	201.9205	166.9801	343
Creep rate (m)	0.3054	0.3147	0.335

The aggregate used is high-quality basalt, while the filler consists of basalt mineral powder. To ensure accurate test results and maintain uniformity in the production process, the aggregate employed for specimen production undergoes screening to achieve different particle sizes. In order to ascertain the performance of these aggregates, various performance indices were tested according to the Technical Specification for Highway Asphalt Pavement Construction (JTG F40-2004). Consequently, aggregates' physical and mechanical performance indices were obtained, as presented in Tables 4–6. The results demonstrate that all utilized aggregate parameters meet the specified technical requirements.

Table 4. Density test results for limestone aggregates.

Serial Number	Particle Size (mm)	Performance Density (g/cm^3)	Bulk Density (g/cm^3)	Dry Apparent Density (g/cm^3)	Water Absorption (%)
1	13.2~9.5	2.673	2.569	2.608	1.53
2	9.5~4.75	2.661	2.572	2.607	1.35
3	4.75~2.36	2.649			
4	2.36~1.18	2.642			
5	1.18~0.6	2.606	—	—	—
6	0.6~0.3	2.592			
7	0.3~0.15	2.586			
8	0.15~0.075	2.615			

Table 5. Test results for crushing value, polishing value, and abrasion value of limestone aggregate.

Aggregate Type	Crushing Value (%)	Polished Stone Value (BPN)	Wear Value (%)
Limestone	15.8	57.4	19.9
Standard requirement	≤28	≥45	≤30

Table 6. Basic technical index of mineral powder.

Test Items		Test Results	Technical Requirements
Apparent relative density (g/cm^3)		2.688	≥2.50
water content (%)		0.4	≤1
Appearance characteristic		No agglomerative caking	—
Hydrophilic coefficient		0.57	<1
Plasticity index (%)		2.2	<4
Granulometric range (%)	<0.6 mm	100	100
	<0.15 mm	95.2	90~100
	<0.075 mm	84.4	75~100

This study's crumb rubber-modified asphalt mixture grading design method primarily refers to the Technical Specification for Highway Asphalt Pavement Construction (JTG F40-2004). Tables 7–9 display the grading of each crumb rubber-modified asphalt mixture, while Figure 2 illustrates the grading curve. The aggregate grading satisfies the requirements outlined in JTG F40-2004, and the asphalt-aggregate ratio for three crumb rubber asphalt mixtures at 30%, 40%, and 50% is 5.22%, 4.79%, and 4.36%, respectively.

Table 7. Mineral aggregate gradation inARHM-13.

Mesh Size (mm)	16	13.2	9.5	4.75	2.36	1.18	0.6	0.3	0.15	0.075
The upper limit of gradation	100	100	71	35	28	23	19	15	12	10
The lower limit of gradation	100	95	62	25	20	15	12	10	8	6
Median of gradation	100	97.5	66.5	30	24	19	15.5	12.5	10	8
Passing rate (%)	100	95.7	67	30.3	24.3	18.5	13.2	10.3	9.1	8.6

Table 8. Mineral aggregate gradation in ARHM-20.

Mesh Size (mm)	26.5	19	16	13.2	9.5	4.75	2.36	1.18	0.6	0.3	0.15	0.075
The upper limit of gradation	100	100	88	76	59	35	27	21	17	13	10	8
The lower limit of gradation	100	90	77	64	47	25	18	14	10	7	5	4
Median of gradation	100	95	82.5	70	53	30	22.5	17.5	13.5	10	7.5	6
Passing rate (%)	100	95.5	84.6	72.8	53.6	29.5	23.1	16.4	10.4	7.2	6.3	6.0

Table 9. Mineral aggregate gradation in ARHM-25.

Mesh Size (mm)	31.5	26.5	19	16	13.2	9.5	4.75	2.36	1.18	0.6	0.3	0.15	0.075
The upper limit of gradation	100	100	82	73	65	53	35	27	20	16	12	9	7
The lower limit of gradation	100	90	70	62	54	42	25	17	12	9	6	4	3
Median of gradation	100	95	76	67.5	59.5	47.5	30	22	16	12.5	9	6.5	5
Passing rate (%)	100	99.4	75.4	68.1	59.8	48.7	29.9	19.0	13.7	9.6	7.2	6.4	5.9

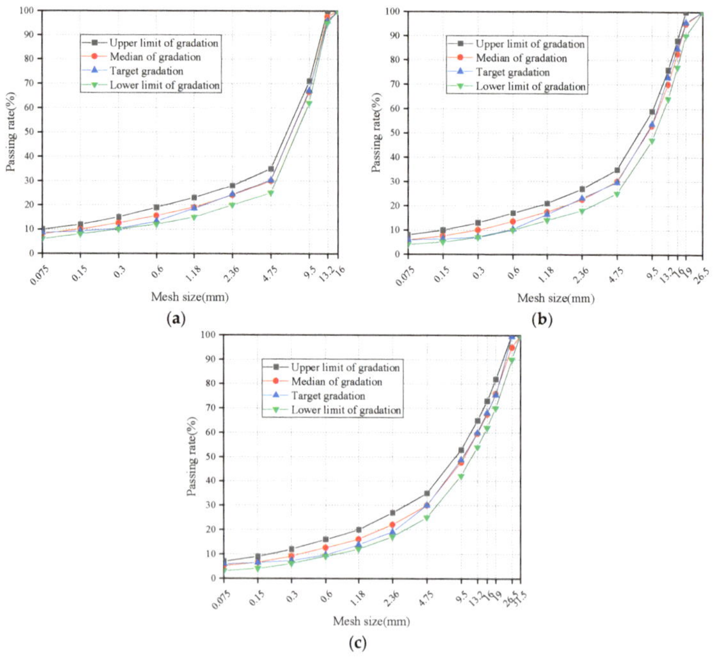

Figure 2. Asphalt mixture mineral grading curve: (**a**) ARHM-13; (**b**) ARHM-20; (**c**) ARHM-25.

The limestone aggregate is placed in a constant-temperature oven at 110 °C until it reaches a consistent weight through drying. The asphalt and aggregate are heated in the oven at 180 °C for over 2 h, following the grading mentioned above design. The crumb rubber-modified asphalt mixture is thoroughly mixed using an indoor mixer, maintaining the optimal oil-to-stone ratio. Specimens were prepared using the rotating compaction method (76-B0252 rotating compactor of Italian CONTROLS). After demudding, cylindrical specimens with a height of 150 mm and a diameter of 100 mm were obtained. The asphalt mixture specimens are shown in Figure 3.

Figure 3. Asphalt mixture specimens.

2.3. *Methods for Calculating Vibroelastic Parameters*

2.3.1. Dynamic Modulus, Phase Angle, and Loss Factor

The dynamic modulus, which is the ratio of stress to strain, characterizes a material's resistance to deformation. The phase angle, representing the phase difference between strain and stress, quantifies the extent of viscoelasticity. The loss factor, expressed as the tangent value of the phase angle, indicates the material's viscoelastic properties. A higher loss factor indicates greater viscosity, while a lower loss factor signifies enhanced elasticity. Calculation methods for dynamic modulus and phase angle are provided below:

$$\sigma_0 = P_i / A \quad (1)$$

where

σ_0—axial stress amplitude (MPa);
P_i—average amplitude of axial test load in the last five loading cycles (N);
A—radial cross-section area of the specimen (mm^2).

$$|E^*| = \sigma_0 / \varepsilon_0 \quad (2)$$

where

$|E^*|$—dynamic modulus (MPa);
ε_0—the axial strain amplitude measured by the extensometer (mm/mm).

$$\varphi = t_i / l_p \times 360 \quad (3)$$

where

φ—phase angle (°);
t_i—the average lag time between peak deformation and peak load in the last five loading cycles (s);
l_p—the average load period of the last five load cycles (s).

2.3.2. Storage Modulus, Loss Modulus, and Rutting Factor

The complex modulus of the crumb rubber-modified asphalt mixture can be expressed by the equation below, where the natural part represents the storage modulus, and the imaginary part denotes the loss modulus. The storage modulus characterizes the elastic behavior of the crumb rubber-modified asphalt mixture, with a higher value indicating better shape retention capability. On the other hand, the loss modulus reflects the viscous behavior of crumb rubber-modified asphalt mixture. A more significant loss modulus implies increased energy consumption during deformation and more incredible difficulty returning to its original shape. The subsequent equation describes the relationship between the complex, storage, and loss modulus.

$$E^* = E' + iE'' \tag{4}$$

$$E' = |E^*| \cos \varphi \tag{5}$$

$$E'' = |E^*| \sin \varphi \tag{6}$$

where

E^*—complex modulus (MPa);
E'—storage modulus (MPa);
E''—loss modulus (MPa);
i—complex unit, $i^2 = -1$.

WITCZAK [44] tested and compared the dynamic modulus under varying loading frequencies and temperatures, both with and without confining pressure. It was concluded that the permanent deformation of asphalt pavement strongly correlates with the unconstrained rutting factor, while this correlation weakens under confining pressure. The symbol k is used to represent the rutting factor.

$$k = |E^*| / \sin \varphi \tag{7}$$

3. Dynamic Modulus Test Results and Analysis

3.1. Influence of Loading Frequency

3.1.1. Dynamic Modulus

The dynamic modulus changes in three different crumb rubber-modified asphalt mixtures with loading frequency under different temperature conditions are shown in Figure 4.

The frequency-dependent behavior of the dynamic moduli for the three crumb rubber-modified asphalt mixtures is evident in Figure 4. As the loading frequency increases, a consistent upward trend can be observed in the dynamic moduli of ARHM-13, ARHM-20, and ARHM-25. However, this increasing trend gradually diminishes with higher loading frequencies. The dynamic modulus is stable when the frequency increases beyond a certain threshold. The asphalt mixture belongs to viscoelastic materials, and its deformation hysteresis is caused by external load stress. As the loading frequency increases, the energy that cannot be released in time gradually increases, increasing the dynamic modulus of the asphalt mixture. When the loading frequency is high, the viscosity characteristics of the asphalt mixture are no longer evident and approach elastic deformation, so the dynamic modulus tends to stabilize.

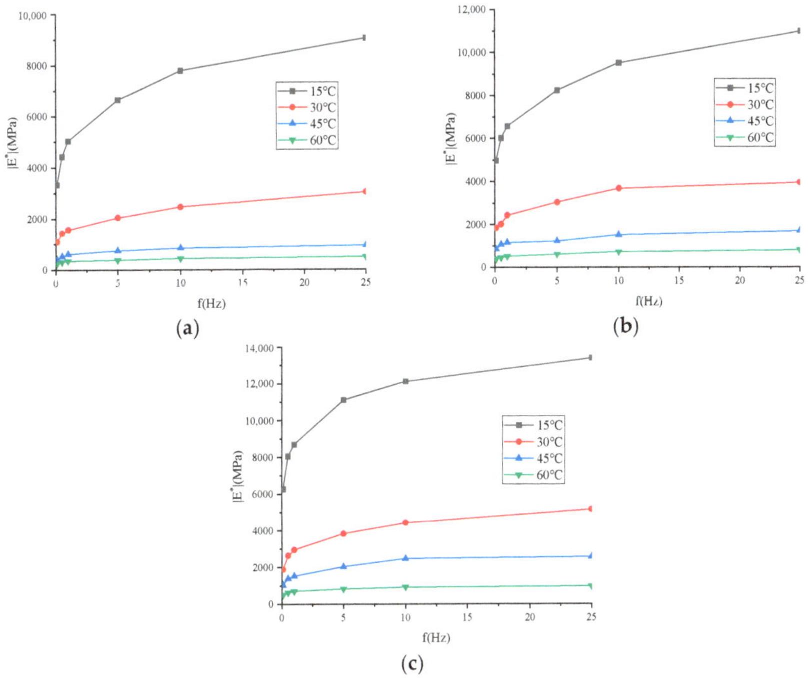

Figure 4. Variation trend of dynamic modulus of asphalt mixture with loading frequency: (**a**) ARHM-13; (**b**) ARHM-20; (**c**) ARHM-25.

3.1.2. Phase Angle

The phase angle changes with the loading frequency of three different crumb rubber-modified asphalt mixtures under different temperature conditions are shown in Figure 5.

As shown in Figure 5, when the temperature is between 15 °C and 30 °C, the phase angle of the three crumb rubber-modified asphalt mixtures decreases with the increase in loading frequency, and the decreasing trend gradually slows down with the growth of loading frequency. When the temperature is between 45 °C and 60 °C, the phase angle of the three kinds of rubber-modified asphalt mixtures increases with the increase in loading frequency, and the decreasing trend gradually slows down with the increase in loading frequency. The phase angle is stable when the frequency increases beyond a certain threshold. The phase angle characterizes the ability of the mixture to resist deformation under load from the perspective of viscosity, reflecting the proportion and degree of influence of viscosity and elastic components in viscoelasticity. Under the conditions of 15 °C and 30 °C, the phase angle decreases with the increase in loading frequency, and the decreasing trend gradually slows down. The formation mechanism of high-temperature stability of asphalt mixtures comes from the bonding properties of the binder and the interlocking effect of mineral aggregate gradation. Under low temperatures and high-frequency loads, the primary performance is elastic. With the increase in loading frequency, the elastic performance is enhanced, resulting in a decrease in phase angle. As the loading frequency increases, the elastic performance tends to stabilize to a certain extent, and the ratio of the elastic part to the viscous part tends to stabilize.

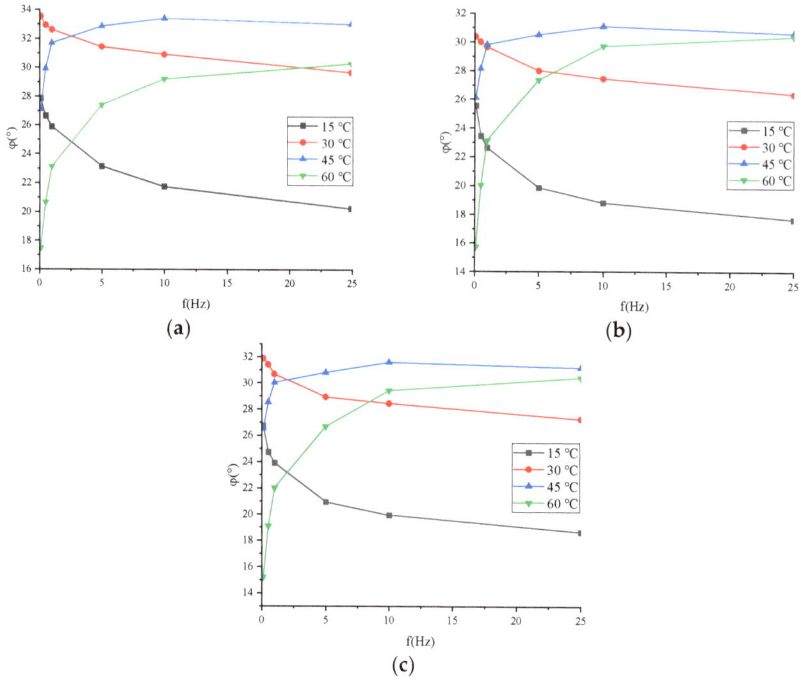

Figure 5. Variation trend of phase angle of asphalt mixtures with loading frequency: (**a**) ARHM-13; (**b**) ARHM-20; (**c**) ARHM-25.

3.1.3. Storage Modulus

The change of storage modulus of three different crumb rubber-modified asphalt mixtures with loading frequency under different temperature conditions is shown in Figure 6.

It can be seen from Figure 6 that the storage moduli of the three crumb rubber-modified asphalt mixtures change regularly with frequency. With the increase in loading frequency, the storage moduli of ARHM-13, ARHM-20, and ARHM-25 all show an increasing trend, and the growing trend gradually slows down with the increase in loading frequency. The transformation law of the storage modulus with loading frequency is the same as that of the dynamic modulus with loading frequency. The storage modulus refers to the amount of energy stored in a material due to elastic (reversible) deformation during deformation, reflecting the magnitude of the material's elasticity. It represents the index of material rebound after deformation. The dynamic modulus refers to the mechanical response of a material under load, which is not only related to the influencing factors under static action but also to the time, magnitude, frequency, and repetition effects of load action, demonstrating a specific stress dependence. The storage modulus reflects the elastic properties of asphalt mixtures, while the dynamic modulus reflects the elastic and viscous properties of asphalt mixtures. When asphalt mixtures exhibit elasticity, the value of the stored modulus tends to align with that of the dynamic modulus and is calculated as the cosine of the dynamic modulus multiplied by the phase angle.

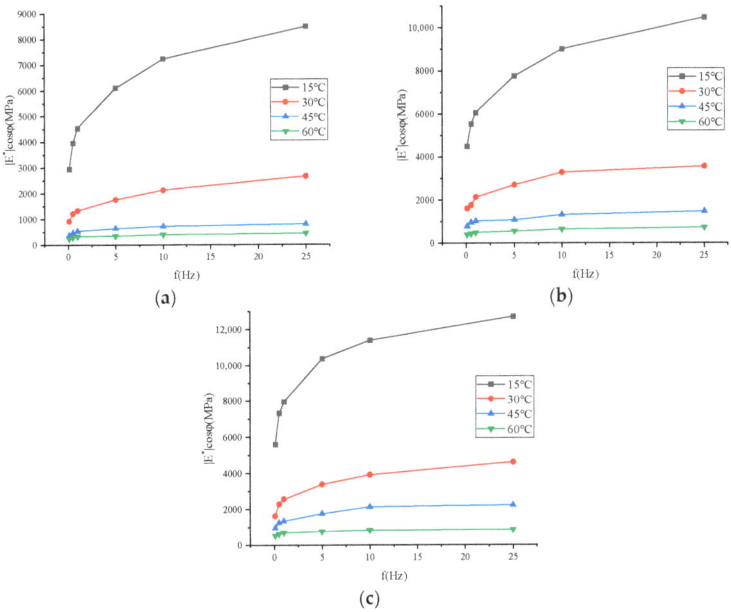

Figure 6. Variation trend of storage moduli of asphalt mixture with loading frequency: (**a**) ARHM-13; (**b**) ARHM-20; (**c**) ARHM-25.

3.1.4. Loss Modulus

The loss modulus changes with the loading frequency of three different crumb rubber-modified asphalt mixtures under different temperature conditions are shown in Figure 7.

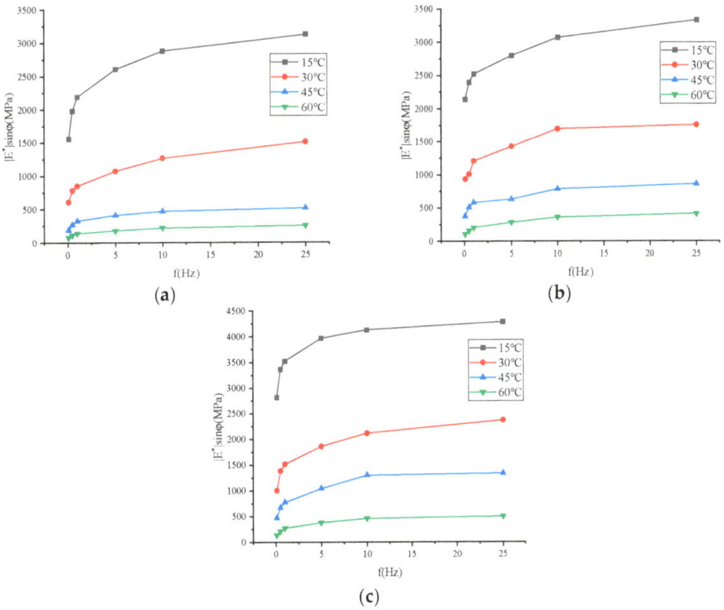

Figure 7. Variation trend of asphalt mixture loss moduli with loading frequency: (**a**) ARHM-13; (**b**) ARHM-20; (**c**) ARHM-25.

It can be seen from Figure 7 that the loss modulus of the three crumb rubber-modified asphalt mixtures changes regularly with the frequency. With the increase in loading frequency, the dynamic modulus shows an increasing trend, which gradually slows down with the growth of loading frequency. The law governing the change in the loss modulus with loading frequency is the same as that of the dynamic and storage moduli with loading frequency. This is because, at low frequencies, the molecular and lattice movements inside the material are relatively slow, the resistance is relatively small, and the loss modulus is low. At high frequencies, molecular and lattice movements become more frequent and complex, leading to increased resistance and a gradual increase in loss modulus. However, when the frequency reaches a particular value, the movement of molecules and lattices begins to be influenced by external constraints and dissipation effects, so the loss modulus will gradually decrease.

3.1.5. Loss Factor

The loss factors of the three different crumb rubber-modified asphalt mixtures under different temperature conditions vary with loading frequency, as shown in Figure 8.

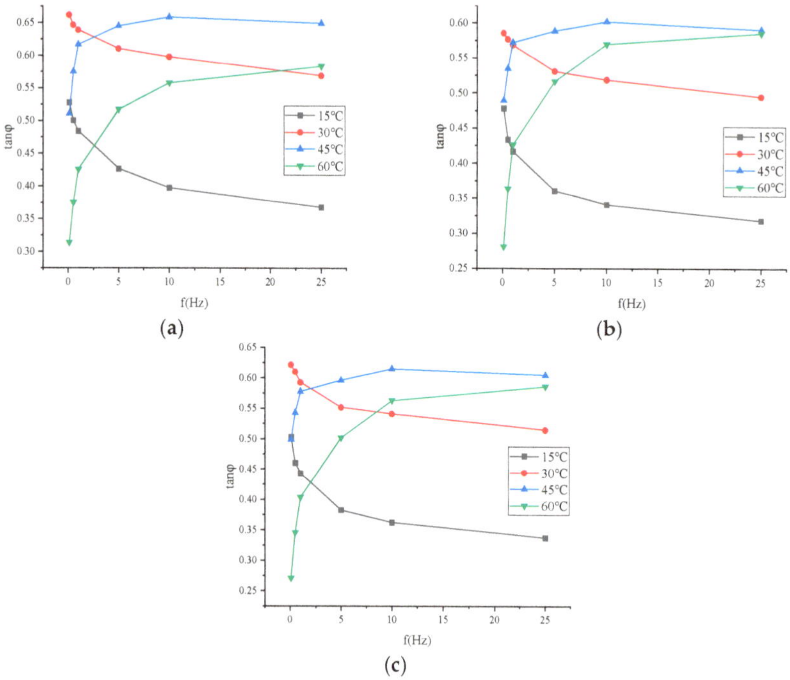

Figure 8. Variation trend of asphalt mixture loss factors with loading frequency: (**a**) ARHM-13; (**b**) ARHM-20; (**c**) ARHM-25.

As shown in Figure 8, when the temperature is between 15 °C and 30 °C, the loss factors of the three crumb rubber-modified asphalt mixtures decrease with the increase in loading frequency, and the decreasing trend gradually slows down with the growth of loading frequency. When the temperature is between 45 °C and 60 °C, the loss factors of the three kinds of rubber-modified asphalt mixtures increase with the increase in loading frequency, and the decreasing trend becomes slower with the growth of loading frequency. The variation pattern of the loss factor with loading frequency aligns with the phase angles. The loss factor is the tangent of the phase angle between the strain and stress period of a viscoelastic material under the action of an alternating force field, which is also equal to the ratio of the loss modulus to the storage modulus of the material. When the temperature is between 45 °C and 60 °C, the loss factor can be calculated from the phase angle obtained from the test.

3.1.6. Rutting Factor

Under different temperature conditions of the three crumb rubber-modified asphalt mixtures, the rutting factors change with loading frequency, as shown in Figure 9.

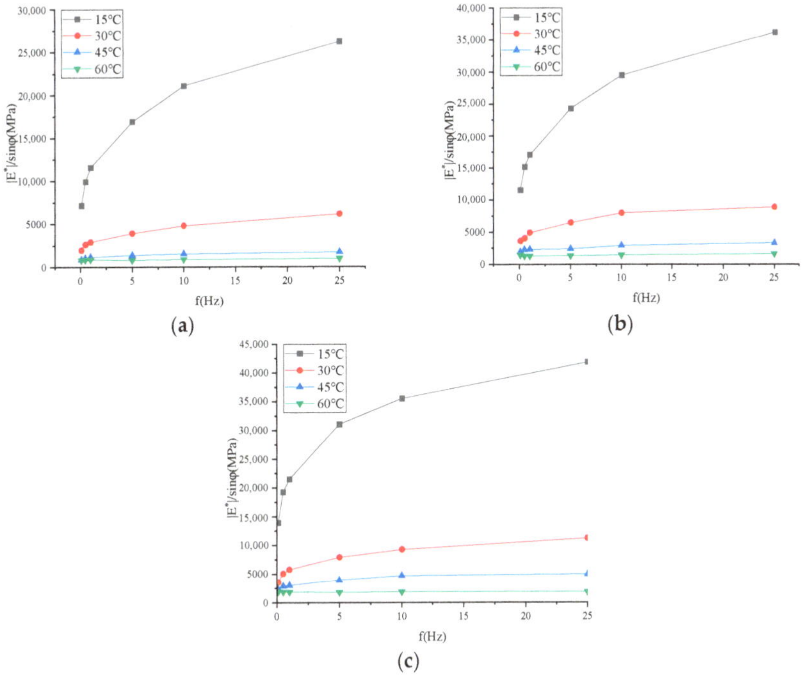

Figure 9. Trends in the rutting factors of asphalt mixtures with loading frequency: (**a**) ARHM-13; (**b**) ARHM-20; (**c**) ARHM-25.

It can be seen from Figure 9 that the rutting factors of the three crumb rubber-modified asphalt mixtures change regularly with frequency. With the increase in loading frequency, the rutting factors all show an increasing trend, which gradually slows down with the growth of loading frequency. The law governing the change in the rutting factor with loading frequency is the same as that of the dynamic modulus, storage modulus, and loss modulus with loading frequency.

3.2. Influence of Temperature

3.2.1. Dynamic Modulus

The dynamic modulus changes with temperature for the three different crumb rubber-modified asphalt mixtures at different frequencies are shown in Figure 10.

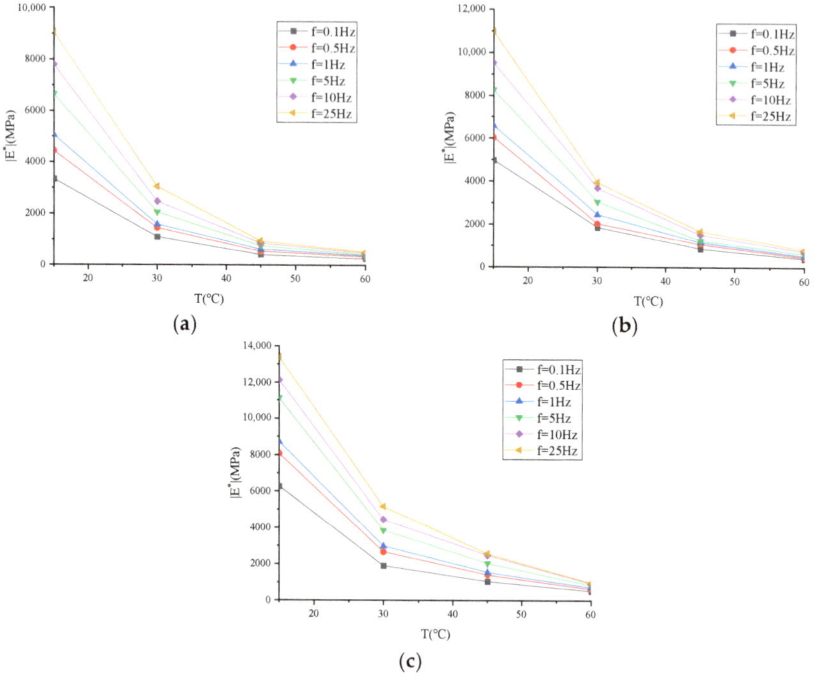

Figure 10. Variation trends of dynamic moduli of asphalt mixtures with temperature: (**a**) ARHM-13; (**b**) ARHM-20; (**c**) ARHM-25.

It can be seen from Figure 10 that the dynamic moduli of the three crumb rubber-modified asphalt mixtures change regularly with temperature. With the increase in the temperature, the dynamic moduli of ARHM-13, ARHM-20, and ARHM-25 all show a decreasing trend, and the decreasing trend gradually slows down with the increase in the temperature. At higher temperatures, the viscosity characteristics of asphalt mixtures increase, resulting in a decrease in the dynamic modulus.

3.2.2. Phase Angle

The results of the phase angle variation with temperature for the three different crumb rubber-modified asphalt mixtures at different frequencies are shown in Figure 11.

It can be seen from Figure 11 that the phase angles of the three crumb rubber-modified asphalt mixtures change regularly with the temperature. With the increase in temperature, the dynamic moduli of ARHM-13, ARHM-20, and ARHM-25 increase first and then decrease. When the frequency is less than or equal to 1 Hz, the phase angle reaches the maximum at 30 °C, and when the frequency is greater than or equal to 5 Hz, the phase angle reaches the maximum at 45 °C. The low-temperature mixture exhibits elasticity, and as the frequency decreases, the phase angle becomes larger. The high-temperature mixture exhibits viscoelasticity, and as the frequency decreases, the phase angle becomes smaller. When the temperature is between high and low, the phase angle increases first and then decreases as the frequency decreases.

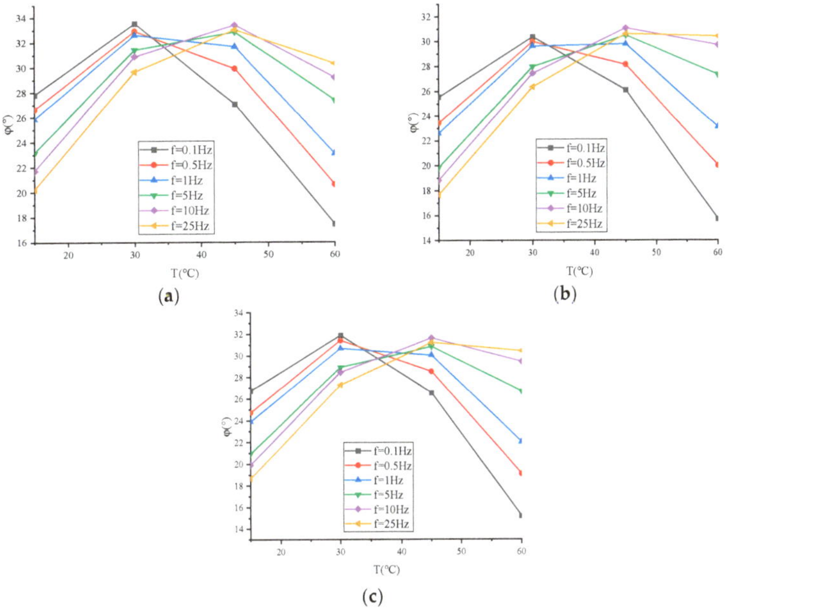

Figure 11. Variation trends in phase angles of asphalt mixtures with temperature: (**a**) ARHM-13; (**b**) ARHM-20; (**c**) ARHM-25.

3.2.3. Storage Modulus

The results of the storage modulus variation with temperature for the three different crumb rubber-modified asphalt mixtures at different frequencies are shown in Figure 12.

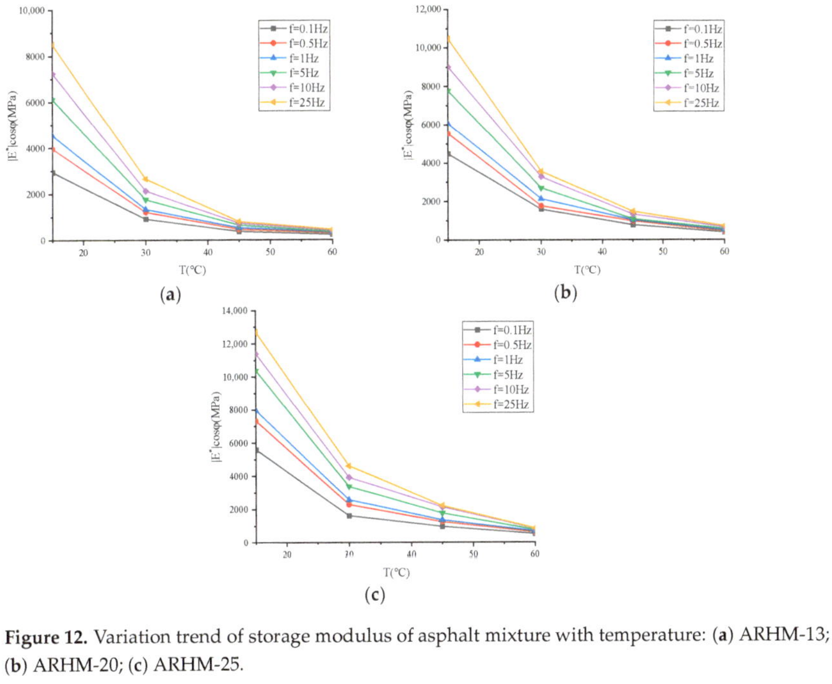

Figure 12. Variation trend of storage modulus of asphalt mixture with temperature: (**a**) ARHM-13; (**b**) ARHM-20; (**c**) ARHM-25.

It can be seen from Figure 12 that the storage moduli of the three crumb rubber-modified asphalt mixtures change regularly with temperature. With the increase in temperature, the storage moduli of ARHM-13, ARHM-20, and ARHM-25 all show a decreasing trend, and the decreasing trend gradually slows down with the increase in temperature. The temperature-dependent transformation law of the storage modulus is identical to that of the dynamic modulus.

3.2.4. Loss Modulus

The loss modulus changes with temperature of the three different crumb rubber-modified asphalt mixtures at different frequencies are shown in Figure 13.

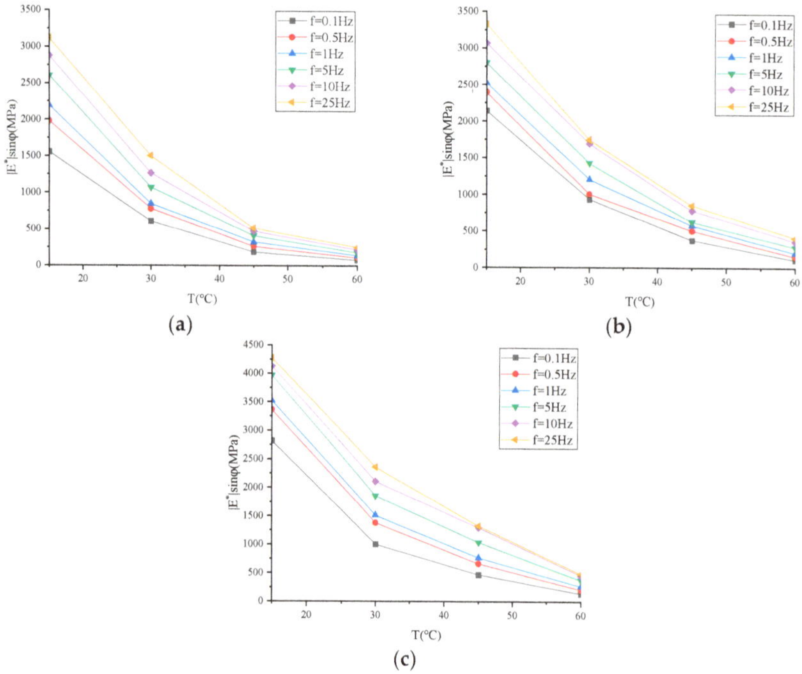

Figure 13. Variation trend of loss moduli of asphalt mixtures with temperature: (**a**) ARHM-13; (**b**) ARHM-20; (**c**) ARHM-25.

It can be seen from Figure 13 that the loss moduli of the three crumb rubber-modified asphalt mixtures change regularly with temperature. With the increase in temperature, the loss moduli of ARHM-13, ARHM-20, and ARHM-25 all show a decreasing trend, and the decreasing trend gradually slows down with the temperature increase. The temperature-dependent transformation law of the storage modulus is consistent with that of both the dynamic and storage moduli.

3.2.5. Loss Factor

The loss factors of the three different crumb rubber-modified asphalt mixtures under different frequency conditions change with temperature, as shown in Figure 14.

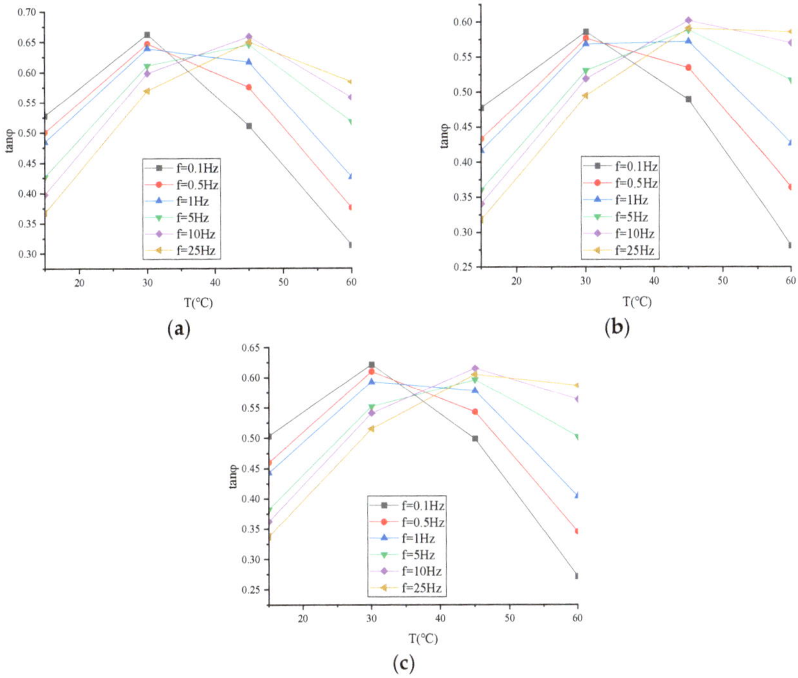

Figure 14. Variation trend of asphalt mixture loss factors with temperature: (**a**) ARHM-13; (**b**) ARHM-20; (**c**) ARHM-25.

It can be seen from Figure 14 that the loss factors of the three crumb rubber-modified asphalt mixtures change regularly with temperature. With the increase in temperature, the loss factors of ARHM-13, ARHM-20 and ARHM-25 show first an increase and then a decrease trend. When the frequency is less than or equal to 1 Hz, the loss factor reaches the maximum at 30 °C, and when the frequency is greater than or equal to 5 Hz, the loss factor reaches the maximum at 45 °C. The temperature-dependent variation in the loss factor aligns with the corresponding change in the phase angle. The loss factor is the tangent of the phase angle between the strain and stress periods of a viscoelastic material under the action of an alternating force field, which is equal to the ratio of the loss modulus to the storage modulus of the material. The low-temperature mixture exhibits elasticity, and the lower the frequency, the greater the loss factor. The high-temperature mixture exhibits viscoelasticity, and the lower the frequency, the smaller the loss factor. When the temperature is withing the range of high and low temperatures, as the frequency decreases, the loss factor first increases and then decreases.

3.2.6. Rutting Factor

The results of the rutting factor changes with temperature for the three different crumb rubber-modified asphalt mixtures at different frequencies are shown in Figure 15.

It can be seen from Figure 15 that the rutting factors of the three crumb rubber-modified asphalt mixtures change regularly with temperature. With the increase in temperature, the rutting factors of ARHM-13, ARHM-20, and ARHM-25 all show a decreasing trend, and the decreasing trend gradually slows down with the increase in temperature. The temperature-dependent variation in the rutting factors aligns with that observed for the dynamic, storage, and loss moduli. The rutting factor is used as an indicator to evaluate the high-temperature performance of asphalt. The higher the rutting factor is, the stronger the asphalt mixture resistance is to permanent deformation high-temperatures; that is, the higher the high-temperature stability of the mixture, the better.

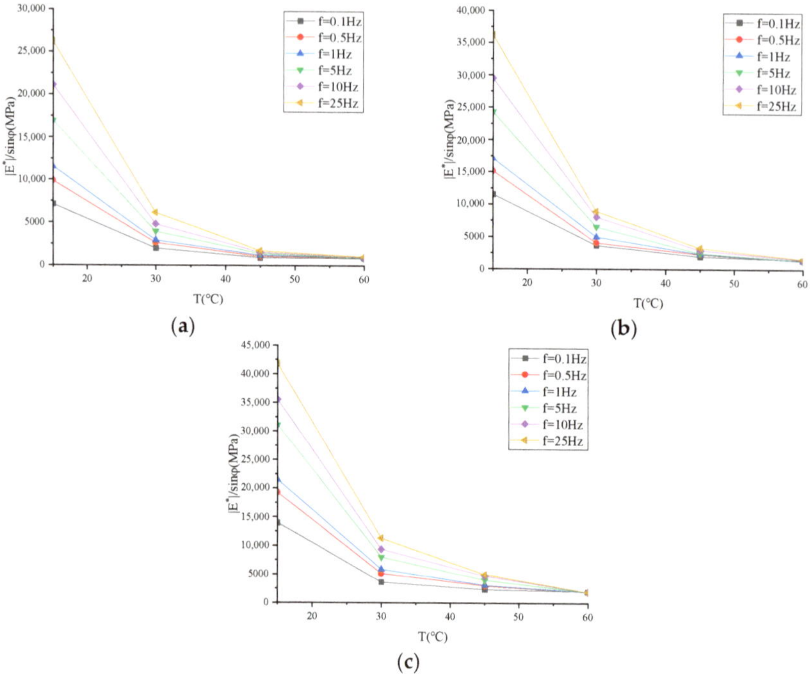

Figure 15. Trend of rutting factors of asphalt mixtures with temperature: (**a**) ARHM-13; (**b**) ARHM-20; (**c**) ARHM-25.

3.3. Influence of Rubber-Modified Asphalt Mixture's Type

3.3.1. Dynamic Modulus

Under four temperature conditions and different frequency conditions, the dynamic modulus results for the three different kinds of rubber-modified asphalt mixtures are shown in Figure 16.

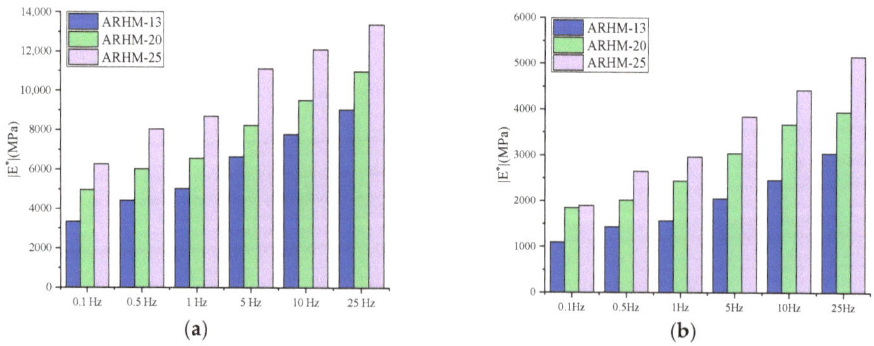

Figure 16. *Cont.*

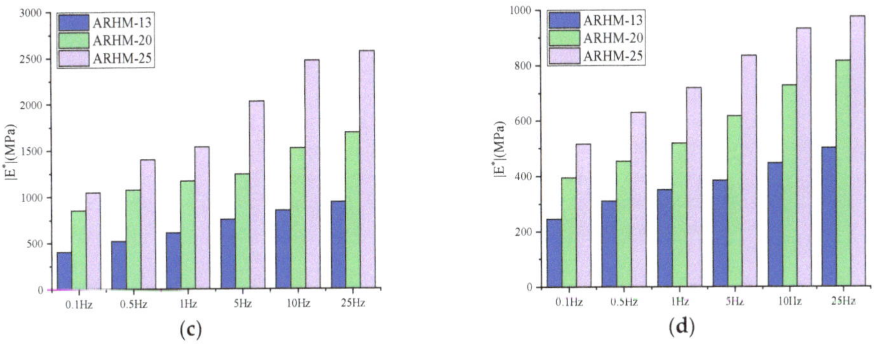

Figure 16. Relationship between type of asphalt mixture and dynamic modulus: (**a**) T = 15 °C; (**b**) T = 30 °C; (**c**) T = 45 °C; (**d**) T = 60 °C.

As can be seen from Figure 16, under different temperature and frequency conditions, the dynamic modulus results for the three different kinds of rubber-modified asphalt mixtures are as follows: ARHM-25 > ARHM-20 > ARHM-13. The experimental results show that as the maximum nominal particle size of the gradation increases, the dynamic modulus of the asphalt mixture is significantly enhanced. Therefore, it is recommended to use coarse-grained asphalt mixture as pavement material in areas and sections with heavy traffic and high temperatures.

3.3.2. Phase Angle

Under four temperature conditions and different frequency conditions, the phase angle results for the three different kinds of rubber-modified asphalt mixtures are shown in Figure 17.

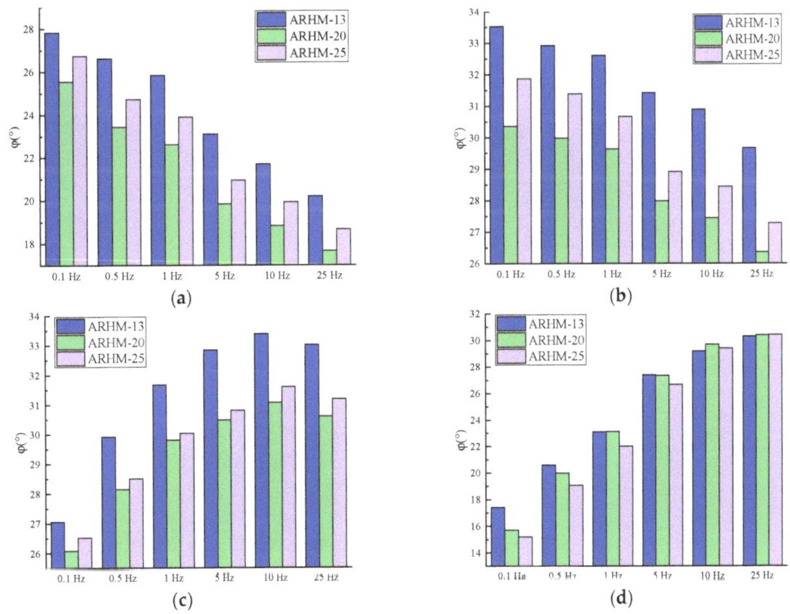

Figure 17. Relationship between asphalt mixture type and phase angle: (**a**) T = 15 °C; (**b**) T = 30 °C; (**c**) T = 45 °C; (**d**) T = 60 °C.

As can be seen from Figure 17, under the condition of T ≤ 45 °C and different frequencies, the phase angle results for the three different kinds of rubber-modified asphalt mixture are as follows: ARHM-13 > ARHM-25 > ARHM-20. Under the conditions of T = 60 °C and f ≤ 5 Hz, the phase angle results for the three different kinds of rubber-modified asphalt mixture are as follows: ARHM-13 > ARHM-20 > ARHM-25. Under the condition of T = 60 °C and f = 10 Hz, the phase angle results for the three different kinds of rubber-modified asphalt mixture are as follows: ARHM-20 > ARHM-25 > ARHM-13. Under the conditions of T = 60 °C and f = 25 Hz, the phase angle results for the three different kinds of rubber-modified asphalt mixture are as follows: ARHM-25 > ARHM-20 > ARHM-13.

3.3.3. Storage Modulus

Under four temperature conditions and different frequency conditions, the storage modulus results for three different kinds of rubber-modified asphalt mixture are shown in Figure 18.

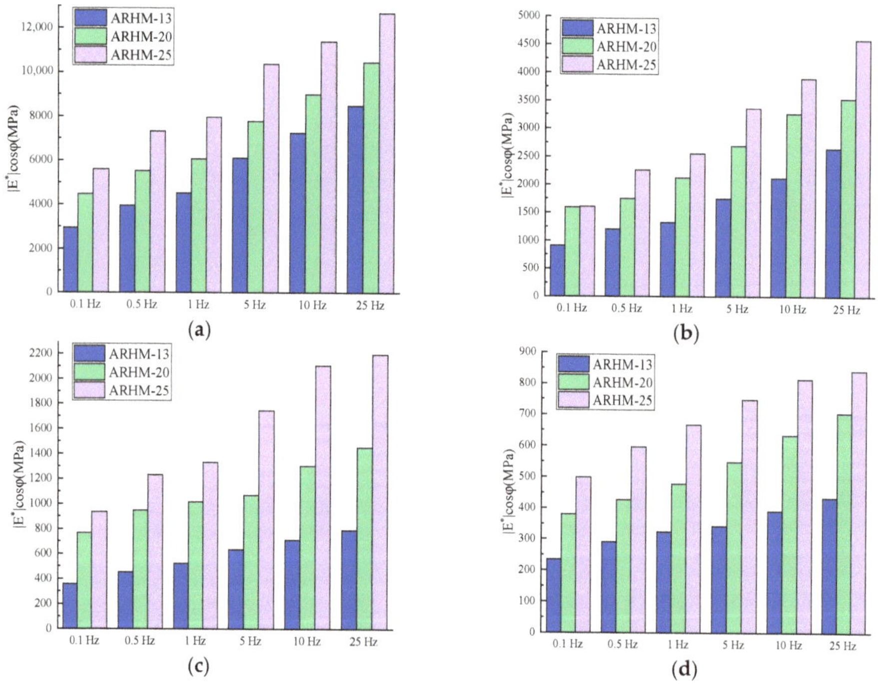

Figure 18. Relationship between asphalt mixture type and storage modulus: (**a**) T = 15 °C; (**b**) T = 30 °C; (**c**) T = 45 °C; (**d**) T = 60 °C.

As can be seen from Figure 18, under different temperature and frequency conditions, the storage modulus results for the three different kinds of rubber-modified asphalt mixture are as follows: ARHM-25 > ARHM-20 > ARHM-13.

3.3.4. Loss Modulus

Under four temperature conditions and different frequency conditions, the loss modulus results for the three different kinds of rubber-modified asphalt mixture are shown in Figure 19.

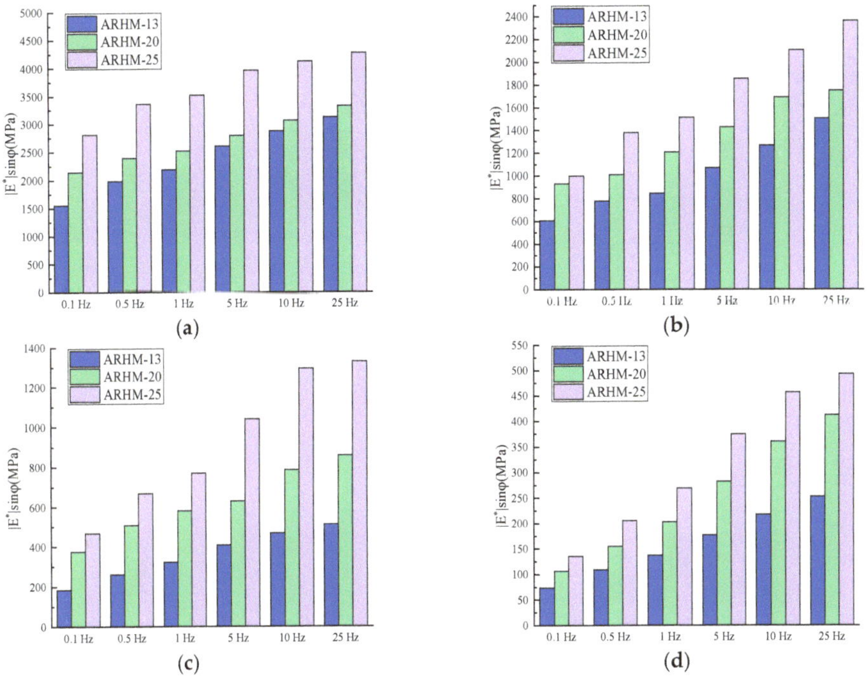

Figure 19. Relationship between asphalt mixture type and loss modulus: (**a**) T = 15 °C; (**b**) T = 30 °C; (**c**) T = 45 °C; (**d**) T = 60 °C.

As can be seen from Figure 19, under different temperature and frequency conditions, the storage modulus results for the three different kinds of rubber-modified asphalt mixture are as follows: ARHM-25 > ARHM-20 > ARHM-13.

3.3.5. Loss Factor

Under four temperature conditions and different frequency conditions, the loss factors of the three different kinds of rubber-modified asphalt mixtures are shown in Figure 20.

As can be seen from Figure 20, under the condition of T = 15 °C and different frequencies, the loss factors of the three different kinds of rubber-modified asphalt mixture are as follows: ARHM-13 > ARHM-25 > ARHM-20. Under the condition of T = 30 °C or 45 °C and different frequencies, the loss factors of the three different kinds of rubber-modified asphalt mixtures are as follows: ARHM-25 > ARHM-20 > ARHM-13. Under T = 60 °C and f $\leq$ 0.5 Hz, the loss factors of the three different kinds of rubber-modified asphalt mixtures are as follows: ARHM-25 > ARHM-13 > ARHM-20. Under T = 60 °C, f = 1 Hz, the loss factors of the three different kinds of rubber-modified asphalt mixtures are as follows: ARHM-20 > ARHM-13 > ARHM-25. Under the condition of T = 60 °C, f = 10 Hz, the loss factors of the three different kinds of rubber-modified asphalt mixture are as follows: ARHM-20 > ARHM-25 > ARHM-13. Under T = 60 °C and f = 25 Hz, the loss factors of the three different kinds of rubber-modified asphalt mixtures are as follows: ARHM-20 > ARHM-13 > ARHM-25.

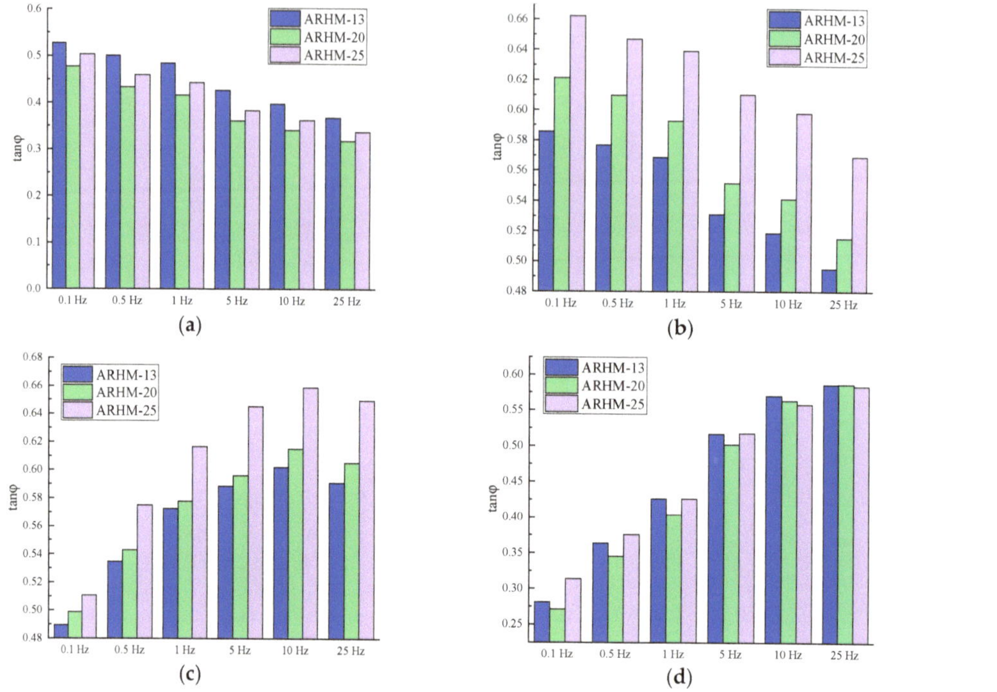

Figure 20. Relationship between asphalt mixture type and loss factor: (**a**) T = 15 °C; (**b**) T = 30 °C; (**c**) T = 45 °C; (**d**) T = 60 °C.

3.3.6. Rutting Factor

Under four temperature conditions and different frequency conditions, the loss factors of the three different kinds of rubber-modified asphalt mixtures are shown in Figure 21.

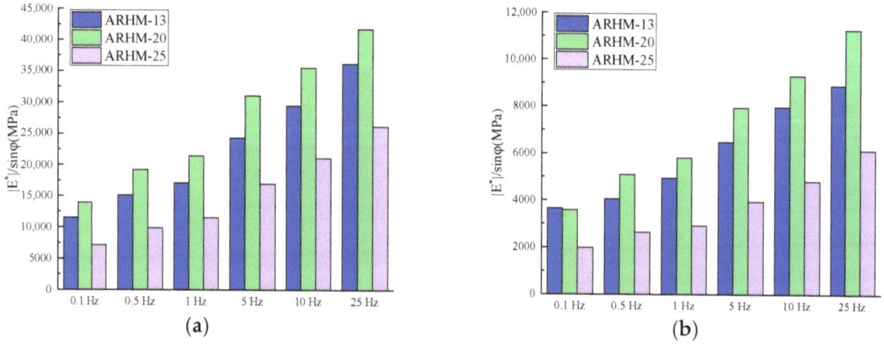

Figure 21. *Cont.*

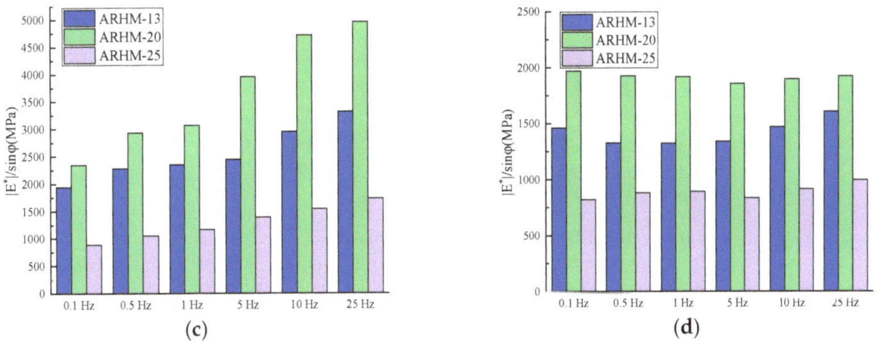

Figure 21. Relationship between asphalt mixture type and rutting factor: (**a**) T = 15 °C; (**b**) T = 30 °C; (**c**) T = 45 °C; (**d**) T = 60 °C.

As can be seen from Figure 21, under the conditions of T = 30 °C and f = 0.1 Hz, the rutting factor results for the three different kinds of rubber-modified asphalt mixtures are as follows: ARHM-13 > ARHM-20 > ARHM-25. Under other temperature and frequency conditions, the rutting factors of the three different types of rubber-modified asphalt mixtures are as follows: ARHM-20 > ARHM-13 > ARHM-25.

4. Conclusions

This study adopted a blending method of equal volume addition of rubber particles, using a step-by-step compaction method with coarse and fine aggregates. Asphalt mixture specimens ARHM-13, ARHM-20, and ARHM-25 were prepared with 30%, 40%, and 50% rubber content, respectively. MTS was selected for dynamic modulus testing under unconfined conditions to analyze the viscoelastic parameters of the rubber-modified asphalt mixtures, the variation pattern under different loading frequencies, and loading temperatures. The main conclusions are as follows:

(1) The relationship between the dynamic, storage, and loss moduli of the three rubber-modified asphalt mixtures under different loading frequencies and temperatures is ARHM-25 > ARHM-20 > ARHM-13. Based on this, it is recommended to use a coarse-grained asphalt mixture as the pavement surface material in areas and sections with heavy traffic and high temperatures.

(2) The relationship between the rutting factors of the three types of rubber-modified asphalt mixtures under different frequency and temperature conditions is ARHM-20 > ARHM-13 > ARHM-25. The modified asphalt mixture exhibits a significantly enhanced delayed elastic recovery ability, enabling partial deformation recovery upon unloading. However, this phenomenon inevitably introduces errors, thereby imposing limitations on the utilization of the rutting factor for evaluating the high-temperature performance of modified asphalt.

(3) The research results of this article indicate that high rubber-content modified asphalt mixtures have excellent performance. In addition, it supports the view that adding 30% to 50% rubber particles to asphalt mixtures is reasonable, pointing out that increasing the gradation will increase the rubber content.

(4) To establish a relationship between the maximum reasonable dosage of rubber particles and the gradation of the asphalt mixture, more indoor test data need to be obtained in the future. Through many experiments, the relationship between the maximum reasonable dosage of rubber particles and the gradation of the asphalt mixture needs to be established.

Author Contributions: Conceptualization: S.L.; methodology: G.Z.; validation: G.Z. and N.Z.; formal analysis: S.L. and G.Z.; investigation: S.L. and N.Z.; resources: N.Z. and G.Z.; data creation: S.L. and G.Z.; writing: N.Z. and G.Z.; writing—review and editing: N.Z. and G.Z.; visualization: N.Z. and G.Z.; supervision: N.Z.; project administration: G.Z.; funding acquisition: S.L. All authors have read and agreed to the published version of the manuscript.

Funding: This work was supported by the National Natural Science Foundation of China (52225806 and 52078063) and the Open Fund of Key Laboratory of Special Environment Road Engineering of Hunan Province (kfj230602).

Data Availability Statement: The data analyzed in this study are subject to the following licenses/restric the restrictions can be received form the first author. Requests to access these datasets should be directed to zgz@stu.csust.edu.cn.

Conflicts of Interest: Authors declare no conflicts of interest.

References

1. Bakhshi, B.; Arabani, M. Numerical Evaluation of Rutting in Rubberized Asphalt Mixture Using Finite Element Modeling Based on Experimental Viscoelastic Properties. *J. Mater. Civ. Eng.* **2018**, *30*, 04018088.1–04018088.11. [CrossRef]
2. Yang, Y.; Li, Y.; Zhang, Z.; Cao, D. Performance Evaluation and Parameter Analysis of Asphalt Mixture Using Simple Performance Test (SPT). *J. China Foreign Highw.* **2013**, *33*, 262–268. [CrossRef]
3. Ji, J.; Suo, Z.; Xu, Y.; Xu, S. Experimental Research on Performance of Warm-recycled Mixture Asphalt with SMA. *China J. Highw. Transp.* **2013**, *26*, 28–33. [CrossRef]
4. Zhu, L. Research on Viscoelastic Properties of Asphalt Mixture. Master's Thesis, Chongqing Jiaotong University, Chongqing, China, 2010.
5. Irfan, M.; Waraich, A.S.; Ahmed, S.; Ali, Y. Characterization of Various Plant-Produced Asphalt Concrete Mixtures Using Dynamic Modulus Test. *Adv. Mater. Sci. Eng.* **2016**, *2016*, 5618427. [CrossRef]
6. Espersson, M. Effect in the High Modulus Asphalt Concrete with the Temperature. *Constr. Build. Mater.* **2014**, *71*, 638–643. [CrossRef]
7. Lee, S.J.; Akisetty, C.K.; Amirkhanian, S.N. The Effect of Crumb Rubber Modifier (CRM) on the Performance Properties of Rubberized Binders in HMA Pavements. *Constr. Build. Mater.* **2008**, *22*, 1368–1376. [CrossRef]
8. Luo, Y.; Li, S.; Tian, G.; Liu, Y. Experimental Study on Freeze-thaw Performance of Rubber-modified Asphalt Mixture. *Transp. Sci. Technol.* **2022**, *8*, 116–120, 126. [CrossRef]
9. Fang, C.; Guo, N.; Sun, Y.; Wang, J. Research on Fatigue Life of Rubber Asphalt Mixture Based on Stiffness Modulus. *Eng. Mech.* **2020**, *37*, 196–204. [CrossRef]
10. Liu, X. Rutting Resistance Evaluation of Asphalt Pavement Modified with Rubber Powder. *China Highw.* **2022**, *618*, 106–109. [CrossRef]
11. Chen, Z.; Liu, B.; Feng, D.; Li, G. Analysis of Factors Influencing the Modulus of Hot-Recycled Asphalt Mixture with High RAP. *Materials* **2023**, *16*, 5280. [CrossRef]
12. Wang, H.; Zhan, S.; Liu, G. The Effects of Asphalt Migration on the Dynamic Modulus of Asphalt Mixture. *Appl. Sci.* **2019**, *9*, 2747. [CrossRef]
13. Ruan, L.; Luo, R.; Hu, X.; Pan, P. Effect of Bell-Shaped Loading and Haversine Loading on the Dynamic Modulus and Resilient Modulus of Asphalt Mixtures. *Constr. Build. Mater.* **2018**, *161*, 124–131. [CrossRef]
14. Han, D.; Yuan, M.; Hu, H. Establishing Prediction Master Curve of Dynamic Modulus of Asphalt Mixture Considering Randomness of Aggregate Morphology. *Constr. Build. Mater.* **2021**, *294*, 123575. [CrossRef]
15. Li, X.-J.; Li, Y.; Li, J.; Shen, Y.; Zhang, C.; Wei, J. Effect of Laboratory Long-Term Oven Aging on the Stiffness of Asphalt Mixtures. *Constr. Build. Mater.* **2020**, *258*, 120252. [CrossRef]
16. Harnaeni, S.R.; Pramesti, F.P.; Budiarto, A.; Setyawan, A.; Khan, M.I.; Sutanto, M.H. Study on Structural Performance of Asphalt Concrete and Hot Rolled Sheet Through Viscoelastic Characterization. *Materials* **2020**, *13*, 1133. [CrossRef] [PubMed]
17. Qin, X.; Ma, L.; Wang, H. Comparison Analysis of Dynamic Modulus of Asphalt Mixture: Indirect Tension and Uniaxial Compression Test. *Transp. A Transp. Sci.* **2019**, *15*, 165–178. [CrossRef]
18. Xu, H.; Fu, X.; Meng, A.; Cui, H.; Tan, Y. Rayleigh Wave Technology for Analysis Dynamic Modulus of Asphalt Mixtures: Feasibility and Application. *Int. J. Pavement Eng.* **2020**, *21*, 1235–1247. [CrossRef]
19. Zhang, Q.; Kang, J.; Zhao, Y. A Predictive Model for Permanent Deformation of Asphalt Mixtures Modified with High Content Rubber Powder/SBS Composites. *J. China Foreign Highw.* **2022**, *42*, 197–202. [CrossRef]
20. Zhou, Z.; Liu, Z.; Wang, X.; Zhang, C. Study on Dynamic Flexural Modulus of Rubber Asphalt Mixture. *J. Highw. Transp. Res. Dev.* **2019**, *36*, 16–21+54.
21. Wang, Z. A Study on Dynamic Creep Tests and Viscoelastic Model of Dry-Blending Crumb Rubber Modified Asphalt Mixtures. *J. Inn. Mong. Univ. (Nat. Sci. Ed.)* **2013**, *44*, 647–653.

22. Kocak, S.; Kutay, M.E. Fatigue Performance Assessment of Recycled Tire Rubber Modified Asphalt Mixtures Using Viscoelastic Continuum Damage Analysis and AASHTOWare Pavement ME Design. *Constr. Build. Mater.* **2020**, *248*, 118658. [CrossRef]
23. Bai, F.; Yang, X.; Zeng, G. A Stochastic Viscoelastic–Viscoplastic Constitutive Model and Its Application to Crumb Rubber Modified Asphalt Mixtures. *Mater. Des.* **2016**, *89*, 802–809. [CrossRef]
24. Wang, G.; Yuan, S.; Qiu, W.; Qin, L.; He, Z. Study on Fatigue Properties of High-volume Rubber-modified Asphalt. *J. Wuhan Univ. Technol. (Transp. Sci. Eng.)* **2023**, *44*, 1–7.
25. Tan, H.; Hu, S.; Liu, B.; Qin, R.; Tong, S.; Ren, S. Study on Viscoelastic Properties of Rubber Modified Asphalt Based on Rheology. *China Civ. Eng. J.* **2017**, *50*, 115–122. [CrossRef]
26. Wang, X.; Wang, G.; Qin, L.; Meng, H.; Wang, X.; Cai, B. Influence of Rubber Asphalt Mortar on Dynamic Modulus of Mixture. *J. Shenzhen Univ. Sci. Eng.* **2019**, *36*, 74–79. [CrossRef]
27. Wang, X.; Wang, G.; Wang, Q.; Gao, Z. Design and Performance Research of High Content Rubber Powder Modified Asphalt SRA-20 Mixture. *Bull. Chin. Ceram. Soc.* **2018**, *37*, 2985–2990. [CrossRef]
28. Zhao, Z.; Xiao, F.; Toraldo, E.; Crispino, M.; Ketabdari, M. Effect of Crumb Rubber and Reclaimed Asphalt Pavement on Viscoelastic Property of Asphalt Mixture. *J. Clean. Prod.* **2023**, *428*, 139422. [CrossRef]
29. Yu, M.; Wu, G.X.; Liu, W.; Wang, X.Y.; Zhang, L. The Shaping Process Research on Crumb Rubber Modified Asphalt Mixture. *Appl. Mech. Mater.* **2012**, *178–181*, 1568–1572. [CrossRef]
30. Yu, M.; Zhao, Z.; Chen, H. Study on Viscoelasticity of Crumb Rubber Modified Asphalt Mixture. *J. Highw. Transp. Res. Dev.* **2023**, *40*, 1–7. [CrossRef]
31. Wang, G.; Cao, D.; Wang, Z.; Qiu, W.; Zhang, S. Comparative Study on Rheological Properties of Asphalt Binder Modified with High Content Crumb Rubber. *J. Highw. Transp. Res. Dev.* **2022**, *39*, 7–14+22. [CrossRef]
32. Ansari, A.H.; Jakarni, F.M.; Muniandy, R.; Hassim, S.; Elahi, Z.; Zair, M.M.B. Mechanical Performance of Cup Lump Rubber Modified Asphalt Mixtures Incorporating Polyphosphoric Acid. *Constr. Build. Mater.* **2023**, *408*, 133625. [CrossRef]
33. Guo, R.; Tang, J.; Gu, J.; Guo, G.; Feng, X. Analysis on the Road Performance of Graphene Composite Rubber Asphalt and Its Mixture. *Case Stud. Constr. Mater.* **2022**, *17*, e01664. [CrossRef]
34. Wang, J.C.; Zeng, X. Influence of Temperature and Pressure on the Dynamic Properties of Rubber-Modified Asphalt Concrete. *J. Mater. Civ. Eng.* **2006**, *18*, 125–131. [CrossRef]
35. Ren, R.; Geng, L.; Sun, J.; Liu, Z. Dynamic Modulus Research of Stabilized Rubber and Plastic Modified Asphalt Mixtures. *J. Shandong Jianzhu Univ.* **2016**, *31*, 307–313, 335. [CrossRef]
36. Kumar, S.A.; Veeraragavan, A. Dynamic Mechanical Characterization of Asphalt Concrete Mixes with Modified Asphalt Binders. *Mater. Sci. Eng. A* **2011**, *528*, 6445–6454. [CrossRef]
37. Roja, K.L.; Aljarrah, M.F.; Sirin, O.; Al-Nuaimi, N.; Masad, E. Rheological, Thermal, and Chemical Evaluation of Asphalt Binders Modified Using Crumb Rubber and Warm-Mix Additive. *J. Mater. Civ. Eng.* **2022**, *34*, 04022049. [CrossRef]
38. Wang, Z.; Guo, N.; Zhao, Y. Prediction of the Dynamic Modulus of Granular Crumb Rubber Modified Asphalt Concrete Based on Micromechanics. In Proceedings of the International Workshop on Energy and Environment in the Development of Sustainable Asphalt Pavements, Xi'an, China, 6 June 2010.
39. Abdulridha, A.A. Effect of Crumb Rubber Size and Content on the Properties of Dry Mixed Rubberized Asphalt Mixtures. Ph.D. Thesis, Universiti Teknologi Malaysia, Skudai, Malaysia, 2014.
40. Tang, Y.; Wang, Y.; Wu, D.; Chen, M.; Pang, L.; Sun, J.; Feng, W.; Wang, X. Exploring Temperature-Resilient Recycled Aggregate Concrete with Waste Rubber: An Experimental and Multi-Objective Optimization Analysis. *Rev. Adv. Mater. Sci.* **2023**, *62*, 20230347. [CrossRef]
41. Zhou, T.; Yu, F.; Li, L.; Dong, Z.; Fini, E.H. Swelling-Degradation Dynamic Evolution Behaviors of Bio-Modified Rubberized Asphalt under Thermal Conditions. *J. Clean. Prod.* **2023**, *426*, 139061. [CrossRef]
42. Karalar, M.; Ozturk, H.; Ozkilic, Y.O. Experimental and Numerical Investigation on Flexural Response of Reinforced Rubberized Concrete Beams Using Waste Tire Rubber. *Steel Compos. Struct.* **2023**, *48*, 43–57. [CrossRef]
43. Beskopylny, A.N.; Shcherban', E.M.; Stel'makh, S.A.; Meskhi, B.; Shilov, A.A.; Varavka, V.; Evtushenko, A.; Özkılıç, Y.O.; Aksoylu, C.; Karalar, M. Composition Component Influence on Concrete Properties with the Additive of Rubber Tree Seed Shells. *Appl. Sci.* **2022**, *12*, 11744. [CrossRef]
44. Witczak, M.W.; Kaloush, K.; Pellinen, T.; Elbasyouny, M.; Quintus, H.V. *Simple Performance Test for Superpave Mix Design*; National Cooperative Highway Research Program; Transportation Research Board: Washington, DC, USA, 2002.

buildings

MDPI

Article

Research on the Modulus Decay Model under a Three-Dimensional Stress State of Asphalt Mixture during Fatigue Damage

Yonghai He [1], Songtao Lv [2,3,4], Ziyang Wang [2,5,*], Huabao Ma [1], Wei Lei [1], Changyu Pu [1], Huilin Meng [1], Nasi Xie [2] and Xinghai Peng [2]

1 Hebei Provincial Communications Planning, Design and Research Institute Co., Ltd., Shijiazhuang 050090, China; yonghaihe@126.com (Y.H.); pushumhb@163.com (H.M.); hbleiwei@163.com (W.L.); puchangyu_1980@163.com (C.P.); smenghl@163.com (H.M.)
2 School of Traffic and Transportation Engineering, Changsha University of Science & Technology, Changsha 410114, China; lst@csust.edu.cn (S.L.); xns@stu.csust.edu.cn (N.X.); pengxinghsi@stu.csust.edu.cn (X.P.)
3 National Engineering Research Center of Highway Maintenance Technology, Changsha University of Science & Technology, Changsha 410114, China
4 National Key Laboratory of Green and Long-Life Road Engineering in Extreme Environment (Changsha) Changsha University of Science & Technology, Changsha 410114, China
5 School of Transportation, Southeast University, Nanjing 211189, China
* Correspondence: wzyang@seu.edu.cn

Abstract: Fatigue damage can develop within asphalt pavement due to the continuous impact of driving loads and natural elements. Understanding the process of asphalt mixtures' fatigue damage is crucial for guiding the design, maintenance, and repair of asphalt pavement. This research aims to establish a model that characterizes the mixtures' modulus decay under a three-dimensional (3-D) stress state. Firstly, asphalt mixes were subjected to direct tensile (DT), indirect tensile (IDT), unconfined compressive (UC) strength and fatigue tests, and the resulting data were analyzed. Then, modulus decay models under DT, IDT, and UC conditions were established, and the modulus decay patterns under the three loading modes were compared and analyzed. Finally, using the fatigue stress strength ratio Δ (a fatigue resistance index for asphalt mixtures that takes into account the impacts of stress state and loading rate), a unified characterization model for asphalt mixes' modulus decay under a 3-D stress state was created. According to the study's findings, asphalt mixes' modulus decay during fatigue damage exhibits obvious nonlinear characteristics. While the asphalt mixes' modulus decay law with various loading modalities is similar under the same conditions, the decay rate may differ. Essentially, the speed of the modulus decay of a certain asphalt mixture primarily depends on the value of Δ during service. A larger Δ indicates a faster modulus decay. This study offers a theoretical foundation for the conversion from material fatigue damage to structural fatigue damage, which is vital for enhancing the asphalt pavements' construction quality and longevity.

Keywords: fatigue damage; asphalt mixture; modulus decay model; three-dimensional stress state; fatigue stress strength ratio

Citation: He, Y.; Lv, S.; Wang, Z.; Ma, H.; Lei, W.; Pu, C.; Meng, H.; Xie, N.; Peng, X. Research on the Modulus Decay Model under a Three-Dimensional Stress State of Asphalt Mixture during Fatigue Damage. *Buildings* **2023**, *13*, 2570. https://doi.org/10.3390/buildings13102570

Academic Editor: Pengfei Liu

Received: 8 September 2023
Revised: 7 October 2023
Accepted: 10 October 2023
Published: 11 October 2023

1. Introduction

1.1. Background

In the discipline of solid mechanics, known as fatigue damage mechanics [1–3], the evolution of damage factors [4], like internal stress, strain, and modulus of materials subjected to repeated loading over time, is primarily studied. Due to factors like vehicular loads and natural environmental conditions, asphalt pavement may suffer from fatigue damage within the structure [5,6]. With the accumulation of damage, the materials and structural performance of the pavement will progressively deteriorate until they eventually fail [7]. Optimizing the design of the pavement structure and carrying out the necessary maintenance procedures throughout its service life is crucial for improving the construction quality and durability of asphalt pavements. Further, investigating the development process of asphalt mixes' fatigue damage is critical for optimizing pavement structure design and making informed maintenance decisions [8,9].

Therefore, numerous indoor experimental studies have been conducted by road researchers on asphalt mixtures [10–15], simulating various service conditions of asphalt pavement through diverse fatigue test methods. The asphalt mixes' fatigue performance was examined in these investigations, and fatigue damage models were suggested for several scenarios.

Specifically, Zheng et al. [10] established a nonlinear fatigue damage equation for asphalt mixes based on the S-N fatigue equation and damage mechanics theory. Zhou et al. [11] simulated the asphalt mixes' fatigue damage process using three-point bending fatigue tests and a nonlinear finite element method, respectively. They predicted asphalt concrete specimens' fatigue life and crack length during unstable fracture. To evaluate the stiffness variations of two types of asphalt mixes at rest brought on by temperature, fatigue damage growth, and healing, Daniel et al. [12] used the impact resonance method. Shan et al. [13] utilized CT scanning technology to capture internal cross-sectional images of asphalt mixtures at various stages of fatigue damage during IDT fatigue testing. Subsequently, they examined the progression of fatigue damage over time. Based on four-point bending tests, Izaks et al. [14] found that developing High Modulus Asphalt Concrete (HMAC) mixtures with high fatigue resistance performance using RA content up to 70% by using polymer-modified bitumen binder is possible. Lv et al. [15] conducted small-scale accelerated pavement tests on asphalt mixtures using the Model Moving Load Simulator (MMLS3) to evaluate their fatigue damage performance.

It should be noted that one crucial mechanical parameter that plays a significant role in determining the asphalt pavement's service life is the modulus of pavement material [16–18]. An essential factor to take into account in pavement design and maintenance is the decay pattern of the modulus, which can serve as a general measure of asphalt pavement's damage law [18–22]. Thus, according to the strain equivalence assumption of viscoelastic damage mechanics theory, some scholars defined the reduction in modulus (stiffness) as a damage variable to study the asphalt mixes' fatigue damage features. For example, Zhou et al. [20] established a damage evolution equation considering the joint impact of fatigue and creep damage effects. Based on the small-scale accelerated pavement test results, Lv et al. [15] established a stiffness decrease equation for asphalt mixes and proposed the corresponding reduction law during fatigue. Lv et al. [21] also analyzed the modulus decay characteristics of asphalt mixtures with various loading circumstances. By utilizing DT fatigue testing, Liu et al. [22] revealed the attenuation law of remaining strength and dynamic modulus. Additionally, they established a coupled fatigue damage evolution model of remaining strength and dynamic modulus.

However, two aspects need to be noted when studying the modulus decay law of asphalt mixtures. Firstly, the modulus of asphalt pavement materials has a specific stress dependence. As asphalt pavement materials are viscoelastic and composed of multiple phases, scales, and components, their modulus is not an inherent property of asphalt mixtures and is influenced by stress states [23,24]. Secondly, the modulus of asphalt pavement materials is influenced by the pavement's structure. Modulus levels and stress states change as a result of the different layers of asphalt mixture that make up the pavement structure.

During the actual service period, asphalt pavement structures experience a 3-D stress condition [25]. Consequently, there may be significant discrepancies between the results of indoor material tests conducted using one particular stress condition and the actual pavement service state, which hinders the precise assessment of the asphalt pavement structures' fatigue performance and impedes scientific maintenance decision-making. Although full-scale accelerated loading tests [26–28] can provide a more realistic simulation of the actual service state of asphalt pavement, their high testing costs and lengthy testing cycles have limited their widespread use. Therefore, an urgent need is to develop a model that can accurately reflect the modulus decay of asphalt mixtures under actual 3-D service conditions through indoor experiments.

This study focuses on understanding the process of fatigue damage in asphalt mixtures under the three-dimensional service condition, which is crucial for guiding the design and maintenance of asphalt pavement structures. On the basis of the fatigue stress strength ratio Δ as a measure of durability against fatigue [29], a unified characterization model for the asphalt mixes' modulus decay under a 3-D stress condition was developed. The findings of this study provide a theoretical foundation for the reasonable conversion of structural fatigue damage from material fatigue damage, allowing for a more realistic characterization of the asphalt pavement structures' fatigue damage patterns through indoor material tests, which is beneficial for improving asphalt pavements' construction quality and service life.

1.2. Objective

This paper aims to establish a model that characterizes the asphalt mixtures' modulus decay under a 3-D stress state during the fatigue damage. Strength and fatigue tests were conducted under the DT, IDT, and UC modes. The primary objectives of this research are as follows:

(1) The relationship between the asphalt mixtures' strength and loading rate was established.
(2) The conventional S-N fatigue equations for asphalt mixtures under the DT, IDT, and UC modes were obtained and analyzed. Additionally, the normalized fatigue equation for asphalt mixtures under the various modes was established based on the fatigue stress strength ratio Δ.
(3) Through stress-controlled fatigue tests, the modulus decay equations of asphalt mixtures under the DT, IDT, and UC modes were developed and compared.
(4) A unified characterization model for the modulus decay of asphalt mixture under the 3-D stress state was established based on the fatigue stress strength ratio Δ.

2. Materials and Methodology

2.1. Materials

2.1.1. Asphalt

Table 1 displays the conventional performance test outcomes of the high-content (30%) rubber-modified asphalt employed in this investigation. The test methods followed the specification JTG E20-2011 [30].

Table 1. Performance test outcomes of the asphalt in this investigation.

Test Items		Test Results	Indicator Requirements	Specifications
Penetration 25 °C, 100 g, 5 s (0.1 mm)		62	60~80	T 0604-2011
Ductility 5 cm/min, 5 °C (cm)		16	≥15	T 0605-2011
Softening point (°C)		72.5	≥55	T 0606-2011
Dynamic viscosity 180 °C (Pa·s)		2.8	1~4	T 0625-2011
Elastic recovery 25 °C (%)		84	≥55	T 0662-2000
48 h softening point difference (°C)		2.0	≤5.5	T 0661-2011
Residue after TFOT (163 °C, 85 min)	Mass loss (%)	−0.29	≤±0.8	T 0610-2011
	Residual penetration ratio 25 °C (%)	77	≥60	T 0604-2011
	Residual ductility 5 °C (cm)	11	≥10	T 0605-2011

2.1.2. Aggregate and Filler

Limestone aggregate and limestone mineral powder filler were utilized in this study. The aggregate density and mechanical index test results are presented in Tables 2 and 3, respectively. Table 4 shows the basic technical indicators of the filler. Each index complies with the JTG F40-2004 specification [31].

Table 2. Density and water absorption of limestone aggregate.

Particle Size (mm)	Apparent Relative Density (g/cm^3)	Gross Volume Relative Density (g/cm^3)	Water Absorption (%)
13.2~9.5	2.692	2.619	0.42
9.5~4.75	2.686	2.562	0.46
4.75~2.36	2.655	2.554	1.06
2.36~1.18	2.632	2.612	1.29
1.18~0.6	2.625	2.607	1.31
0.6~0.3	2.618	2.596	1.34
0.3~0.15	2.602	2.588	1.53
0.15~0.075	2.618	2.605	1.58

Table 3. Technical indexes of limestone aggregate.

Test Items	Crushing Value (%)	Polishing Value (BPN)	Abrasion Value (%)
Results	15.8	57.4	19.9
Technical indicators	≤28	≥45	≤30

Table 4. Technical indexes of filler.

Test Items		Test Results	Technical Indicators
Apparent relative density (g/cm^3)		2.688	≥2.50
Water content (%)		0.4	≤1
Apparent property		No agglomeration	—
Hydrophilic coefficient		0.57	<1
Plasticity index (%)		2.2	<4
Particle size (%)	<0.6 mm	100	100
	<0.15 mm	95.2	90~100
	<0.075 mm	84.4	75~100

2.1.3. Mix Design

The SAC-13 graded rubber asphalt mixture [32–34] was utilized in this study. The design principle of the SAC mix is to form a skeleton with coarse aggregates and fill the voids in the structure with fine aggregates and asphalt. SAC asphalt mixture has excellent high-temperature deformation resistance, water damage resistance, and slip resistance [35]. Additionally, it has advantages such as small segregation, easy compaction, and asphalt conservation [36].

The grading design for SAC-13 [35,36] was conducted, which included a maximum particle size of 13.2 mm, a control percent passing of 30% for the 4.75 mm sieve, and 8% for the 0.075 mm sieve. The grading curve is plotted as shown in Figure 1, and the final asphalt mixture asphalt–aggregate ratio is 5.5%. Table 5 displays the Marshall performance indicators of the SAC-13 mixes at the optimal asphalt–aggregate ratio.

2.2. Experiment Approach

This study tested asphalt mixtures for strength and fatigue using the DT, IDT, and UC modes. The strength test adopts two methods: stress and displacement control. The fatigue test determines a sample's fatigue life by using a stress-control mode and counting the number of loading cycles till a specimen fails, either marked by obvious fracture cracking or deformation. Strength and fatigue tests were conducted on an MTS testing machine. LVDT sensors are utilized during the fatigue test to detect real-time specimen displacement, allowing for the analysis of modulus changes throughout the process. Testing is consistently conducted at 15 ± 1 °C, and the specimen should be insulated at that temperature for 4 h or more before the test.

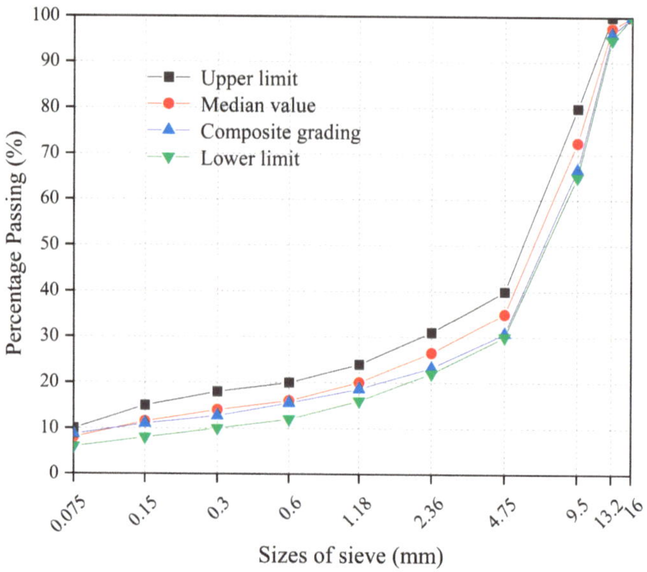

Figure 1. SAC-13 gradation curve.

Table 5. Marshall performance indicators.

Asphalt–Aggregate Ratio (%)	Bulk Density (g/cm^3)	Volume of Air Voids VV (%)	Voids in Mineral Aggregate VMA (%)	Voids Filled with Asphalt VFA (%)	Marshall Stability MS (kN)	Flow Value FL (mm)	Dynamic Stability DS (time/mm)
5.5	2.512	3.91	13.89	71.78	12.71	3.9	9581

2.2.1. Specimen Preparation

The DT samples were created following the guidelines of specification JTG E20-2011 [30]. Initially, a rectangular specimen (length and width of 300 mm, thickness of 50 mm) was molded using an Asphalt Mixer Wheel Grinding Machine. Subsequently, it was divided into beam samples (length of 250 mm, width and height of 50 mm). The asphalt mixture beam specimen is fixed between two circular iron plates with joints using epoxy stone dry-hanging adhesive, which has a higher tensile strength than the asphalt mixture and causes minimal deformation, ensuring accurate test results. Two displacement sensors are symmetrically attached on either side during fatigue testing of the asphalt mixture beam specimen. Refer to Figure 2a for the DT specimen.

The SGC rotary compactor was utilized to create cylindrical IDT specimens with a height and diameter of 100 ± 2 mm. The compaction process involved a vertical loading pressure of 600 ± 18 kPa, an effective internal rotation angle of 1.16 ± 0.02°, and a compaction speed of 30 ± 0.5 r/min. Afterward, the specimens were cut into 60 ± 2 mm height and 100 ± 2 mm diameter specimens for indirect tensile testing, as displayed in Figure 2b.

The UC samples, with a height and diameter of 100 ± 2 mm, were produced using the SGC rotary compactor. This method is consistent with the indirect stretching method discussed earlier. The unconfined compressive specimen is shown in Figure 2c.

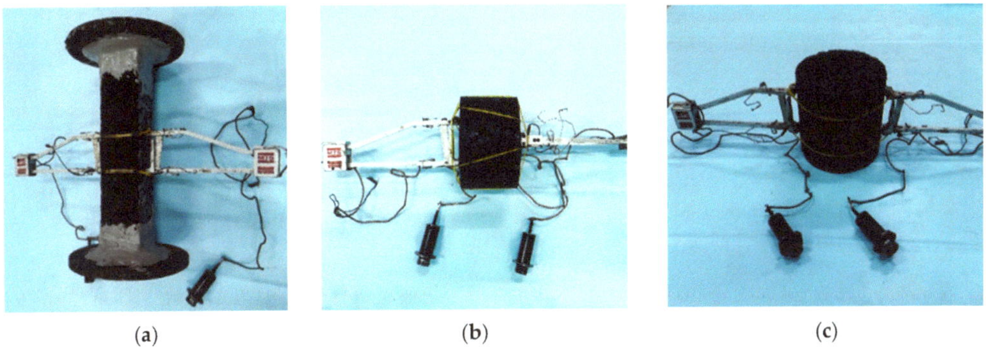

Figure 2. Specimen preparation. (a) DT specimen. (b) IDT specimen. (c) UC specimen.

2.2.2. Strength Test Approach

The MTS testing machines were employed to conduct the strength tests. For the stress-control mode DT, IDT, and UC strength tests, eight loading rates ranging from 0.02 MPa/s to 6 MPa/s were utilized. Before the fatigue test, displacement-control mode strength tests were performed. During the displacement-control mode, the loading rates for the DT, IDT, and UC strength tests were maintained at 5 mm/min, 50 mm/min, and 2 mm/min, respectively, following JTG E20-2011 standards [30]. Each group of tests consisted of three parallel specimens, and the maximum load (N) was recorded upon specimen failure. The DT, IDT, and UC strength tests are shown in Figure 3.

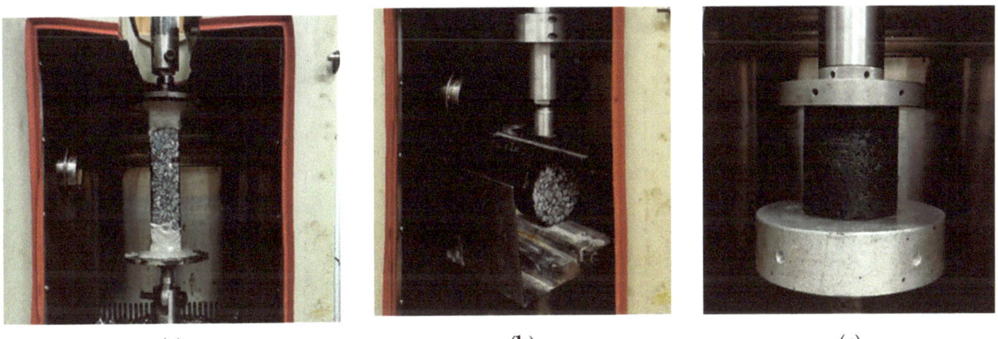

Figure 3. Strength test of asphalt mixture. (a) DT strength test. (b) IDT strength test. (c) UC strength test.

The formula for calculating DT strength:

$$S_{DT} = \frac{P}{A} = \frac{P}{ab} \tag{1}$$

where S_{DT} represents DT strength (MPa); P is the maximum load (N) at the point of specimen failure; a, b represents the length and width of the cross section ($a = b = 50$ mm); A is the cross-sectional area (mm^2).

The formula for calculating IDT strength:

$$S_{IDT} = 0.006287\frac{P}{h} \tag{2}$$

where S_{IDT} represents IDT strength (MPa); h denotes specimen height (mm); P is the maximum test load (N).

The formula for calculating UC strength:

$$S_{UC} = \frac{P}{A} = \frac{P}{\pi D^2/4} \tag{3}$$

where S_{UC} represents UC strength (MPa); P is the maximum load (N) at the point of specimen failure; D denotes specimen diameter (mm); A is the cross-sectional area (mm^2).

2.2.3. Fatigue Test Approach

In this study, the selection of stress levels for the DT, IDT, and UC fatigue tests was based on the strength (static load strength) under displacement-control mode, respectively. Moreover, the selected stress level should not exceed the static load strength. Four specimens were tested in parallel for fatigue at each stress level, with a loading frequency of 10 Hz and a continuous haversine wave loading waveform. Refer to Figure 4 for the fatigue test schematic diagram.

(**a**) (**b**) (**c**)

Figure 4. Fatigue test of asphalt mixture. (**a**) DT fatigue test. (**b**) IDT fatigue test. (**c**) UC fatigue test.

2.3. Derivation of Modulus Decay Equation

2.3.1. Damage Variables Based on Modulus

When subjected to the factors such as forces, temperature, and loading speed, microdefects can form within a material. These defects can nucleate, expand, and converge, ultimately leading to material failure [7]. Variations in macroscopic physical variables like modulus, stress, and strength frequently accompany this process of material deterioration. Specific physical quantities can be defined as damage variables to study the microscopic changes within materials during the damage process, and corresponding fatigue damage models can be established to describe the damage process [4].

In 1963, the mechanist Rabotnov used the following damage factor D to describe the damage and study the constitutive equations of metal materials [37].

$$D = 1 - \frac{A'}{A} \tag{4}$$

where A' is the effective bearing area; A is the nominal area, which is the area in the undamaged state.

The fatigue test of the mixes allows for continuous measurement of the modulus, enabling the asphalt mixes' modulus decay during the test to be used as a damage variable.

In this study, a damage factor formula based on the modulus was defined to describe damage in fatigue tests, as shown below.

$$D = 1 - \frac{E}{E_0} \tag{5}$$

where D is the damage variable of the specimen; E is the modulus of the material after fatigue; E_0 is the initial modulus of the specimen.

2.3.2. Nonlinear Fatigue Damage Equation for Asphalt Mixes

Creating a fatigue damage model is crucial in studying material fatigue damage, as it links the number of load actions to the resulting material damage. On the one hand, materials' nonlinear fatigue damage law is related to their initial damage state, and the amount of damage that occurs at different damage stages, even under the same load state, is not the same. On the other hand, the nonlinear fatigue damage law of materials varies with different load levels during cyclic loading. The most intuitive manifestation of the nonlinearity of material fatigue damage is that the image describing the material fatigue damage model is nonlinear. Equation (6) is a commonly used nonlinear fatigue damage model considering the influence of stress amplitude.

$$\frac{dD}{dN} = (\frac{\sigma}{2M})^{\alpha}(1-D)^{-(\alpha+\gamma)} = (\frac{1}{2M})^{\alpha}(\frac{\alpha}{1-D})^{\alpha}(1-D)^{-\gamma} \tag{6}$$

By integrating Equation (6), the fatigue damage variable formula can be derived as follows:

$$D(N) = 1 - \left(1 - \frac{N}{N_f}\right)^{\frac{1}{1+\alpha+\gamma}} \tag{7}$$

Correspondingly, Equation (8) is the expression of the material fatigue equation:

$$N_f = \frac{1}{1+\alpha+\gamma}(\frac{\sigma}{2M})^{-\sigma} \tag{8}$$

where α, γ and M are the material parameters linked to the average stress, temperature, and stress amplitude; D represents the damage variable; N stands for the number of cycles of load in the fatigue test; N_f represents the fatigue life of the material.

2.3.3. Modulus Decay Equation

Fatigue tests allow for continuous modulus measurement without affecting asphalt mixture performance. The modulus steadily decreases as internal damage accumulates within the material, making it an effective measure for describing damage in asphalt mixtures. The simultaneous Equations (5) and (7) are as follows:

$$\begin{cases} D = 1 - \frac{E}{E_0} \\ D = 1 - \left(1 - \frac{N}{N_f}\right)^{\frac{1}{1+\alpha+\gamma}} \end{cases} \tag{9}$$

The modulus decay equation can be obtained by eliminating the damage variable D in Equation (9).

$$\frac{E}{E_0} = \left(1 - \frac{N}{N_f}\right)^{\frac{1}{1+\alpha+\gamma}} \tag{10}$$

Let $m = \frac{1}{1+\alpha+\gamma}$, then:

$$\frac{E}{E_0} = (1 - \frac{N}{N_f})^{m} \tag{11}$$

where m is the modulus decay equation parameter, which can be fitted through fatigue tests; E is the effective modulus; E_0 is the initial modulus; N represents the number of load cycles in the fatigue test; N_f denotes the fatigue life.

3. Results and Discussion

3.1. Asphalt Mixture Strength and Fatigue Test Results Analysis

3.1.1. Analysis of Strength Test Results

Displacement-controlled strength tests for asphalt mixture were performed implementing the procedure in Section 2.2.2, and the experimental results are shown in Table 6.

Table 6. Displacement-controlled strength test results.

Loading Modes	Displacement-Controlled Strength (MPa)	Mean Strength (MPa)	Coefficient of Variation (%)
DT	1.043/1.138/1.032	1.071	5.442
IDT	2.192/2.064/2.284	2.180	5.068
UC	7.732/7.029/7.583	7.448	4.974

A stress-control mode was adopted to perform the DT, IDT, and UC strength tests at various loading rates to investigate the impact of the loading rate on asphalt mix strength. The experiment outcomes, presented in Table 7, provide insights into the viscoelastic behavior of asphalt mixtures.

Table 7. Strength test results by stress-controlled mode.

Loading Rate (MPa/s)	DT		IDT		UC	
	Mean Strength (MPa)	Coefficient of Variation (%)	Mean Strength (MPa)	Coefficient of Variation (%)	Mean Strength (MPa)	Coefficient of Variation (%)
0.02	0.928	6.819	0.931	4.988	6.525	9.982
0.05	1.162	7.066	1.132	8.896	7.036	5.849
0.1	1.537	4.988	1.185	7.494	8.315	5.430
0.5	1.987	5.125	1.859	9.116	10.303	4.566
1.0	2.155	5.882	2.113	5.100	12.348	5.670
2.0	2.439	6.534	2.214	9.853	12.712	5.225
4.0	3.319	7.115	2.358	7.303	14.594	8.024
6.0	3.407	8.310	2.915	5.585	14.978	5.837

Equation (12) shows that the strength and loading rate were fitted using a power function equation. The curves are displayed in Figure 5.

$$S = Av^B \tag{12}$$

where S is the strength of asphalt mixture; A and B are the equation parameters; v is the loading rate.

Figure 5 demonstrates that the loading rate positively affects the strength of asphalt mixtures for all stress states, but the pace of increase reduces with higher loading rates. Unconfined compressive mode yields much higher strength values than the direct and indirect tensile modes, while the direct and indirect tensile modes produce similar strength values at the same loading rate.

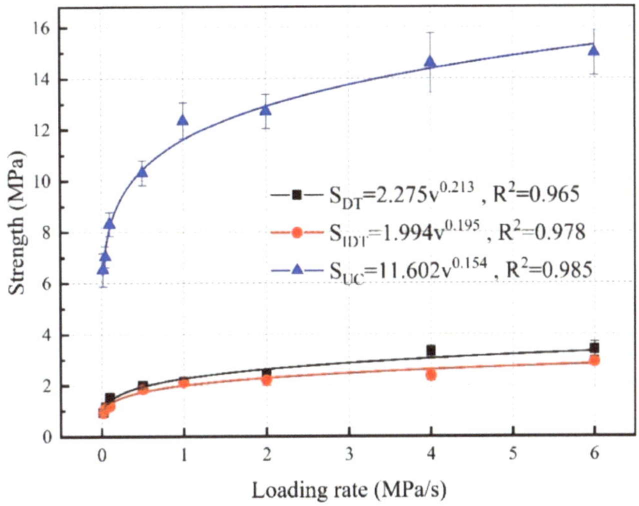

Figure 5. Fitting results of strength and loading rate.

3.1.2. Analysis of Fatigue Test Results

Fatigue experiments were performed on asphalt mixtures by applying the experimental procedure outlined in Section 2.2.3. Before the fatigue test, a displacement-control mode strength test was conducted, and the outcomes are presented in Table 6. On this basis, five stress levels ranging from 0.2 MPa to 1 MPa were chosen for the DT and IDT fatigue tests, while UC fatigue experiments were performed at stress levels of 1.5 MPa, 2 MPa, 2.5 MPa, 3 MPa, and 3.5 MPa. The fatigue test data are summarized in Table 8.

Table 8. Fatigue test results of asphalt mixture.

DT				IDT				UC			
σ	t_n	N_f	CV	σ	t_n	N_f	CV	σ	t_n	N_f	CV
0.2	0.187	934,042	10.350	0.2	0.092	699,521	9.137	1.5	0.134	722,354	12.562
0.4	0.373	61,249	16.107	0.4	0.183	21,498	13.737	2.0	0.201	280,096	15.050
0.6	0.560	19,905	15.983	0.6	0.275	5115	8.491	2.5	0.269	41,052	8.834
0.8	0.747	7889	21.118	0.8	0.367	1205	14.842	3.0	0.336	11,884	11.396
1.0	0.934	3503	13.354	1.0	0.459	519	11.666	3.5	0.403	6268	9.785

In the table: σ is the stress level (MPa); t_n is the nominal stress ratio; N_f is the mean fatigue life (cycles); CV is the coefficient of variation (%).

Equation (13) is the S-N fatigue model based on phenomenological methods [38,39].

$$N_f = k(\frac{1}{t_n})^c \quad (13)$$

Equation (14) is the formula for calculating t_n.

$$t_n = \sigma/\overline{S} \quad (14)$$

where N_f is the fatigue life; k and c are the regression parameters; t_n is the nominal stress ratio; $\overline{S}$ is the static load strength; σ is the stress levels of fatigue test.

Figure 6 illustrates the fitting of the S-N fatigue formulas for asphalt mixtures under various stress states (loading modes) using Equation (12), founded on the fatigue experiment data obtained from the DT, IDT, and UC modes.

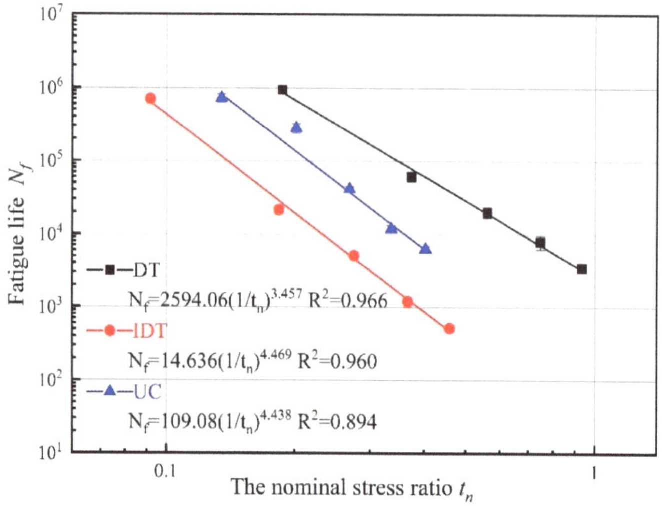

Figure 6. S-N fatigue equations under different stress states.

Upon examining Figure 6, it is apparent that the S-N equation of asphalt concrete displays noteworthy discrepancies when subjected to the different loading modes. It is evident in the considerable variation in the fatigue equation parameter k values obtained from the three methods, resulting in a substantial separation between the fatigue equation curves under the DT, IDT, and UC modes. The fundamental reason for this phenomenon is that the nominal stress ratio t_n, which serves as a fatigue resistance indicator, fails to account for the impact of the stress state and loading rate on asphalt mixtures as viscoelastic materials. Hence, to accurately assess the fatigue and damage properties of the mixtures, it is imperative to employ a fatigue resistance index that considers the material's 3-D stress condition and loading rate. This approach can significantly mitigate the influence of the experimental loading modes, specimen geometry, and size.

3.2. The Modulus Decay Law of Asphalt Mix with Various Stress States

3.2.1. Fitting Results of Modulus Decay Equation with Various Stress States

This study defines the modulus' attenuation as a fatigue damage variable, and the focus of this section is to investigate asphalt mixes' modulus decay equation under various stress conditions during fatigue testing. A modulus value can be calculated in each load cycle of the fatigue test. The calculation method for modulus E is as follows.

$$E = \frac{\sigma}{\varepsilon_{ei}} \quad (15)$$

where σ is the stress level of fatigue test; ε_{ei} is the strain; E is the modulus of asphalt mixture.

The Equation (11) derived in Section 2.3.3 was used to fit the modulus ratio E/E_0 and fatigue life ratio N/N_f. The fitted modulus decay model parameter m of DT, IDT, and UC are summarized in Tables 9–11. Examples of modulus decay curves (one of the four parallel UC experiments) are shown in Figure 7.

Table 9. Modulus decay equation parameters m fitting results of DT.

Stress Levels σ (MPa)	Nominal Stress Ratio t_n	m Value	Mean of m Values	Coefficient of Variation (%)
0.2	0.187	0.418/0.389/0.395/0.456	0.415	7.345
0.4	0.373	0.499/0.443/0.524/0.501	0.492	6.997
0.6	0.560	0.574/0.527/0.541/0.634	0.569	8.354
0.8	0.747	0.598/0.568/0.563/0.607	0.584	3.752
1.0	0.934	0.655/0.617/0.621/0.673	0.642	4.226

Table 10. Modulus decay equation parameters m fitting results of IDT.

Stress Levels σ (MPa)	Nominal Stress Ratio t_n	m Value	Mean of m Values	Coefficient of Variation (%)
0.2	0.092	0.446/0.432/0.488/0.431	0.449	5.960
0.4	0.183	0.575/0.524/0.548/0.625	0.568	7.606
0.6	0.275	0.602/0.615/0.641/0.553	0.603	6.097
0.8	0.367	0.626/0.593/0.577/0.668	0.616	6.551
1.0	0.459	0.669/0.636/0.659/0.696	0.665	3.726

Table 11. Modulus decay equation parameters m fitting results of UC.

Stress Levels σ (MPa)	Nominal Stress Ratio t_n	m Value	Mean of m Values	Coefficient of Variation (%)
1.5	0.134	0.397/0.436/0.395/0.389	0.404	5.281
2.0	0.201	0.423/0.415/0.436/0.438	0.428	2.547
2.5	0.269	0.465/0.431/0.446/0.522	0.466	8.550
3.0	0.336	0.495/0.488/0.521/0.514	0.504	3.093
3.5	0.403	0.509/0.547/0.486/0.520	0.516	4.916

Figure 7 shows that the parameter m in the modulus decay equation positively correlates with the stress level. It means that as the stress level increases, the dynamic modulus decays faster.

3.2.2. Comparative Analysis of Modulus Decay Law under Different Stress States

Because the stress levels during fatigue testing differ between the UC mode and DT/IDT modes, the modulus decay equations for the three modes cannot be directly compared. To address this, the nominal stress ratio was obtained by dividing the stress levels equally by the static load strength for dimensionless processing. As a result, the modulus decay equations under the DT, IDT, and UC modes could be compared by examining the parameter m under the same nominal stress ratio. The m values and nominal stress ratios t_n for each mode were fitted using a power function, and the resulting fitting curves are displayed in Figure 8.

Figure 8 displays that the parameter m in the mixes' modulus decay equation rises as the nominal stress ratio t_n increases. The corresponding fitting equations are presented below.

$$\text{DT}: \ m = 0.648{t_n}^{0.267}, \ R^2 = 0.979 \tag{16}$$

$$\text{IDT}: \ m = 0.790{t_n}^{0.221}, \ R^2 = 0.936 \tag{17}$$

$$\text{UC}: \ m = 0.656{t_n}^{0.310}, \ R^2 = 0.970 \tag{18}$$

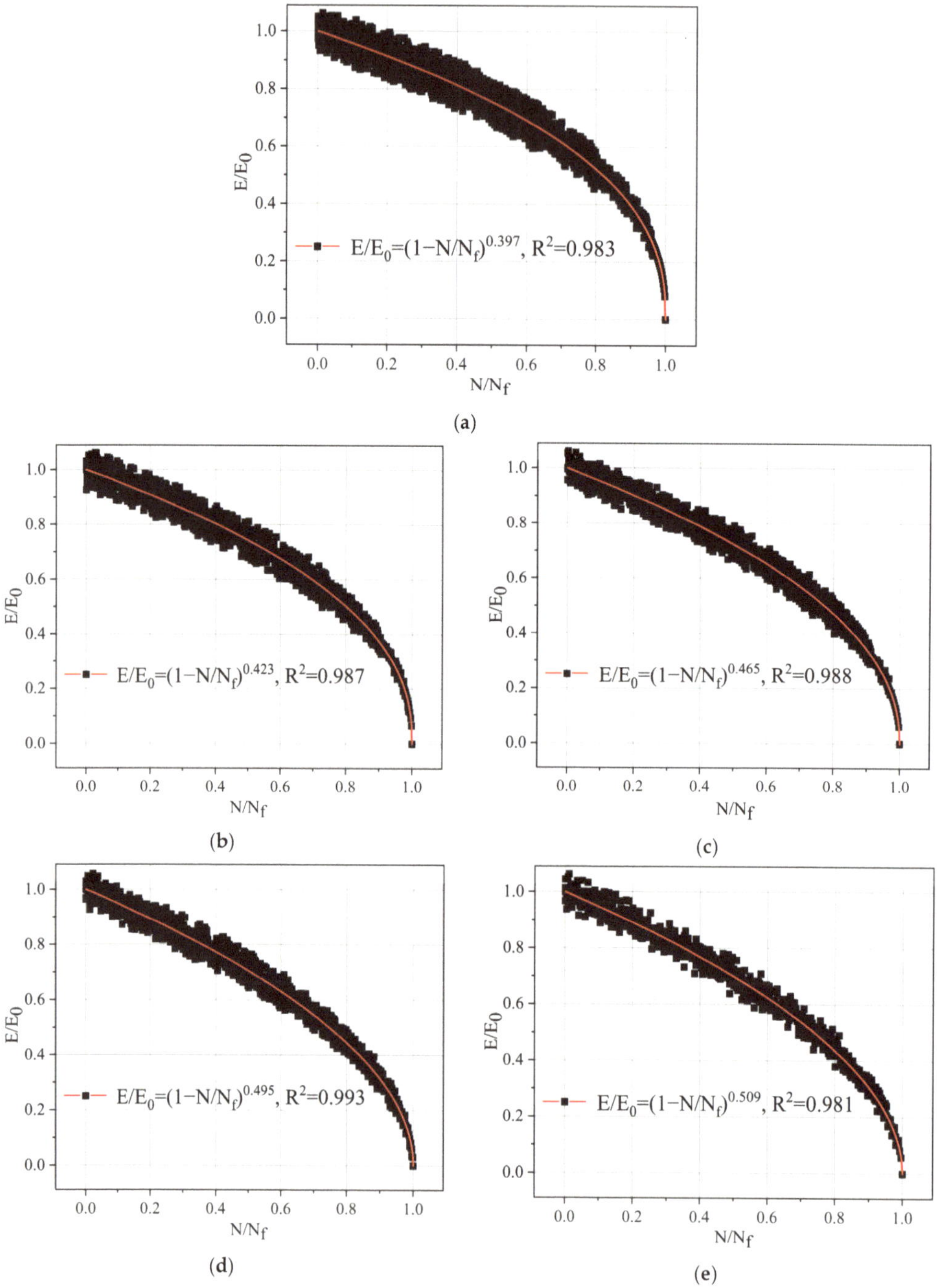

Figure 7. Modulus decay curves at different stress levels under UC mode. (**a**) 1.5 MPa. (**b**) 2.0 MPa. (**c**) 2.5 MPa. (**d**) 3.0 MPa. (**e**) 3.5 MPa.

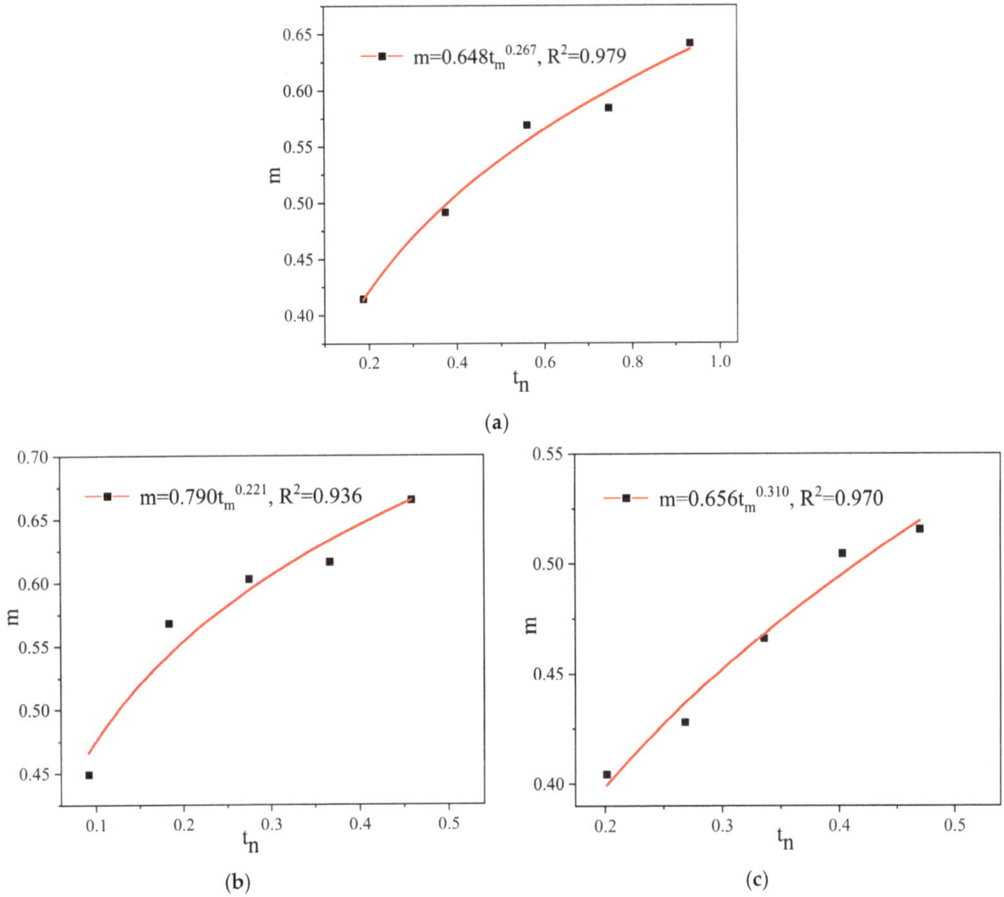

Figure 8. Fitting curves of parameter m and t_n under the different loading modes. (**a**) DT. (**b**) IDT. (**c**) UC.

In an effort to contrast and analyze the modulus decay patterns with various stress states, it is essential to ensure that the fatigue test load levels remain consistent, i.e., the nominal stress ratio t_n should be the same. Taking $t_n = 0.2$, the modulus decay equations for the DT, IDT, and UC modes are:

$$\text{DT}: \; E/E_0 = (1 - N/N_f)^{0.422} \tag{19}$$

$$\text{IDT}: \; E/E_0 = (1 - N/N_f)^{0.554} \tag{20}$$

$$\text{UC}: \; E/E_0 = (1 - N/N_f)^{0.398} \tag{21}$$

The modulus decay curves were plotted according to Equations (19)–(21), as shown in Figure 9.

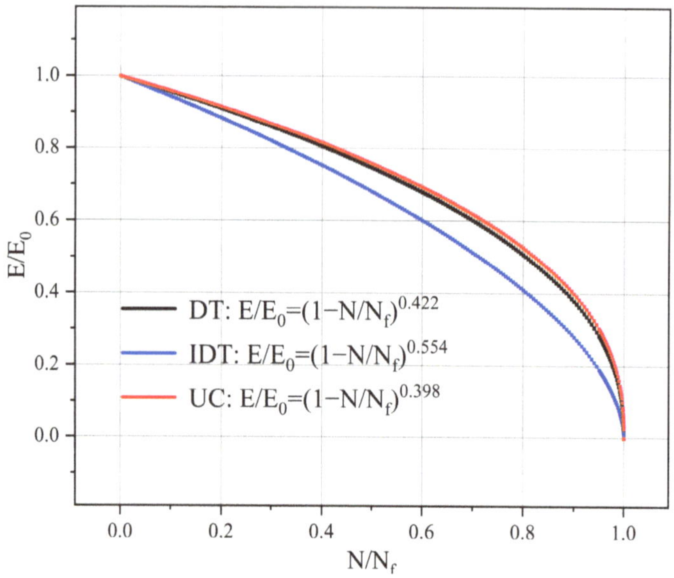

Figure 9. Modulus decay curves under the different loading modes when t_n = 0.2.

Figure 9 suggests that the modulus decay patterns are generally similar under the different loading modes despite differences in the stress states. The decay of the modulus first occurs in the stable decay stage, followed by the rapid decay stage. However, the m value of the mixes' modulus decay curve varies under the different loading modes, even when subjected to the same nominal stress ratio. The modulus decay rate from fast to slow is IDT > DT > UC.

This result is likely because the t_n does not consider how the loading rate and stress condition affect asphalt mixtures. To address this, the next step in this study will be to establish a unified model for describing asphalt mixes' modulus decay pattern under a 3-D stress condition, using the fatigue stress strength ratio Δ as a fatigue resistance index.

3.3. Unified Characterization Model for Modulus Decay of Asphalt Mixture under the 3-D Stress State

3.3.1. Fatigue Stress Strength Ratio and Normalized Fatigue Equation

The stress condition inside asphalt pavement is highly complex under actual driving loads. Asphalt mixtures' strength and fatigue behavior are affected by the stress state and loading rate. Traditional phenomenological fatigue analysis methods fail to consider these factors, resulting in an unscientific approach. To overcome this, Lv et al. [29,39] proposed a novel approach for analyzing the fatigue characteristic of asphalt mix, which accounts for the actual pavement service conditions and utilizes the idea of a 3-D stress state strength yield surface. This compensates for the limitations of traditional fatigue performance analysis methods that do not consider the impacts of stress conditions and loading rates. The fatigue stress strength ratio Δ is served as a fatigue resistance index, with Equation (22) defining its formula.

$$\Delta = \frac{\sigma_{ei}}{\sigma_{es}} \tag{22}$$

where σ_{ei} represents the initial equivalent stress, while σ_{es} represents the equivalent stress at the point of fatigue failure.

The formula for equivalent stress is:

$$\sigma_e = \sqrt{\left[(\sigma_1 - \sigma_2)^2 + (\sigma_2 - \sigma_3)^2 + (\sigma_3 - \sigma_1)^2\right]/2} \tag{23}$$

where σ_1 is the first principal stress; σ_2 is the second principal stress; σ_3 is the third principal stress.

Equation (22) shows the ratio of the stress level to material resistance during asphalt mixture fatigue testing. Using the ratio Δ, which considers the stress state and loading rate, a normalized fatigue equation is established as follows.

$$N_f = \left(\frac{1}{\Delta}\right)^{\gamma} \quad (24)$$

where N_f represents fatigue life; γ represents the regression parameter.

The initial equivalent stress σ_{ei} of the fatigue test and the equivalent stress during fatigue failure σ_{es} can be calculated according to Tables 12 and 13, respectively.

Table 12. Calculation method of σ_{ei}.

Loading Modes	Stress Levels σ_i (MPa)	σ_1	σ_2	σ_3	σ_e
DT	σ_d	σ_d	0	0	σ_d
IDT	σ_t	σ_t	0	$-3\sigma_t$	$\sqrt{13}\sigma_t$
UC	σ_c	0	0	$-\sigma_c$	σ_c

Table 13. Calculation method of σ_{es}.

Loading Modes	Stress-Controlled S_i	σ_1	σ_2	σ_3	σ_{es}
DT	S_d	S_d	0	0	S_d
IDT	S_t	S_t	0	$-3S_t$	$\sqrt{13}S_t$
UC	S_c	0	0	$-S_c$	S_c

In this study, the fatigue tests apply a haversine waveform load. The loading rate v can be calculated by the test's loading frequency f (loading cycle T) and stress level σ.

$$v = \frac{\sigma}{T/2} = 2f\sigma \quad (25)$$

Based on Equation (24) and the power function equations established in Section 3.1.1 for the strength and loading rate of asphalt mixtures in the DT, IDT, and UC modes, the ultimate strength at various stress levels can be determined. Thus, σ_{es} can be calculated. Table 14 displays the calculated Δ values.

Based on the data of fatigue life N_f (in Section 3.1.2) and the fatigue stress strength ratio, the normalized fatigue equation for various stress conditions can be fitted by Equation (24). The resulting curve for the asphalt mix is displayed in Figure 10.

Figure 10 illustrates that the use of the fatigue stress strength ratio Δ has successfully achieved the uniform description of asphalt mixtures' fatigue behavior with the various stress states. As a fatigue resistance indicator, the ratio Δ considers the impact of the loading mode and rate on asphalt mixtures and correlates with the stress state and vehicle speed within a pavement structure. Based on the ratio Δ, this study established a modulus decay model for asphalt mixes under a 3-D stress condition.

3.3.2. Establishment of the Modulus Decay Model under the 3-D Stress State

Section 3.2 developed the modulus decay equations of asphalt mixes under various stress conditions corresponding to the different loading modes. In Section 3.3.1, the fatigue stress strength ratios were calculated for the DT, IDT, and UC modes at different stress levels. Table 15 displays the values of the modulus decay equation parameter m and the fatigue stress strength ratio Δ. m and Δ are fitted using a power function equation, as shown in Figure 11.

Table 14. The calculated Δ values.

Loading Modes	Stress Level σ (Mpa)	Loading Rate v (MPa/s)	Ultimate Strength S (Mpa)	σ_{ei}	σ_{es}	Δ
DT	0.2	4	3.056	0.2	3.056	0.065
	0.4	8	3.543	0.4	3.543	0.113
	0.6	12	3.862	0.6	3.862	0.155
	0.8	16	4.106	0.8	4.106	0.195
	1	20	4.306	1	4.306	0.232
IDT	0.2	4	2.613	0.721	9.421	0.077
	0.4	8	2.991	1.442	10.784	0.134
	0.6	12	3.237	2.163	11.672	0.185
	0.8	16	3.424	2.884	12.345	0.234
	1	20	3.576	3.606	12.894	0.280
UC	1.5	30	19.589	1.5	19.589	0.077
	2	40	20.476	2	20.476	0.098
	2.5	50	21.192	2.5	21.192	0.118
	3	60	21.796	3	21.796	0.138
	3.5	70	22.319	3.5	22.319	0.157

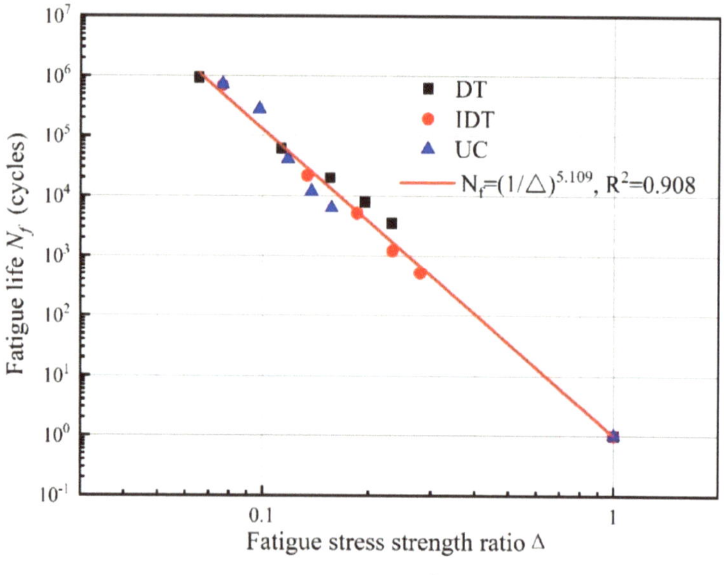

Figure 10. The normalized fatigue equation curve of asphalt mixture.

By observing Figure 11, it is evident that the parameter m in the modulus decay equation increases as the ratio Δ increases. The power function relationship equation between m and Δ, obtained by fitting the data, is presented below.

$$m = \Delta^{0.328}, R^2 = 0.910 \tag{26}$$

Therefore, a modulus decay model of asphalt mixture under the 3-D stress condition, Equation (27), was established by uniting Equations (11) and (26).

Table 15. The values of m and Δ.

Loading Modes	Stress Level σ (Mpa)	Fatigue Stress Strength Ratio Δ	Modulus Decay Equation Parameter m
DT	0.2	0.065	0.415
	0.4	0.113	0.492
	0.6	0.155	0.569
	0.8	0.195	0.584
	1	0.232	0.642
IDT	0.2	0.077	0.449
	0.4	0.134	0.568
	0.6	0.185	0.603
	0.8	0.234	0.616
	1	0.280	0.665
UC	1.5	0.077	0.404
	2	0.098	0.428
	2.5	0.118	0.466
	3	0.138	0.504
	3.5	0.157	0.516

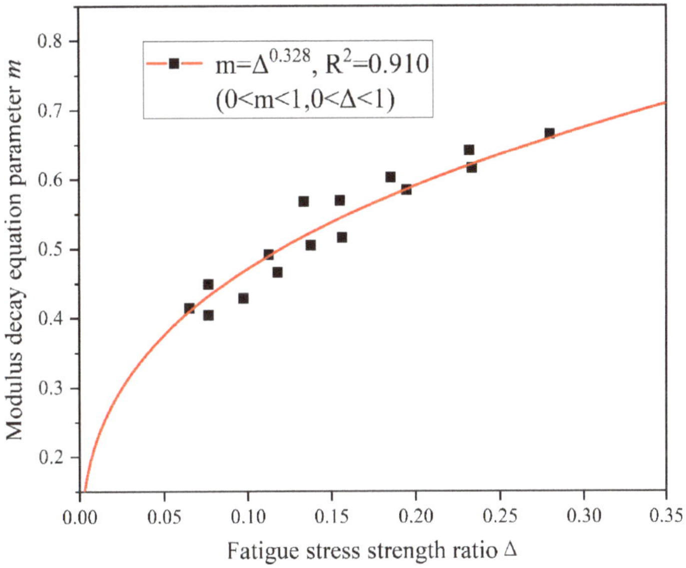

Figure 11. The fitting results of m and Δ.

$$\begin{cases} \frac{E}{E_0} = \left(1 - \frac{N}{N_f}\right)^m \\ m = \Delta^n,\ (n = 0.328,\ 0 < m < 1,\ 0 < \Delta < 1) \end{cases} \tag{27}$$

where E represents the modulus during fatigue, while E_0 is its initial value; N represents the number of fatigue load cycles, while N_f represents the fatigue life; m is a parameter of the modulus decay equation that can be determined through fatigue tests; Δ is the fatigue stress strength ratio, which accounts for the effects of the stress state and loading rate, and represents the ratio of the stress level to material resistance during fatigue testing of asphalt mixtures; n is the regression parameter.

Figure 12 shows the modulus decay surface under the 3-D stress state, which was drawn using MATLAB R2018a based on Equation (27). The surface function is $Z = (1-y)^{x^{0.328}}$, $x \in (0,1)$, $y \in (0,1)$.

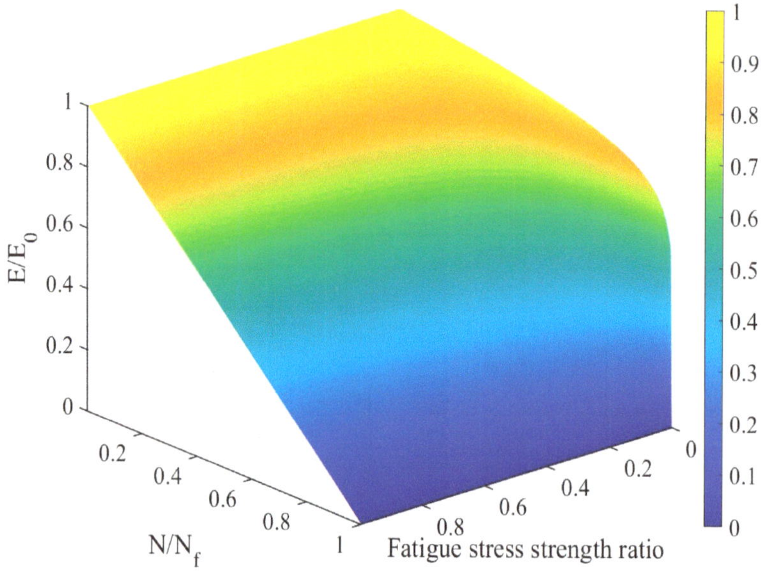

Figure 12. Modulus decay curved surface under the three-dimensional stress state.

Figure 12 illustrates that the modulus decay model (27) can describe the modulus decay behavior under any service state (represented by fatigue stress strength ratio Δ). In asphalt pavement structures, the fatigue stress strength ratio at a given point continuously changes during service. Analyzing the variation law of the fatigue stress strength ratio at a particular point in the asphalt pavement structure under different fatigue life ratios, and using Equation (27), a structural modulus decay model can be established for that point in the asphalt pavement structure. Theoretically, the modulus decay curve of an asphalt pavement structure should be a curve on the asphalt mixes' modulus decay curved surface under the 3-D stress condition. As a result, the relationship between material modulus decay and structural modulus decay has been established.

This study establishes a theoretical foundation for the reasonable conversion of structural fatigue damage from material fatigue damage, making it possible to characterize the fatigue damage pattern of asphalt pavement structural layers through indoor experiments.

4. Conclusions

This study investigated the fatigue damage properties of asphalt mixtures under the DT, IDT, and UC modes. Firstly, the strength and fatigue experiment results were analyzed, followed by establishing the modulus decay models under the different loading modes. The modulus decay patterns were then compared and analyzed. Using the fatigue stress strength ratio Δ as a measure of fatigue resistance, a unified description model for the modulus decay of asphalt mixtures under the 3-D stress condition was established. In summary, the study provides the following conclusions.

(1) The power function relationships between asphalt mixtures' strength and loading rate v are as follows.

$$S_{DT} = 2.275v^{0.213},\ R^2 = 0.965$$

$$S_{IDT} = 1.994v^{0.195},\ R^2 = 0.978$$

$$S_{UC} = 11.602v^{0.154},\ R^2 = 0.985$$

The strength of the asphalt mixture rises with the loading rate during testing. However, under various loading approaches, the asphalt mixes' strength varies even at the same loading rate.

(2) The conventional S-N fatigue equations of asphalt mix under the DT, IDT, and UC methods are as follows.

$$\text{DT}:\ N_f = 2594.06(1/t_n)^{3.457},\ R^2 = 0.966$$

$$\text{IDT}:\ N_f = 14.636(1/t_n)^{4.469},\ R^2 = 0.960$$

$$\text{UC}:\ N_f = 109.08(1/t_n)^{4.438},\ R^2 = 0.894$$

The S-N fatigue equation of the mix exhibits significant differences under the DT, IDT, and UC modes. This phenomenon can be attributed to the fact that the nominal stress ratio t_n, as a fatigue resistance indicator, does not account for the impacts of the stress conditions and loading rates on asphalt mixtures as viscoelastic materials.

(3) The modulus decay patterns are generally similar across the different loading modes. Modulus decay occurs first in the stable decay stage and then in the rapid decay stage. However, the parameter m of the modulus decay curve varies across the different loading modes under the same nominal stress ratio t_n, with the modulus decay being fastest for IDT, followed by the DT and UC modes in descending order.

(4) Using the fatigue stress strength ratio Δ, the fatigue features of asphalt mixtures with various loading conditions can be uniformly described. The normalized fatigue equation for asphalt mixtures under different stress conditions was established.

$$N_f = (1/\Delta)^{5.109},\ R^2 = 0.908$$

The fatigue resistance index Δ considers the impact of the loading mode and rate on asphalt mixtures. It can correspond to the stress condition within the pavement structure and vehicle driving speed.

(5) A unified characterization model for the modulus decay of asphalt mixture under the 3-D stress state was established based on the ratio Δ.

$$\begin{cases} \frac{E}{E_0} = (1 - \frac{N}{N_f})^m \\ m = \Delta^n,\ (n = 0.328,\ 0 < m < 1,\ 0 < \Delta < 1) \end{cases}$$

This model comprehensively characterizes the asphalt mixes' modulus decay law under various service conditions, providing a reference to understanding the fatigue damage behavior of asphalt pavement structures. The modulus decay curve of an asphalt pavement structure should theoretically be a curve in the modulus decay curved surface of asphalt mixture under the 3-D stress condition, thus establishing a relationship between material modulus decay and structural modulus decay.

Author Contributions: Conceptualization, Y.H., S.L. and Z.W.; methodology, S.L. and Z.W.; software, Y.H., S.L., Z.W. and H.M. (Huabao Ma); formal analysis, W.L., C.P. and H.M. (Huilin Meng); investigation, H.M. (Huilin Meng) and N.X.; resources, H.M. (Huabao Ma) and X.P.; data curation, Y.H.; writing—original draft, Y.H., S.L., Z.W., H.M. (Huabao Ma), W.L., C.P., H.M. (Huilin Meng) and N.X.; writing—review and editing, Y.H., S.L., Z.W. and X.P.; visualization, W.L. and C.P.; supervision, X.P.; project administration, Y.H.; funding acquisition, Y.H. and S.L. All authors have read and agreed to the published version of the manuscript.

Funding: The research was funded by the National Science Fund for Distinguished Young Scholars (No. 52225806), the National Natural Science Foundation of China (No. 52078063), the Science and Technology Project of Hebei Provincial Department of Transportation (No. TH-201918), and the Postgraduate Scientific Research Innovation Project of Changsha University of Science and Technology (No. CXCLY2022012 and No. CLSJCX22016).

Data Availability Statement: Some or all data, models, or code generated or used during the study are available from the corresponding author by request.

Conflicts of Interest: The authors declare no conflict of interest.

References

1. Ge, Z.-S.; Huang, X.-M. Prediction of asphalt mixtures fatigue properties using damage mechanics. *J. Traffic Transp. Eng.* **2003**, *3*, 40–42.
2. Chow, C.L.; Wei, Y. A model of continuum damage mechanics for fatigue failure. *Int. J. Fract.* **1991**, *50*, 301–316. [CrossRef]
3. Ritchie, R.O.; Gilbert, C.J.; McNaney, J.M. Mechanics and mechanisms of fatigue damage and crack growth in advanced materials. *Int. J. Solids Struct.* **2000**, *37*, 311–329. [CrossRef]
4. Xiao, J.Q.; Ding, D.X.; Jiang, F.L.; Xu, G. Fatigue damage variable and evolution of rock subjected to cyclic loading. *Int. J. Rock Mech. Min. Sci.* **2010**, *47*, 461–468. [CrossRef]
5. Yang, S.; Park, H.; Baek, C.; Li, M.E. Fatigue Cracking Characteristics of Asphalt Pavement Structure under Aging and Moisture Damage. *Sustainability* **2023**, *15*, 4815. [CrossRef]
6. Le, X.Q.; Nguyen, M.L.; Hornych, P.; Nguyen, Q.T. Analysis of LVE behaviour and fatigue damage evolution of asphalt pavements with different interface conditions in an accelerated full-scale experiment. *Int. J. Pavement Eng.* **2022**. [CrossRef]
7. Moreno-Navarro, F.; Rubio-Gámez, M.C. A review of fatigue damage in bituminous mixtures: Understanding the phenomenon from a new perspective. *Constr. Build. Mater.* **2016**, *113*, 927–938. [CrossRef]
8. Liu, G.; Jia, Y.; Yang, T.; Du, H.; Zhang, J.; Zhao, Y. Fatigue performance evaluation of asphalt mixtures based on energy-controlled loading mode. *Constr. Build. Mater.* **2017**, *157*, 348–356. [CrossRef]
9. Yao, Y.L.; Ren, Y.; Chen, S.F. Preventive maintenance timing of asphalt pavement. *J. Chang. Univ.* **2006**, *26*, 3438.
10. Zheng, J.; Lu, S. Nonlinear Fatigue Damage Model for Asphalt Mixtures. *China J. Highw. Transp.* **2009**, *22*, 21–28.
11. Zhou, Z.G.; Zhang, Q.P.; Yuan, X.X. Fatigue damage analysis of aspahlt mixture specimen during cyclic bending test. *J. Cent. South Univ. Sci. Technol.* **2011**, *42*, 1743–1751.
12. Daniel, J.S.; Kim, Y.R. Laboratory Evaluation of Fatigue Damage and Healing of Asphalt Mixtures. *J. Mater. Civ. Eng.* **2001**, *13*, 434–440. [CrossRef]
13. Shan, L.; Yang, H.; Guo, F.; Li, Z. Fatigue damage evolution in asphalt mixture based on X-ray CT images. *Constr. Build. Mater.* **2022**, *358*, 129242. [CrossRef]
14. Izaks, R.; Rathore, M.; Haritonovs, V.; Zaumanis, M. Performance properties of high modulus asphalt concrete containing high reclaimed asphalt content and polymer modified binder. *Int. J. Pavement Eng.* **2022**, *23*, 2255–2264. [CrossRef]
15. Lv, S.; Hu, L.; Xia, C.; Wang, X.; Borges Cabrera, M.; Guo, S.; Chen, J. Development of fatigue damage model of asphalt mixtures based on small-scale accelerated pavement test. *Constr. Build. Mater.* **2020**, *260*, 119930. [CrossRef]
16. Rezazadeh Eidgahee, D.; Jahangir, H.; Solatifar, N.; Fakharian, P.; Rezaeemanesh, M. Data-driven estimation models of asphalt mixtures dynamic modulus using ANN, GP and combinatorial GMDH approaches. *Neural Comput. Appl.* **2022**, *34*, 17289–17314. [CrossRef]
17. Vestena, P.M.; Schuster, S.L.; de Almeida, P.O., Jr.; Faccin, C.; Specht, L.P.; da Silva Pereira, D. Dynamic modulus master curve construction of asphalt mixtures: Error analysis in different models and field scenarios. *Constr. Build. Mater.* **2021**, *301*, 124343. [CrossRef]
18. Wahengbam, R.D.; Rajbongshi, P. An approach for dynamic stiffness evaluation in asphalt concrete. *Constr. Build. Mater.* **2015**, *96*, 541–549. [CrossRef]
19. Lv, S.; Yuan, J.; Liu, C.; Wang, J.; Li, J.; Zheng, J. Investigation of the fatigue modulus decay in cement stabilized base material by considering the difference between compressive and tensile modulus. *Constr. Build. Mater.* **2019**, *223*, 491–502. [CrossRef]
20. Zhu, H.Z.; Yan, H.; Tang, B.M. Damage Model of Interaction Between Fatigue and Creep for Asphalt Mixture. *China J. Highw. Transp.* **2011**, *24*, 15–20.
21. Lv, S.T.; Fan, X.Y.; Xia, C.D.; Zheng, J.L.; Chen, D.; You, L.Y. Characteristics of Moduli Decay for the Asphalt Mixture under Different Loading Conditions. *Appl. Sci.* **2018**, *8*, 840. [CrossRef]
22. Liu, H.; Yang, X.; Xia, C.; Zheng, J.; Huang, T.; Lv, S. Nonlinear Fatigue Damage Model of Asphalt Mixture Based on Dynamic Modulus and Residual Strength Decay. *Materials* **2019**, *12*, 2236. [CrossRef]
23. Zhu, J.; Ahmed, A.; Said, S.; Dinegdae, Y.; Lu, X. Experimental analysis and predictive modelling of linear viscoelastic response of asphalt mixture under dynamic shear loading. *Constr. Build. Mater.* **2022**, *328*, 127095. [CrossRef]
24. Zhang, J.; Bao, L. Determination of asphalt mixture's viscoelastic constitutive parameters for pavement response analysis using dynamic modulus transformation. *Constr. Build. Mater.* **2022**, *315*, 125729. [CrossRef]
25. Wang, H.; Al-Qadi, I.L. Near-Surface Pavement Failure Under Multiaxial Stress State in Thick Asphalt Pavement. *Transp. Res. Rec.* **2010**, *2154*, 91–99. [CrossRef]
26. Wang, W.; Zhao, K.; Li, J.; Luo, R.; Wang, L. Characterization of dynamic response of asphalt pavement in dry and saturated conditions using the full-scale accelerated loading test. *Constr. Build. Mater.* **2021**, *312*, 125355. [CrossRef]

27. Liu, Z.; Gu, X. Performance evaluation of full-scale accelerated pavement using NDT and laboratory tests: A case study in Jiangsu, China. *Case Stud. Constr. Mater.* **2023**, *18*, e02083. [CrossRef]
28. Ahmed, A.W.; Erlingsson, S. Numerical validation of viscoelastic responses of a pavement structure in a full-scale accelerated pavement test. *Int. J. Pavement Eng.* **2017**, *18*, 47–59. [CrossRef]
29. Lv, S.; Wang, Z.; Zhu, X.; Yuan, J.; Peng, X. Research on strength and fatigue properties of asphalt mixture with different gradation curves. *Constr. Build. Mater.* **2023**, *364*, 129872. [CrossRef]
30. *JTG E20-2011*; Standard Test Methods of Asphalt and Asphalts Mixtures for Highway Engineering. Renmin Communication Press: Beijing, China, 2011.
31. *JTG F40-2004*; Technical Spcifications for Construction of Highway Asphalt Pavements. Renmin Communication Press: Beijing, China, 2004.
32. Wang, F.; Li, Y.; Ma, G.; Ren, L. Pavement Performance and Application of New SAC13 Asphalt Mixture. *Road Mach. Constr. Mech.* **2009**, *26*, 39–42.
33. Huang, Y.; Wang, X.D.; Liu, Z.H.; Zhou, X.Y.; Zhang, L. Experimental study on dynamic modulus of SAC-13 asphalt mixture with trapezoid beam test. *J. Highw. Transp. Res. Dev.* **2017**, *34*, 7–14.
34. Xue, Z. Pavement Performance Study on Long Life Pavement of SAC-13. *Highway* **2013**, *58*, 40–44.
35. Huang, W.; Huang, M.; Zheng, M.; Li, Y. Air void variation of asphalt rubber mixture grade SAC-13. *J. Tongji Univ.* **2012**, *40*, 685–690.
36. Wang, F.Y.; Sha, Q.L.; Zhang, Y.; Dai, W.T. Application of aggregate gradation test methods of VCA_(DRF) and VCA_(AC) for SAC13. *J. Jilin Univ.* **2007**, *7*, 538–543.
37. Bai, J.W.; Zhao, Y.R.; Shi, J.; He, X.Y. Damage degradation model of aeolian sand concrete under freeze-thaw cycles based on macro-microscopic perspective. *Constr. Build. Mater.* **2022**, *327*, 126885. [CrossRef]
38. Fernández-Canteli, A.; Castillo, E.; Blasón, S. A methodology for phenomenological analysis of cumulative damage processes. Application to fatigue and fracture phenomena. *Int. J. Fatigue* **2021**, *150*, 106311. [CrossRef]
39. Wang, Z.; Lu, W.; Liu, K.; Lv, S.; Peng, X.; Yang, S.; Ding, S. Research on failure strength master curve and fatigue performance of asphalt mixture containing high-proportion reclaimed asphalt pavement. *Constr. Build. Mater.* **2023**, *370*, 130537. [CrossRef]

buildings

MDPI

Article

Rheological Properties and Microscopic Morphology Evaluation of UHMWPE-Modified Corn Stover Oil Bio-Asphalt

Jian Li [1], Chuanxi Luo [1,2,*], Jixing Jie [1,2] and Haobin Cui [3]

1 Xiaoning Institute of Roadway Engineering, Guangzhou 510006, China; m15738523861@163.com (J.L.); 13609192658@163.com (J.J.)
2 School of Civil Engineering and Transportation, South China University of Technology, Guangzhou 510641, China
3 School of Traffic and Transportation Engineering, Changsha University of Science & Technology, Changsha 410114, China; 18547093308@163.com
* Correspondence: a940680158@163.com

Abstract: In order to promote the efficient utilization of bio-oil, corn stover oil and petroleum asphalt were used to prepare bio-asphalt. UHMWPE was adopted to strengthen the high-temperature properties of bio-asphalt. UHMWPE-modified corn stover oil asphalt was prepared. Rheological and microscopic tests were carried out to study its performance. The softening point and viscosity of the modified asphalt were enhanced with a rise in the UHMWPE dosage. Its ductility and deformation ability increased at 5 °C. An MSCR test suggested that the asphalt's creep recovery ability and anti-rutting properties decreased at a high stress level. Meanwhile, the low-temperature rheological behavior of UHMWPE-modified corn stover oil asphalt was superior to that of neat asphalt. The corn stover oil ameliorated the asphalt's low-temperature properties but weakened its high-temperature behavior. The optimal preparation schemes for UHMWPE-modified corn stover oil asphalt were recommended through a comprehensive analysis of the properties. The recommended dosage of UHMWPE was 3–4%, while the corn stover oil dosage was 5%. However, when the dosage of corn stover oil was 10%, the recommended dosage of UHMWPE was 4%. UHMWPE powder was melted and cross-linked with neat asphalt during high-temperature preparation, demonstrating that UHMWPE can enhance the properties of asphalt. The UHMWPE polymer macromolecules can be dispersed in corn stover oil. UHMWPE can form a compact and robust network structure with asphalt. The feasible application of corn stover oil in road engineering was verified, which provides efficient solutions for waste utilization. This study will contribute to the sustainable development of road construction.

Keywords: UHMWPE; biomass materials; rheological properties; morphology characteristics; preparation scheme

Citation: Li, J.; Luo, C.; Jie, J.; Cui, H. Rheological Properties and Microscopic Morphology Evaluation of UHMWPE-Modified Corn Stover Oil Bio-Asphalt. *Buildings* **2023**, *13*, 2167. https://doi.org/10.3390/buildings13092167

Academic Editor: Pengfei Liu

Received: 18 July 2023
Revised: 13 August 2023
Accepted: 21 August 2023
Published: 26 August 2023

1. Introduction

During its service period, an asphalt pavement will suffer from vehicle load, changing temperature, and moisture, which aggravate the damage to the pavement structure. Structural damage will shorten the service time of an asphalt pavement, seriously affecting road traffic capacity, leading to the waste of resources, and increasing maintenance project costs. In order to satisfy the demands of an increase in traffic volume and the development of green transportation, it is necessary to improve pavement performance and ensure the development of green pavement materials. Meanwhile, bio-asphalt is widely used in modified asphalt due to its vast source and low price. The development of bio-asphalt can contribute to the development of green transportation. This study will explore the application of ultra-high-molecular-weight polyethylene (UHMWPE) polymer in modified bio-asphalt and provide a novel scheme for the modification of petroleum asphalt.

1.1. Polymer-Material-Modified Asphalt

In the 1930s, Britain began to use rubber-modified asphalt in large quantities. Since 1945, polymer-modified asphalt has been regarded as an effective method to improve asphalt pavement performance. Recently, polymer-modified asphalt has been widely used in road construction. Polymers include styrene–butadiene–styrene (SBS), rubber powder, polyphosphoric acids (PPA), etc. Polymers have remarkable effects on improving asphalt, such as high-temperature stability, low-temperature crack resistance, and ultraviolet aging resistance [1,2].

Viscione [3] used recycled plastics to modify asphalt (RPMA). Furthermore, laboratory tests revealed that the polymers could improve the mechanical properties of the asphalt mixture. The cracking resistance and rutting resistance of RPMA were obviously strengthened. Gabriel [4] summarized a polymeric-waste-modified asphalt mixture. It is beneficial as it raised the deformation resistance, fatigue cracking, and low-temperature cracking behavior of an asphalt pavement [5]. Mashaan [6] suggested that the optimal dosage of waste plastic is 6–8% in modified asphalt, which enhances fatigue, aging, and temperature stability behavior. The efficient utilization of waste plastics in road engineering can minimize the amount of waste plastics and effectively protect the environment. Waste tires can be shredded into rubber powder. Rubber powder has been successfully used in modified asphalt. Dibyendu [7] reported that a rubber-powder-modified asphalt mixture has excellent durability and is more resistant to moisture failure. Rubber powder can ameliorate the tensile strength of the asphalt mixture by 11.7%. Yan [8] studied the degradation within high-dosage polymer-modified asphalt (HDPMA). The results showed that the polymer appears to have noticeable degradation during the aging process. HDPMA is more sensitive to short-term aging at high temperatures. Polymer degradation increases the stiffness of asphalt, decreases its elastic recovery ability, and dramatically affects the phase angle. The polymer could significantly increase the rheological behavior and anti-aging properties of HDPMA. Sulfur improves the viscosity and storage stability of HDPMA [9]. However, excessive polymer has a limited effect on asphalt modification [10]. Esraa R. Al-Gurah [11] investigated polymer-modified asphalt binder (PMAB), adding different fillers. The application of PMBA remarkably improves the properties of asphalt mixtures. The rutting depth of the PMBA mixture decreased by 48% compared with the contrast test, which improved the deformation resistance of the pavement. Vamegh [12] investigated the performance of polypropylene- and styrene butadiene rubber (SBR)-modified asphalt. He reported that SBR could ameliorate the moisture behavior of asphalt, which decreased asphalt spalling by 50%. The combined effect of polypropylene and SBR could dramatically improve the deformation resistance of the asphalt mixture. This successful modification scheme can reduce the modification cost. Polymers can also be used as a rejuvenator to recover the performance of recycled asphalt (RA). Daryoosh [13] reported that a polymer and rejuvenator could improve the properties of aged asphalt. The low-temperature PG of recycled asphalt was upgraded to −22 °C. The polymer greatly enhanced the fatigue behavior of aged asphalt by 300% compared with neat asphalt. Ren [14] evaluated the properties of polymer-modified recycled asphalt (RMA). SBS and rubber could raise the high-temperature anti-deformation performance, aging resistance behavior, and viscosity of recycled asphalt (RA). The RA and PMA could strengthen the asphalt against fatigue cracking [15].

All in all, due to the favorable modification effects of various polymers, they are widely used in modified asphalt. Polyethylene is one of the most widely used plastic products. Polyethylene accounts for one-third of global plastic production. According to the division method of Philips Petroleum Company, polyethylene with a molecular weight of more than 1.5 million is called ultra-high-molecular-weight polyethylene [16,17]. UHMWPE is a kind of thermoplastic engineering material with a moderate price and excellent performance due to its extremely high molecular weight [18]. UHMWPE has incomparable comprehensive properties, such as abrasion resistance, impact resistance, low-temperature resistance, corrosion resistance, and low density compared with ordinary polymers. UHMWPE has been widely used in the textile, papermaking, and chemical industries, and medical treatment [19,20]. The polymer has good compatibility with asphalt. However, there is little research on UHMWPE application in petroleum asphalt [21]. This study broadens the utilization of UHMWPE in engineering.

1.2. Bio-Asphalt

Bio-oil can be produced from various plants using pyrolysis technology [22]. As a renewable material, bio-oil has better application advantages. It's functional groups are similar to the functional groups of petroleum asphalt, and they have outstanding compatibility [23]. Bio-asphalt is a blend of bio-oil and petroleum asphalt. The poor high-temperature performance of bio-asphalt is a crucial factor limiting its promotion. In order to promote the efficient application of bio-asphalt, there are numerous studies about modified bio-asphalt [24,25].

Deng [26] used sludge bio-oil to modify asphalt and conducted tests to investigate the intermolecular interactions between sludge bio-oil and asphalt. Sludge bio-oil's main constituent is esterification, with a dosage of 75.94%. Their FTIR results indicated that no new functional groups appeared in the bio-asphalt, and the sludge bio-oil was blended with asphalt. The sludge bio-oil could influence the colloidal structure of the asphalt, which ameliorated its anti-creaking properties. Abdulnaser [27] investigated the application feasibility of crude palm oil (CPO). Their research demonstrated that palm oil could enhance mixtures' fatigue and low-temperature properties. It was reported that the bio-asphalt containing 5% CPO had the same complex shear modulus as the neat asphalt. He also conducted other studies about nano-silica (NS)-modified CPO bio-asphalt [28]. NS could raise the rutting index of CPO bio-asphalt, which compensated for the drawback of the high-temperature performance of bio-asphalt. At shear stress levels of 0.1 kPa and 3.2 kPa, the recovery rate of this modified asphalt increased by 27.7% and 8.8% compared to neat asphalt. Heny [29] extracted the bio-oil from sugarcane using pyrolysis technology. Their study indicated that the penetration of the bio-asphalt was consistent with that of 70# asphalt (penetration between 60 and 70). Lei [30] introduced a novel preparation method of rubber-modified bio-asphalt (RMBA). The rubber enhanced the anti-stripping behavior of asphalt by about three times. Meanwhile, as a closely connected twist band, bio-oil could enable the modified asphalt to be better adsorbed on the aggregate surface. RMBA has great elastic recovery ability at high temperatures. Furthermore, it also has satisfactory storage stability and temperature sensitivity [31]. The rubber powder formed a skeletal structure in RMPA that dramatically enhanced the rutting resistance of the mixture. Jiang [32] developed the utilization of resin bio-oil. Resins bio-oils can positively strengthen bio-asphalt's elastic recovery behavior and high-temperature stability. Furthermore, it is recommended that the best replacement amount of resin bio-oil is 30% of the asphalt. Zheng [33] studied 4,4′-diphenylmethane diisocyanate (MDI)-modified wood-oil asphalt. The MDI and wood oil changed the colloid type to a gel structure, enhancing the deflection resistance. The chemical reaction between MDI and asphalt leads to the enhancement of high-temperature properties. Yang [34] promoted the utilization of bio-asphalt modified with PPA and SBS. Bio-oil could facilitate the compatibility of the SBS modifier with asphalt. The chemical reaction between PPA and asphalt formed a network structure, which ameliorated the high-temperature behavior of the bio-asphalt [23]. Zhang [35] summarized the preparation and application of bio-asphalt. He pointed out that most types of bio-oil can reduce the asphaltene dosage and strengthen the low-temperature behavior of asphalt. The low viscosity of bio-asphalt leads to lower energy consumption during the compaction of pavement construction [36]. Applying bio-asphalt in road engineering can promote the effective utilization of waste and reduce costs.

In summary, various bio-asphalts have the obvious drawback of poor high-temperature behavior. It is essential to make up for this defect of bio-asphalt. Common polymers, such as SBS, PPA, MDI, and SBR, are used in modified bio-asphalt. The emergence of nanomaterials has also promoted the utilization of bio-asphalt.

1.3. Study Objectives

The objectives of this study were to promote the efficient application of corn stover oil. However, the poor high-temperature performance of bio-asphalt has limited its promotion. So, UHMWPE was adopted to strengthen the high-temperature properties of corn stover oil bio-asphalt. The UHMWPE-modified corn stover oil asphalt (UCA) was prepared. Some laboratory tests were conducted to study its properties. The following objectives were established:

- To promote the efficient application of corn stover oil bio-asphalt.
- To evaluate the high- and low-temperature characteristics and microscopic morphology of UCA.
- To determine the dosage of UHMWPE and CSO to prepare the UCA.

2. Raw Materials and Test Methods

2.1. Materials

SK neat asphalt (SK-NA) was selected in this research study, produced by Qingdao Rishi Jitong Company (Qingdao, China). The technical indicators of SK-NA are displayed in Table 1. The corn straw oil (CSO) selected in this study belongs to one kind of bio-oil. It has a wide range of sources, a low price, and a promising foundation for application [37]. Its technical indicators are shown in Table 2. Ultra-high-molecular-weight Polyethylene (UHMWPE) has excellent serviceability due to its super high molecular weight. UHMWPE is a three-million mesh of powder polymers. Furthermore, the technical indicators of UHMWPE are shown in Table 3.

Table 1. Technical indicators of SK-NA.

Indicator		Value	Requirements
Penetration at 25 °C (0.1 mm)		71.2	60~80
Ductility at 15 °C (cm)		121.6	≥100
Softening point (°C)		48.2	≥46
Flash point (°C)		312	≥260
After RTFO-aging	Mass loss (%)	0.028	≤±0.8
	Penetration ratio (%)	62.6	≥61
	Ductility at 10 °C (cm)	7.3	≥6

Table 2. Technical indicators of CSO.

Indicator	Value
Flash point (°C)	267
Density at 15 °C (g/cm^3)	0.937
Ash dosage (%)	0.064
Aliphatic acid dosage (%)	67.3
Alcohols dosage (%)	4.68

Table 3. Technical indicators of UHMWPE.

Indicator	Value
Density at 15 °C (g/cm^3)	0.986
Molecular weight	≥1.5 million
Appearance	White-colored powder
Water-absorbing capacity (%)	≤0.01
Melting point (°C)	134

2.2. Preparation of UHMWPE-Modified CSO Bio-Asphalt

The SK-NA and CSO were heated to form a flowing liquid in an oven at 135 °C. Firstly, a high-speed shear was utilized to prepare the CSO bio-asphalt. The dosages of CSO were 5% and 10% of the mass of the SK neat asphalt. The shear rate was 2000 rpm. The shear time was 30 min. The temperature was maintained at 135 °C during the preparation of the CSO bio-asphalt. Secondly, UHMWPE was added to the CSO bio-asphalt, and its dosages were 2%, 3%, and 4% of the mass of the CSO bio-asphalt. In this step, the shear rate was 3000 rpm, the shear time was 60 min, and the shear temperature was 150 °C. Then, the UHMWPE-modified CSO bio-asphalt (UCA) was prepared. The diffident types of modified asphalt are displayed in Table 4. The SK-NA was the contrast test in this study and reflects the modification effects of UCA.

Table 4. The various types of modified asphalt.

Dosage of UHMWPE and CSO	Dosage of CSO Is 5%	Dosage of CSO Is 10%
Dosage of UHMWPE is 2%	UCA 2–5	UCA 2–10
Dosage of UHMWPE is 3%	UCA 3–5	UCA 3–10
Dosage of UHMWPE is 4%	UCA 4–5	UCA 4–10

2.3. Test Methods

A test chart of this research study is shown in Figure 1. The tests were divided into three main parts: conventional performance tests, rheological performance tests, and microscopic characteristics tests. The conventional and rheological performance tests were conducted according to the SHRP specifications. The ductility test was conducted in a water bath at 5 °C. The extension rate of the ductility test was 5 ± 0.25 cm/min. This study tested the Brookfield rotational viscosity of asphalt at 135 °C. A dynamic shear rheometer conducted temperature sweep (TS) and multiple stress creep recovery (MSCR) tests. The TS test was performed on the original asphalt sample, and the test temperature was 42–80 °C. The MSCR test was performed on the short-term-aged asphalt sample. The bending beam rheometer (BBR) test proceeded at −18 °C and −24 °C. Scanning electron microscope (SEM) and fluorescence microscope (FM) tests were carried out to reflect the morphology characteristics of the materials.

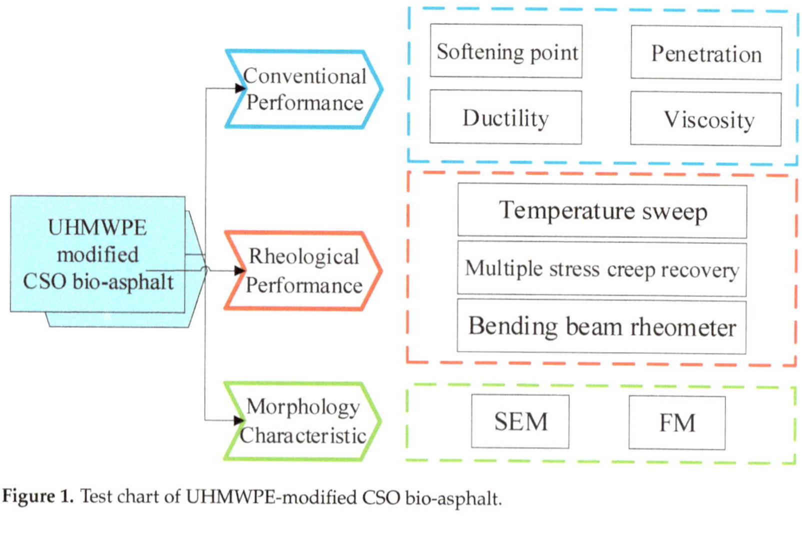

Figure 1. Test chart of UHMWPE-modified CSO bio-asphalt.

3. Results and Discussion

3.1. Conventional Performance of Modified Asphalt

3.1.1. Softening Point

Figure 2 summarizes the softening point and segregation test results of SK-NA and UCA. The softening point of SK-NA is 48.2 °C. The softening point dramatically rose when 5% CSO and a 2–4% dosage of UHMWPE were all blended with SK-NA. This was mainly because UHMWPE enhanced the high-temperature performance of asphalt. However, when the CSO dosage was 10%, the softening points of UCA were dramatically reduced. This was mainly because 10% CSO softened the asphalt and decreased the softening point. The CSO could have a negative effect on the high-temperature performance of asphalt. As detailed in Figure 2a, UCA 4–5 had a maximum value of 59.3 °C. Compared with SK-NA, the softening point of UCA 4–5 was increased by 23.03%. Unfortunately, the softening point of UCA 2–10 was merely 42.5 °C, which was reduced by 11.83% compared with SK-NA. The enhancements of the softening point caused by a 1% dosage of UHMWPE are marked in Figure 2a. It can be noted from Figure 2a that the softening point of UCA was positively enhanced with the increase in the UHMWPE blending amount. It is summarized that UHMWPE can improve the high-temperature property of UCA. When UHMWPE was blended in a certain amount, the softening points of UCA2–10, UCA 3–10, and UCA 4–10 declined by 14.49%, 11.38%, and 11.64%, respectively, compared with UCA 2–5, UCA 3–5, and UCA 4–5. It can be concluded that CSO has the most dramatic impact on the high-temperature property of UCA.

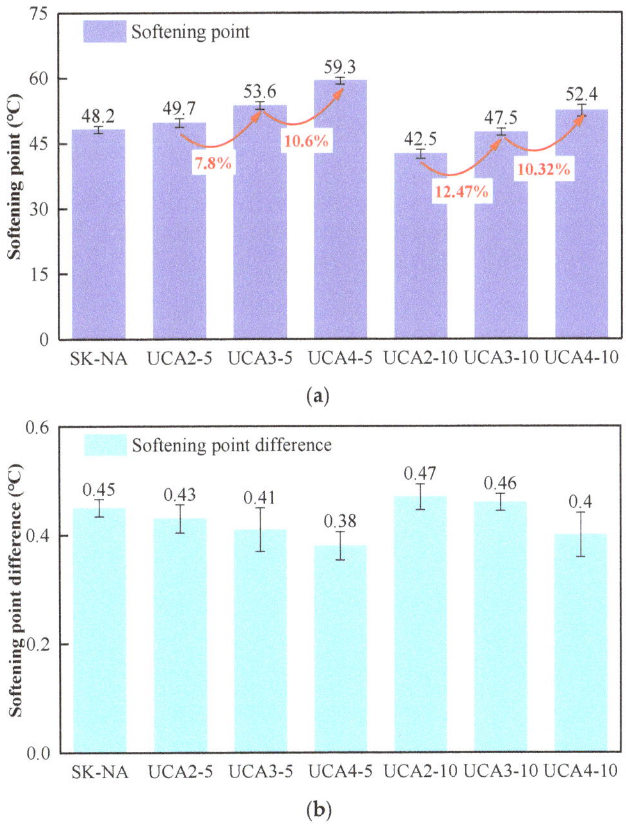

Figure 2. Softening point and segregation test results. (**a**) Softening point; (**b**) segregation test.

Considering the results in Figure 2a, the UHMWPE dosage should be higher than 2% when the CSO dosage is 5%, and the UHMWPE dosage should be higher than 3% when the CSO dosage is 10%. In this sense, the softening point of UCA is greater than that of SK-NA.

Stability is an essential factor that influences the performance of polymer-modified asphalt. The segregation tests were conducted to reveal the stability of UCA. Moreover, the softening point difference was utilized as an evaluation index. The stability test results are displayed in Figure 2b. As displayed in Figure 2b, the softening point difference of SK-NA was 0.45 °C. With the increases in UHMWPE dosage, the softening point difference of UCA gradually decreased. When the CSO dosage was 5%, the softening point difference of UCA was lower than that of SK-NA. However, the softening point difference of UCA increased when the CSO dosage was increased from 5% to 10%. Compared to SK-NA, the softening point differences of UCA 2–10 and UCA 3–10 were increased by 0.2 °C and 0.1 °C. The segregation test results indicate that the UCA maintained good stability.

3.1.2. Penetration

Penetration is a routine index for assessing the consistency of asphalt, in which a higher value means the binder is softer. Figure 3 presents the penetration of SK-NA and UCA. The penetration was tested at 25 °C, near the average service temperature. As detailed in Figure 3, there was a strong correlation between the penetration of UCA and the dosage amount of UHMWPE. The penetration decreased as more UHMWPE was blended with the UCA, indicating that UHMWPE could ameliorate the CSO bio-asphalt stiffness. By contrast with SK-NA, the penetration of UCA 2–5, UCA 3–5, and UCA 4–5 increased by 5.48%, −5.62%, and −15.17%, respectively. There was a noticeable rise in the penetration of UCA as the CSO dosage was changed from 5% to 10%. This illustrates that CSO can soften the UCA and reduce its consistency.

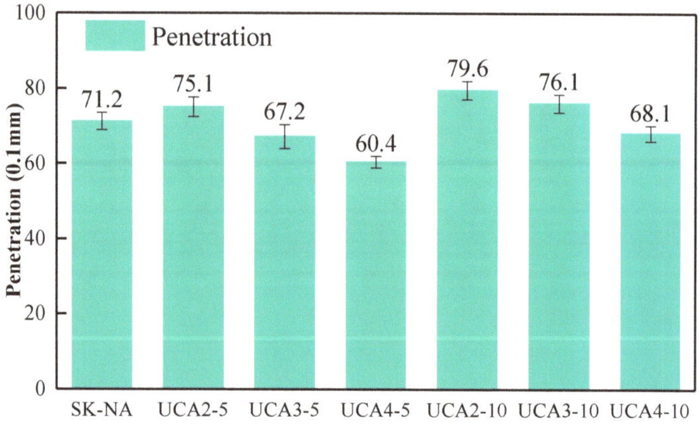

Figure 3. Penetration of UCA and SK-NA.

We can see that the penetration grade of UCA has not changed and still belongs to 70# (penetration between 60 and 80) asphalt. Therefore, in terms of penetration, we suggest that the dosage of CSO and UHMWPE in UCA should be 5–10% and 2–4%, respectively.

3.1.3. Ductility

A ductility test measures the ability of asphalt to be stretched or drawn without breaking. This test was conducted to measure the tensile properties and evaluate the compatibility of the binder. The ductility results are given in Figure 4. It is worth noting that the ductility of UCA was significantly improved. Compared to the results of Figure 4, the ductility of UCA 2–5, UCA 3–5, and UCA 4–5 increased by 45.4%, 66.98%, and 105.4%, respectively. Moreover, when the CSO dosage increased to 10%, the ductility of UCA continued to increase. Significantly, the ductility of UCA 4–10 was enhanced by 155.87%

compared with SK-NA. The UCA's ductility increase will reduce the possibility of cracking of the asphalt pavement at low in-service temperatures. This is mainly due to the dual roles of CSO and UHMWPE. On the one hand, CSO could increase the oil content in UCA. On the other hand, UHMWPE swelled in SK-NA during the preparation at high temperatures, thus strengthening the toughness and resistance to deformation of UCA. UHMWPE and CSO significantly increase the tensile properties of UCA.

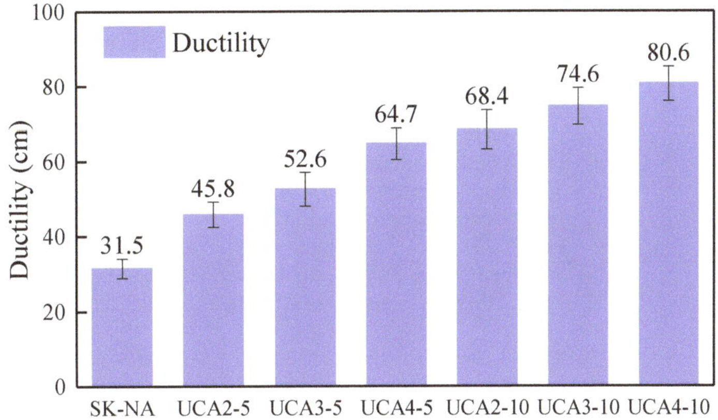

Figure 4. Ductility of UCA and SK-NA.

3.1.4. Viscosity

The asphalt viscosity at 135 °C is near the mixing and laydown temperature for HMA. Viscosity is vital for HMA construction. It can determine the mixing and compaction range of HMA. In terms of performance, the viscosity reflects the high-temperature properties of the asphalt. The viscosity experiment results are given in Figure 5. As seen from Figure 5, when more UHMWPE was blended into the CSO bio-asphalt, the viscosity of UCA gradually increased. Compared to the UCA 2–5 viscosity, that of UCA 3–5 and UCA 4–5 increased by 9.07% and 16.37%, respectively. When a 10% dosage of CSO was blended with asphalt, the viscosity was obviously reduced. The UCA 2–10 viscosity was decreased by 14.8% compared with UCA 2–5.

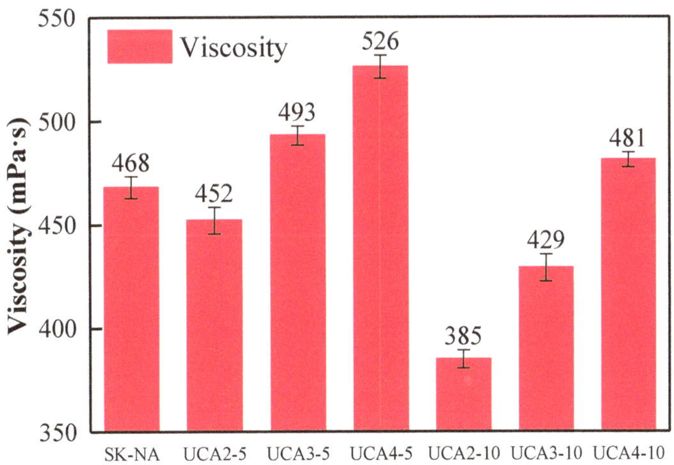

Figure 5. Viscosity test at 135 °C.

As reported in Figure 5, merely three kinds of UCA exceeded SK-NA viscosity. They are UCA 3–5, UCA 4–5, and UCA 4–10, respectively. Therefore, considering the results of Figure 5 separately, it is suggested that when the dosage of CSO is 5%, the UHMWPE dosage is 3–4%. It is recommended that the dosage of CSO is 10% and the dosage of UHMWPE is 4%.

3.1.5. Summary of Conventional Performance

The softening point and viscosity of UCA are enhanced with a UHMWPE dosage increase. Its ductility and deformation ability increase at 5 °C. Because of the composite modification of CSO and UHMWPE, UCA's high- and low-temperature properties are satisfactorily strengthened. Asphalt performance improves because of the light fraction contained in CSO, which increases the oil content in UCA and ameliorates its tensile properties. Meanwhile, UHMWPE is melted in CSO bio-asphalt, and UHMWPE can form a compact and robust network structure in UCA [21]. These lead to dramatic stiffness and toughness enhancements in UCA. The dosages of UHMWPE and CSO in UCA are listed in Table 5 based on conventional performance. In order to make sure the UCA properties were better than SK-NA, the UHMWPE dosage was 3–4%, while the CSO dosage was 5%. The UHMWPE dosage was 4% when the CSO dosage was 10%.

Table 5. The preparation scheme of UCA is based on conventional performance.

Performance Index	Dosage of CSO Is 5%	Dosage of CSO Is 10%
Softening point	2–4%	3–4%
Penetration	2–4%	2–4%
Ductility	2–4%	2–4%
Viscosity	3–4%	4%
Dosage of UHMWPE	3–4%	4%

3.2. *Rheological Performance of Modified Asphalt*

3.2.1. TS Test

In a high-temperature environment, asphalt is a viscous fluid. The higher the temperature, the lower the viscosity, and the asphalt emerges with superior fluidity [38]. A TS test can reflect the rheological properties of asphalt. The rutting index ($G*/\sin\delta$) of the TS test results reveals the high-temperature performance of UCA. The bigger $G*/\sin\delta$, the inferior flow performance of asphalt, indicating satisfactory deformation resistance. Figure 6 lists the TS test results. As detailed in Figure 6, the $G*/\sin\delta$ and temperature are fitted for the asphalt, and the R^2 is 0.99. This exhibits good relevance between $G*/\sin\delta$ and temperature.

We can note that the $G*/\sin\delta$ of UCA rose, while the UHMWPE dosage increased. When the TS temperature was 64 °C, the $G*/\sin\delta$ values of UCA 2–5, UCA 3–5, and UCA 4–5 were 1.564 kPa, 2.674 kPa, and 3.861 kPa, which increased by −11.25%, 51.72%, and 119.07% compared to those of SK-NA. UCA 2–5's high-temperature performance was inferior to that of SK-NA. However, the $G*/\sin\delta$ values of UCA 3–5 and UCA 4–5 were superior to the those of the contrast test, proving that UHMWPE could strengthen the shear deformation resistance of the CSO bio-asphalt. When the CSO dosage of UCA rose to 10%, the 64 °C $G*/\sin\delta$ values of UCA 2–10, UCA 3–10, and UCA 4–10 were 1.2348 kPa, 1.6842 kPa, and 2.9945 kPa, which increased by −29.94%, −4.44%, and 69.91% compared to those of SK-NA. The rutting index of UCA 3–10 was the same as that of the contrast test. Furthermore, the high-temperature properties of UCA 4–10 were outstanding. This also proved that the UHMWPE dosage must be higher than 3% when the CSO dosage is 10%. Comparing the data in Figure 6a,b, it is not difficult to find that the UCA decreased with the increase in CSO dosage. The CSO could overcome the high-temperature rheological behavior.

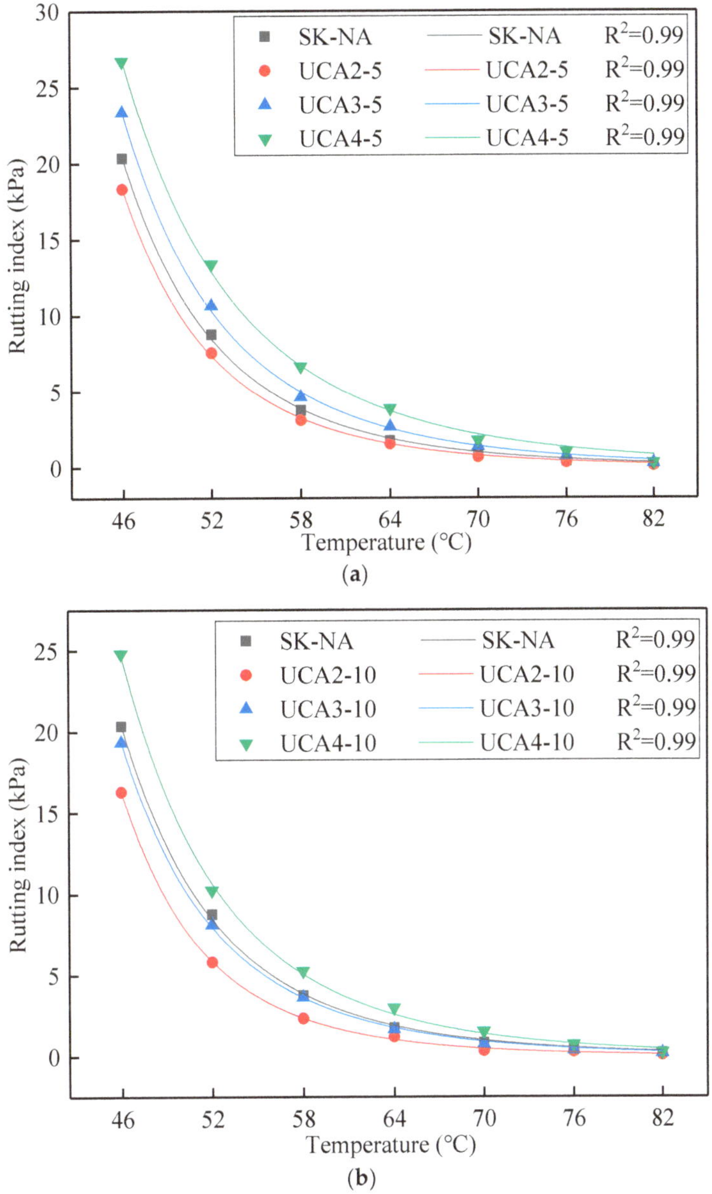

Figure 6. TS test results: (**a**) CSO dosage 5%; (**b**) CSO dosage 10%.

The above TS analysis recommends that the UHMWPE dosage be higher than 3% when the CSO dosage is 5–10%. UCA 3–5, UCA 4–5, and UCA 4–10 have superior high-temperature rheological behavior, indicating that CSO and UHMWPE jointly provide favorable modification. This is mainly due to UHMWPE, as a polymer, having good compatibility and forming a stable network structure with CSO bio-asphalt [16]. The strengthening of the UCA microstructure resulted in a satisfactory amelioration in its performance.

3.2.2. MSCR Test

The MSCR test evaluates the rheological behavior of asphalt by applying repeated cycle loads to aged asphalt specimens. MSCR tests were performed at 58 °C. Recovery rate (R) and Jnr are crucial evaluation indexes of this test, and precisely reflect the service state of an asphalt pavement under actual load [39]. R reflects the recoverability of the asphalt binder. Jnr reflects the resistance against rutting deformation of the asphalt binder. Generally speaking, the larger the R demonstrated, the stronger the elastic recovery behavior of the asphalt sample, and the smaller the Jnr, the better the rutting deformation resistance of the asphalt sample. The MSCR test results are exhibited in Figure 7.

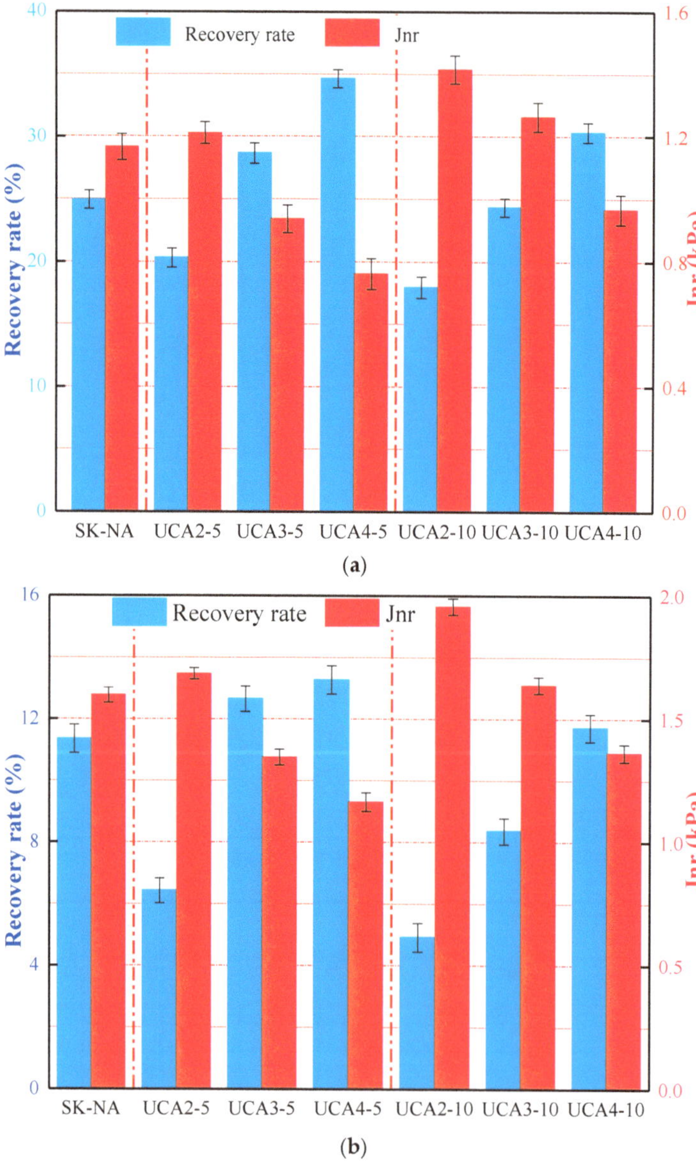

Figure 7. MSCR test result: (**a**) 0.1 kPa; (**b**) 3.2 kPa.

As detailed in Figure 7, whatever the stress level, it can be noted that increasing the UHMWPE dosage or decreasing the dosage of CSO will decrease the high-temperature behavior of UCA.

Under stress of 0.1 kPa, in comparison to SK-NA, the R values of the UCA 2–5, UCA 3–5, and UCA 4–5 samples containing 5% CSO increased by −18.56%, 14.96%, and 38.93%, respectively. Under stress of 3.2 kPa, the R containing 5% CSO increased by −43.34%, 11.45%, and 16.91%, respectively. Under stress of 0.1 kPa, the Jnr values of UCA 2–5, UCA 3–5, and UCA 4–5 were 1.212 kPa, 0.937 kPa, and 0.761 kPa, which decreased by −2.86%, 19.7%, and 34.79% compared to the contrast test. Under stress of 3.2 kPa, Jnr reduced by −5.51%, 15.43%, and 27.07%, respectively. The above analysis suggests that the creep recovery ability and anti-rutting properties of UCA decrease at high stress levels. We can conclude that the UHMWPE dosage should be 3–4% when the CSO dosage is 5%.

When the CSO dosage is 10%, the changing trend of MSCR results in a UCA similar to that with a CSO dosage of 5%. In order to ensure that the MSCR results of UCA containing 10% CSO are superior to those of the contrast test, it is recommended that the dosage of UHMWPE is 4%.

Analyzing the TS and MSCR test results of UCA, it can be concluded that the UHMWPE dosage is 3–4%, and the CSO dosage is 5%. When the CSO dosage is increased to 10%, the UHMWPE dosage is 4%. Only in this way can UCA exhibit excellent high-temperature rheological properties.

3.2.3. BBR Test

When the temperature drops suddenly or in a very low-temperature environment, asphalt will generate temperature stress, leading to cracking on the pavement surface. This will affect the pavement's service state and lifespan [40]. The BBR test can determine the low-temperature cracking resistance of asphalt. The stiffness modulus (S) and m-value are taken as evaluation criteria. Generally speaking, the S must be less than 300 MPa, and the m-value must be higher than 0.3. The BBR results are given in Figure 8.

As listed in Figure 8a, S gradually rose with the increase in UHMWPE dosage. However, there was an inevitable reduction in S when the CSO dosage increased to 10%. The higher dosage of UHMWPE made the asphalt more susceptible to temperature stress. This indicated that UHMWPE would affect the low-temperature crack resistance of UCA. CSO contains more light substances and oil, which can effectively reduce the generation of temperature stress. At the same BBR test temperature, the S of the modified asphalt was lower than that of the SK-NA. It can be seen in Figure 8b that the m-value of UCA was higher than 0.3 at −18 °C. When the BBR test temperature was decreased to −24 °C, only the m-value of UCA 2–10 met the specification.

As shown in Figure 8, when the BBR test temperature was −18 °C, the BBR results of SK-NA and UCA results satisfied the requirements. When the test temperature was decreased to −24 °C, the S values of UCA was all higher than 300 MPa, and the m-value was lower than 0.3. It is worth mentioning that the S and m-values of UCA were better than those of the contrast test. One can conclude that the low-temperature rheological behavior of UCA is superior to that of SK-NA. A high dosage of CSO will ameliorate the low-temperature behavior of UCA. The CSO will weaken the high-temperature behavior to a certain extent. Considering the BBR test results alone, it is appropriate that the UHMWPE dosage is 2–4% when the CSO dosage is 5–10%.

3.3. *Optimal Preparation Scheme of UCA*

The recommended UHMWPE and CSO dosages for each test are summarized in Table 6. We considered UCA's conventional and rheological behavior and determined the dosage of UHMWPE and CSO in UCA. As detailed in Table 6, each test's recommended amounts were considered. In order to make sure the UCA properties were better than those of the SK-NA, the preparation scheme of UCA was obtained. When the dosage of CSO is

5%, the recommended dosage of UHMWPE is 3–4%. However, when the dosage of CSO is 10%, the recommended dosage of UHMWPE is 4%.

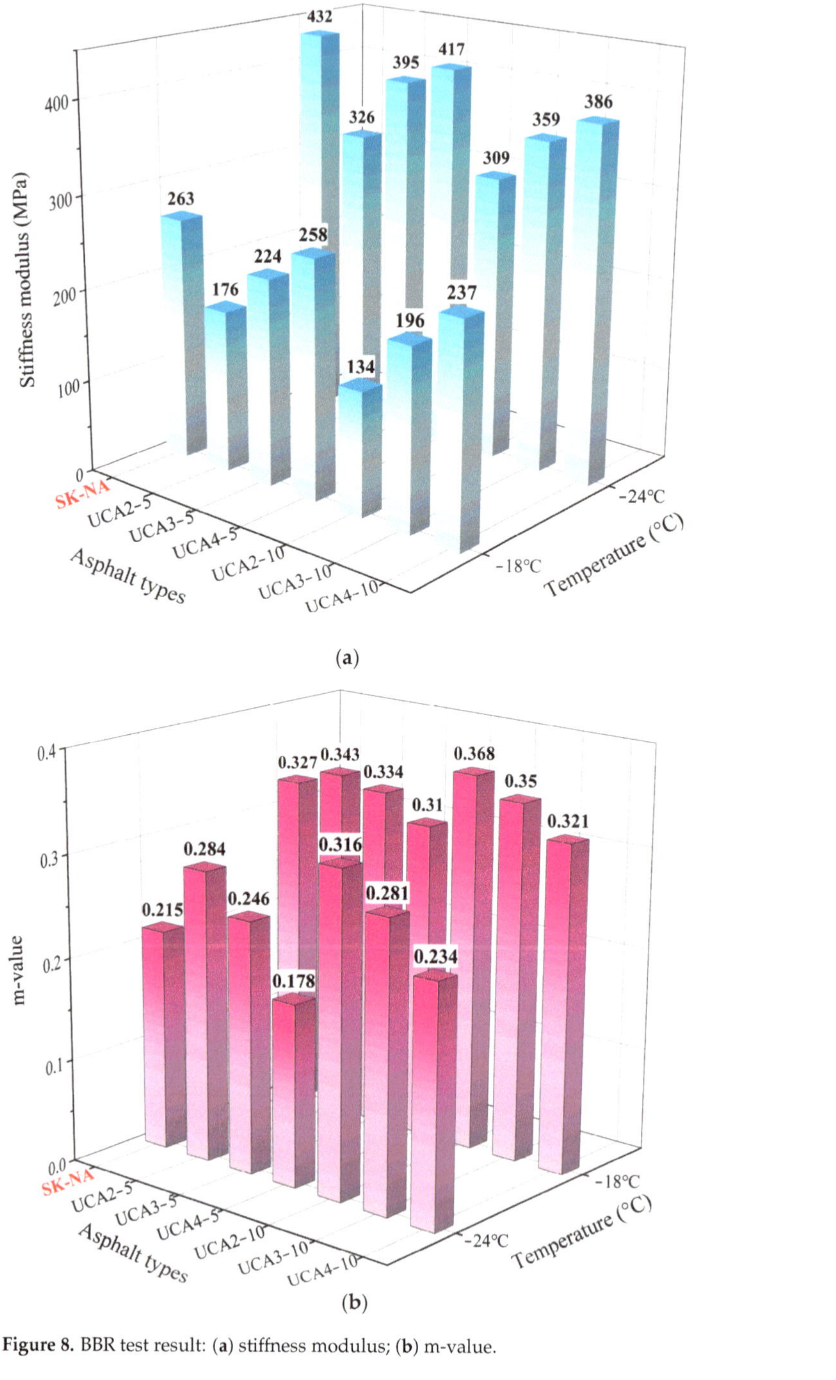

Figure 8. BBR test result: (**a**) stiffness modulus; (**b**) m-value.

Table 6. The preparation scheme of UCA.

Tests	Dosage of CSO Is 5%	Dosage of CSO Is 10%
Conventional tests	3–4%	4%
TS test	3–4%	3–4%
MSCR test	3–4%	4%
BBR test	2–4%	2–4%
Dosage of UHMWPE	3–4%	4%

3.4. Morphology Characteristics

3.4.1. SEM

We used a scanning electron microscope to scan the different types of asphalt and observed the micro-properties. The SEM images of the asphalt are visible in Figure 9.

As seen in Figure 9a, the microstructure surface of SK-NA did not present wrinkles. After blending CSO in SK-NA, smooth wrinkles appeared in the CSO bio-asphalt, as displayed in Figure 9b. Meanwhile, its SEM image became darker, which was caused by CSO [41]. The cross-linked molecular structure of the UHMWPE-modified SK asphalt can be seen in Figure 9c. UHMWPE powder was melted and cross-linked with SK-NA during high-temperature preparation to shape the stable cross-linked structure. This demonstrates that UHMWPE could influence the properties of asphalt. However, the cross-linked structure changed after blending UHMWPE with CSO bio-asphalt, as presented in Figure 9d. This phenomenon may be due to the adsorption of CSO by the reticular structure.

3.4.2. FM

Poor solubility occurs between polymers and asphalt, which results in drawbacks correlated with the properties of the modified asphalt. It is necessary to investigate the compatibility of UHMWPE and CSO in asphalt. The FM images exhibit the morphology between the modifier and SK-NA. Moreover, the FM images of SK-NA, CSO bio-asphalt, UHMWPE-modified SK-NA, and UCA are exhibited in Figure 10.

The SK-NA and CSO bio-asphalt did not appear as fluorescent substances [41]. So, Figure 10a was used to display their microscopic morphology. Figure 10b shows the UHMWPE-modified SK-NA morphology. Many fluorescent substances can be seen in Figure 10b. Compared with Figure 10a, it is presumed that the UHMWPE polymer exhibits fluorescence characteristics. The morphology of UCA is given in Figure 10c. It can be noted that many fluorescent spots appear, and they are evenly distributed. This is because the compositions of CSO and SK-NA are similar, and polymer macromolecules can be dispersed in CSO. UHMWPE appears in UCA as a small-molecular-weight solvent. UHMWPE and CSO change the morphology of asphalt, which leads to the improvement of its performance.

3.5. Economic Analysis

In order to compare the economic benefits of bio-asphalt, the raw material costs of the modified asphalt were calculated. The prices of SK-NA, CSO, and UHMWPE are 5000 CNY/ton, 1000 CNY/ton, and 15,000 CNY/ton. The raw material costs of UCA are displayed in Table 7.

As displayed in Table 7, the raw material costs of UCA 2–10 and UCA 3–10 were lower than that of SK-NA. With the increase in CSO dosage, there was a significant decrease in the cost of UCA. The high price of UHMWPE increased the cost of UCA. UCA 4–5 had the highest price of 5201.4 CNY/ton. Compared to SK-NA, the UCA 4–5 prices increased by 4.028%. UCA 3–5, UCA 4–5, and UCA 4–10 were slightly more expensive than SK-NA, but their conventional and rheological properties were significantly improved. Meanwhile, CSO was used for modified asphalt. The efficient utilization of waste resources generates positive environmental benefits.

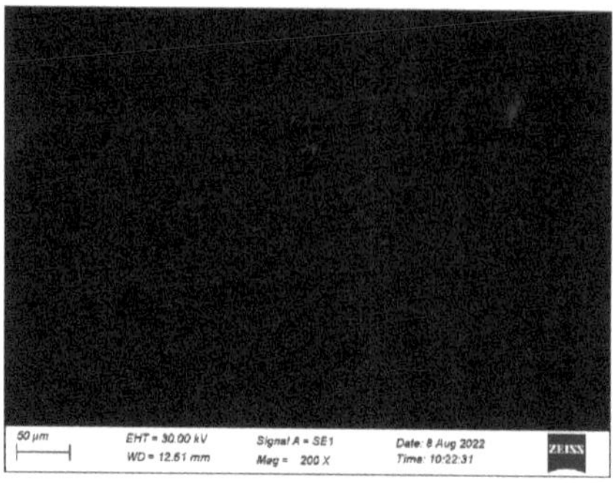

(**a**)

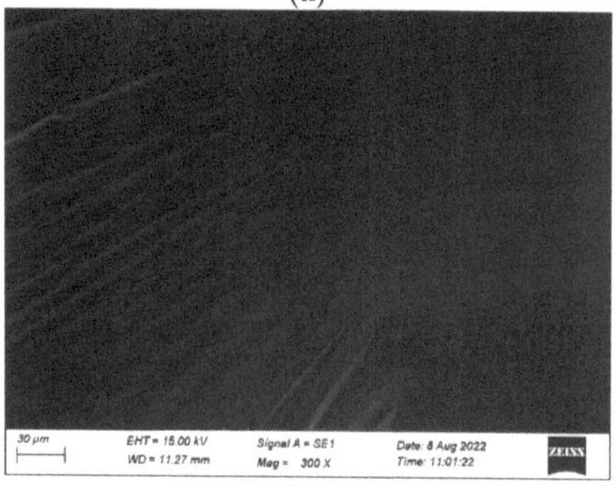

(**b**)

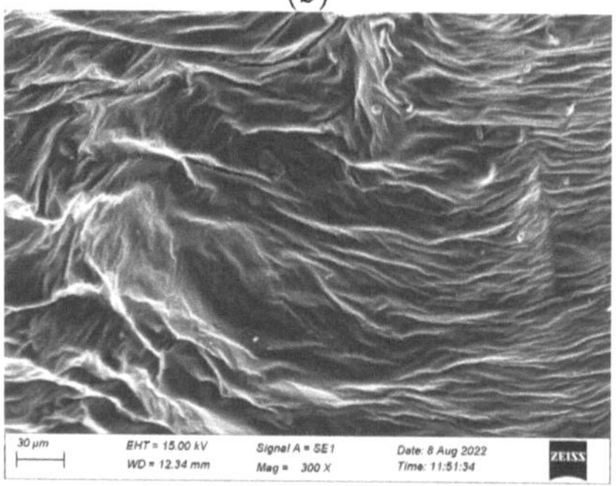

(**c**)

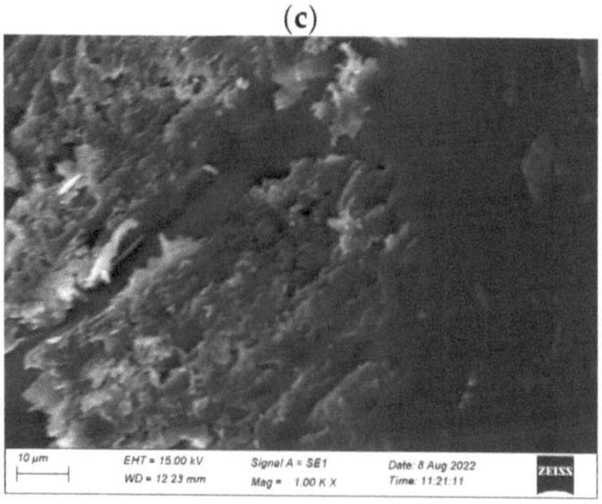

(**d**)

Figure 9. SEM images: (**a**) SK-NA; (**b**) CSO bio-asphalt; (**c**) UHMWPE-modified SK asphalt; (**d**) UCA.

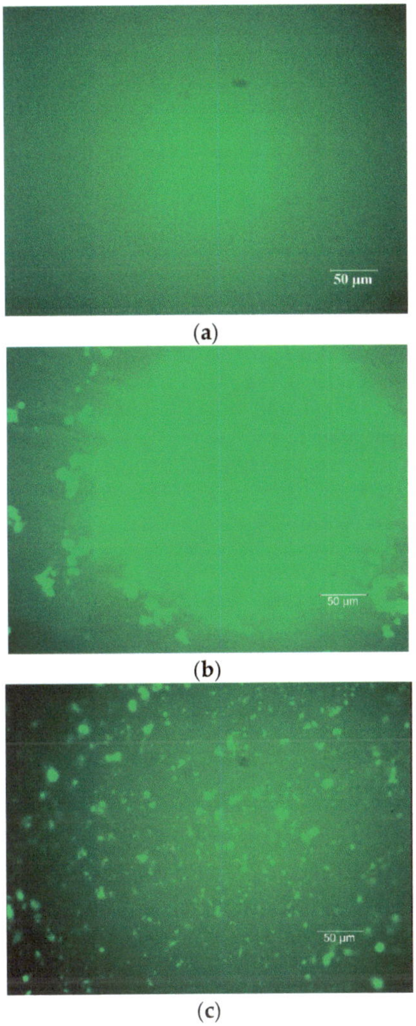

Figure 10. FM images: (a) SK-NA and CSO bio-asphalt; (b) UHMWPE-modified SK-NA; (c) UCA.

Table 7. The raw material costs of UCA.

Items	Raw Material Costs (CNY/ton)	Compared to SK-NA (CNY/ton)
SK-NA	5000	-
UCA 2–5	5009.3	9.3
UCA 3–5	5106.3	106.3
UCA 4–5	5201.4	201.4
UCA 2–10	4839.5	−160.5
UCA 3–10	4938.2	−61.5
UCA 4–10	5034.9	34.9

4. Conclusions

In order to develop the utilization of CSO and UHMWPE in road engineering, UCA was prepared in this paper. We conducted a series of laboratory tests to study its performance. These experiments included conventional property, rheological behavior, and morphology characteristic tests. We comprehensively analyzed these properties and recommended the optimal preparation scheme for UCA. In summary, we draw the following conclusions.

(1) The softening point and viscosity of UCA are enhanced with an increase in the UHMWPE dosage. Its ductility and deformation ability increase at 5 °C. The high- and low-temperature properties of UCA are obviously strengthened. Asphalt performance improves because of the light fraction contained in CSO, which increases the oil content in UCA and ameliorates its tensile properties. UHMWPE can form a compact and robust network structure in UCA. These lead to dramatic stiffness and toughness enhancements in UCA.

(2) UCA 3–5, UCA 4–5, and UCA 4–10 have superior high-temperature rheological behavior, indicating that CSO and UHMWPE jointly provide favorable modification. This is mainly due to UHMWPE, as a polymer, having good compatibility and forming a stable network structure with CSO bio-asphalt. The MSCR test suggests that UCA's creep recovery ability and anti-rutting properties decrease at a high-stress level. The low-temperature rheological behavior of UCA is superior to that of SK-NA. The high dosage of CSO will ameliorate the low-temperature behavior of UCA but will weaken its high-temperature behavior to a certain extent.

(3) The recommended UHMWPE and CSO dosages for each test are summarized. When the dosage of CSO is 5%, the recommended dosage of UHMWPE is 3–4%. However, when the dosage of CSO is 10%, the recommended dosage of UHMWPE is 4%.

(4) UHMWPE powder is melted and cross-linked with SK-NA during high-temperature preparation to shape the stable cross-linked structure. After blending UHMWPE with CSO bio-asphalt, the cross-linked structure is changed. This phenomenon may be due to the adsorption of CSO by the reticular structure. The UHMWPE polymer exhibits fluorescence characteristics. The UHMWPE polymer macromolecules can be dispersed in CSO. UHMWPE appears in UCA as a small-molecular-weight solvent. UHMWPE and CSO change the morphology of asphalt, which leads to the improvement of its performance.

Author Contributions: J.L.: methodology, conceptualization, writing—reviewing and editing, validation, formal analysis. C.L.: methodology, formal analysis, investigation, supervision, writing—reviewing and editing. J.J.: software, validation, formal analysis. H.C.: visualization, writing—reviewing and editing. All authors have read and agreed to the published version of the manuscript.

Funding: This research received no external funding.

Data Availability Statement: Some or all data, models, and code generated or used during this study are available from the corresponding author upon request.

Conflicts of Interest: The authors declare no conflict of interest.

References

1. Liu, H.; Zhang, Z.; Xie, J.; Gui, Z.; Li, N.; Xu, Y. Analysis of OMMT strengthened UV aging-resistance of Sasobit/SBS modified asphalt: Its preparation, characterization and mechanism. *J. Clean. Prod.* **2021**, *315*, 128139. [CrossRef]
2. Gupta, A.; Lastra-Gonzalez, P.; Rodriguez-Hernandez, J.; González González, M.; Castro-Fresno, D. Critical assessment of new polymer-modified bitumen for porous asphalt mixtures. *Constr. Build. Mater.* **2021**, *307*, 124957. [CrossRef]
3. Viscione, N.; Veropalumbo, R.; Oreto, C.; Biancardo, S.A.; Abbondati, F.; Russo, F. Additional procedures for characterizing the performance of recycled polymer modified asphalt mixtures. *Measurement* **2022**, *187*, 110238. [CrossRef]
4. Duarte, G.M.; Faxina, A.L. Asphalt concrete mixtures modified with polymeric waste by the wet and dry processes: A literature review. *Constr. Build. Mater.* **2021**, *312*, 125408. [CrossRef]

5. Sahebzamani, H.; Zia Alavi, M.; Farzaneh, O.; Moniri, A. Laboratory and field investigation of the effect of polymerized pellets on the fatigue and low-temperature performance of asphalt mixtures. *Constr. Build. Mater.* **2022**, *323*, 126527. [CrossRef]
6. Mashaan, N.S.; Chegenizadeh, A.; Nikraz, H.; Rezagholilou, A. Investigating the engineering properties of asphalt binder modified with waste plastic polymer. *Ain Shams Eng. J.* **2021**, *12*, 1569–1574. [CrossRef]
7. Paul, D.; Pal, M.; Majumdar, K.; Suresh, M. Effect of rubber polymeric materials on moisture susceptibility of asphalt mixtures: Optimization and evaluation study. *Mater. Today Proc.* **2022**, *57*, 454–459. [CrossRef]
8. Yan, C.; Huang, W.; Ma, J.; Xu, J.; Lv, Q.; Lin, P. Characterizing the SBS polymer degradation within high content polymer modified asphalt using ATR-FTIR. *Constr. Build. Mater.* **2020**, *233*, 117708. [CrossRef]
9. Wang, S.; Huang, W.; Kang, A. Laboratory evaluation of the properties of high-cured crumb rubber modified asphalt containing sulfur and polymer after the oxidative aging procedure. *Constr. Build. Mater.* **2021**, *304*, 124611. [CrossRef]
10. Tang, N.; Huang, W.; Hao, G. Effect of aging on morphology, rheology, and chemical properties of highly polymer modified asphalt binders. *Constr. Build. Mater.* **2021**, *281*, 122595. [CrossRef]
11. Al-Gurah, E.R.; Al-Humeidawi, B.H. Assessment of performance of hot mix asphalt contained various types of mineral fillers and newly polymer modified bitumen. *Mater. Today Proc.* **2021**, *80*, 3877–3886. [CrossRef]
12. Vamegh, M.; Ameri, M.; Chavoshian Naeni, S.F. Experimental investigation of effect of PP/SBR polymer blends on the moisture resistance and rutting performance of asphalt mixtures. *Constr. Build. Mater.* **2020**, *253*, 119197. [CrossRef]
13. Daryaee, D.; Ameri, M.; Mansourkhaki, A. Utilizing of waste polymer modified bitumen in combination with rejuvenator in high reclaimed asphalt pavement mixtures. *Constr. Build. Mater.* **2020**, *235*, 117516. [CrossRef]
14. Ren, S.; Liu, X.; Wang, H.; Fan, W.; Erkens, S. Evaluation of rheological behaviors and anti-aging properties of recycled asphalts using low-viscosity asphalt and polymers. *J. Clean. Prod.* **2020**, *253*, 120048. [CrossRef]
15. Al-Ghurabi, S.B.; Al-Humeidawi, B.H. Evaluation of performance of hot mix asphalt contained various sizes of Reclaim asphalt pavement and polymer modified Bitumen. *Mater. Today Proc.* **2021**, *80*, 2905–2910. [CrossRef]
16. Sauer, W.L.; Weaver, K.D.; Beals, N.B. Fatigue performance of ultra-high-molecular-weight polyethylene: Effect of gamma radiation sterilization. *Biomaterials* **1996**, *17*, 1929–1935. [CrossRef] [PubMed]
17. Chagas, N.P.d.S.; Aguiar, V.d.O.; Garcia Filho, F.d.C.; Figueiredo, A.B.-H.d.S.; Monteiro, S.N.; Huaman, N.R.C.; Marques, M.d.F.V. Ballistic performance of boron carbide nanoparticles reinforced ultra-high molecular weight polyethylene (UHMWPE). *J. Mater. Res. Technol.* **2022**, *17*, 1799–1811. [CrossRef]
18. Cha, J.-H.; Kumar Sarath Kumar, S.; Noh, J.-E.; Choi, J.-S.; Kim, Y.; Kim, C.-G. Ultra-high-molecular-weight polyethylene/hydrogen-rich benzoxazine composite with improved interlaminar shear strength for cosmic radiation shielding and space environment applications. *Compos. Struct.* **2022**, *300*, 116157. [CrossRef]
19. Alaferdov, A.V.; Lebedev, O.V.; Roggero, U.F.S.; Hernandez-Figueroa, H.E.; Nista, S.V.G.; Trindade, G.M.; Danilov, Y.A.; Ozerin, A.N.; Moshkalev, S.A. Highly conductive nanographite/ultra-high-molecular-weight polyethylene composite. *Results Mater.* **2022**, *15*, 100298. [CrossRef]
20. Yang, H.; Yilmaz, G.; Jiang, J.; Xie, J.; Langstraat, T.; Chu, R.; van Es, M.; Garg, P.; Turng, L.-S. Pelletizing ultra-high molecular weight polyethylene (UHMWPE) powders with a novel tapered die and addition of high density polyethylene (HDPE): Processing, morphology, and properties. *Polymer* **2022**, *256*, 125171. [CrossRef]
21. Hossein Hamedi, G.; Ghalandari Shamami, K.; Mazhari Pakenari, M. Effect of ultra-high-molecular-weight polyethylene on the performance characteristics of hot mix asphalt. *Constr. Build. Mater.* **2020**, *258*, 119729. [CrossRef]
22. Ju, Z.; Ge, D.; Wu, Z.; Xue, Y.; Lv, S.; Li, Y.; Fan, X. The performance evaluation of high content bio-asphalt modified with polyphosphoric acid. *Constr. Build. Mater.* **2022**, *361*, 129593. [CrossRef]
23. Lv, S.; Ju, Z.; Peng, X.; Cabrera, M.B.; Wang, Z.; Jiang, Y.; Long, H. Preparation and Performance of Polyphosphoric Acid/Bio-oil Composite Modified Asphalt Containing a High Content Bio-oil. *J. Mater. Civ. Eng.* **2021**, *34*, 04021484. [CrossRef]
24. Wang, F.; Huang, T.; Xin, G.; Mu, M.; Shen, Q. Study on Conventional and Rheological Properties of Corn Stalk Bioasphalt/PPA Composite Modified Asphalt. *Adv. Civ. Eng.* **2021**, *2021*, 7928189. [CrossRef]
25. Wen, H.; Bhusal, S.; Wen, B. Laboratory Evaluation of Waste Cooking Oil-Based Bioasphalt as an Alternative Binder for Hot Mix Asphalt. *J. Mater. Civ. Eng.* **2013**, *25*, 1432–1437. [CrossRef]
26. Deng, M.; Cao, X.; Li, Z.; Li, X.; Yang, X.; Tang, B. Investigating properties and intermolecular interactions of sludge bio-oil modified asphalt. *J. Mol. Liq.* **2022**, *360*, 119415. [CrossRef]
27. Al-Sabaeei, A.M.; Napiah, M.B.; Sutanto, M.H.; Rahmad, S.; Yusoff, N.I.M.; Alaloul, W.S. Determination of rheological properties of bio-asphalt binders through experimental and multilayer feed-forward neural network methods. *Ain Shams Eng. J.* **2021**, *12*, 3485–3493. [CrossRef]
28. Al-Sabaeei, A.M.; Napiah, M.; Sutanto, M.; Alaloul, W.; Md Yusoff, N.I.; Imran Khan, M.; Modibbo Saeed, S. Physicochemical, rheological and microstructural properties of Nano-Silica modified Bio-Asphalt. *Constr. Build. Mater.* **2021**, *297*, 123772. [CrossRef]
29. Dewajani, H.; Zamrudy, W.; Irfin, Z.; Ningtyas, D.; Mujibur Ridlo, N. Utilization of Indonesian sugarcane bagasse into bio asphalt through pyrolysis process using zeolite-based catalyst. *Mater. Today Proc.* **2023**, *87*, 383–389. [CrossRef]
30. Lyu, L.; Pei, J.; Hu, D.; Sun, G.; Fini, E.H. Bio-modified rubberized asphalt binder: A clean, sustainable approach to recycle rubber into construction. *J. Clean. Prod.* **2022**, *345*, 131151. [CrossRef]

31. Lyu, L.; Mikhailenko, P.; Piao, Z.; Fini, E.H.; Pei, J.; Poulikakos, L.D. Unraveling the modification mechanisms of waste bio-oils and crumb rubber on asphalt binder based on microscopy and chemo-rheology. *Resour. Conserv. Recycl.* **2022**, *185*, 106447. [CrossRef]
32. Jiang, X.; Li, P.; Ding, Z.; Yue, L.; Li, H.; Bing, H.; Zhang, J. Physical, chemical and rheological investigation and optimization design of asphalt binders partially replaced by bio-based resins. *Constr. Build. Mater.* **2022**, *350*, 128845. [CrossRef]
33. Zheng, W.; Wang, H.; You, Z.; Shao, L.; Golroo, A.; Chen, Y. Mechanism and rheological characterization of MDI modified Wood-Based Bio-Oil asphalt. *Constr. Build. Mater.* **2021**, *309*, 125113. [CrossRef]
34. Yang, X.; Liu, G.; Rong, H.; Meng, Y.; Peng, C.; Pan, M.; Ning, Z.; Wang, G. Investigation on mechanism and rheological properties of Bio-asphalt/PPA/SBS modified asphalt. *Constr. Build. Mater.* **2022**, *347*, 128599. [CrossRef]
35. Zhang, X.; Zhang, K.; Wu, C.; Liu, K.; Jiang, K. Preparation of bio-oil and its application in asphalt modification and rejuvenation: A review of the properties, practical application and life cycle assessment. *Constr. Build. Mater.* **2020**, *262*, 120528. [CrossRef]
36. Zhang, Z.; Fang, Y.; Yang, J.; Li, X. A comprehensive review of bio-oil, bio-binder and bio-asphalt materials: Their source, composition, preparation and performance. *J. Traffic Transp. Eng. (Engl. Ed.)* **2022**, *9*, 151–166. [CrossRef]
37. Wang, L.; Si, B.; Han, X.; Yi, W.; Li, Z.; Zhang, A. Study on the effect of red mud and its component oxides on the composition of bio-oil derived from corn stover catalytic pyrolysis. *Ind. Crops Prod.* **2022**, *184*, 114973. [CrossRef]
38. Padhan, R.K.; Sreeram, A. Enhancement of storage stability and rheological properties of polyethylene (PE) modified asphalt using cross linking and reactive polymer based additives. *Constr. Build. Mater.* **2018**, *188*, 772–780. [CrossRef]
39. Lv, S.; Liu, J.; Peng, X.; Liu, H.; Hu, L.; Yuan, J.; Wang, J. Rheological and microscopic characteristics of bio-oil recycled asphalt. *J. Clean. Prod.* **2021**, *295*, 126449. [CrossRef]
40. Chaihad, N.; Karnjanakom, S.; Abudula, A.; Guan, G. Zeolite-based cracking catalysts for bio-oil upgrading: A critical review. *Resour. Chem. Mater.* **2022**, *1*, 167–183. [CrossRef]
41. Huang, T.; He, H.; Zhang, P.; Lv, S.; Jiang, H.; Liu, H.; Peng, X. Laboratory investigation on performance and mechanism of polyphosphoric acid modified bio-asphalt. *J. Clean. Prod.* **2022**, *333*, 130104. [CrossRef]

MDPI AG
Grosspeteranlage 5
4052 Basel
Switzerland
Tel.: +41 61 683 77 34

Buildings Editorial Office
E-mail: buildings@mdpi.com
www.mdpi.com/journal/buildings

www.ingramcontent.com/pod-product-compliance
Lightning Source LLC
Chambersburg PA
CBHW041419210926
51850CB00013B/3317
* 9 7 8 3 7 2 5 8 3 2 0 4 0 *